Chemistry Calculations for Beginners

With decades of combined experience as science teachers at both school and undergraduate levels, the authors have recognised that one of the greatest challenges faced by students studying chemistry is grasping the complexity of the numerous numerical problems found in most parts of the subject. This text is crafted to provide a clear and accessible pathway to overcoming this challenge by assisting students, especially novices or those with minimal knowledge of the subject, in performing chemistry calculations. The content covers fundamental calculations crucial to understanding the principles of chemistry, making it an invaluable tool for students aiming to excel in their studies.

Key features

- Designed with a student-friendly approach, including detailed explanation of chemical concepts underlying each type of calculation, step-by-step explanations, alternative methods for solving problems, numerous practice exercises, answers to practice exercises and appendices.
- The book is tailored to suit various curricula, ensuring relevance for a diverse audience.
- Encompasses a wide range of calculations, offering students a thorough understanding of essential chemistry concepts.
- Serves as an excellent resource for exam preparation and equips students with skills applicable to future scientific endeavours. Employs straightforward language to ensure ease of understanding for beginners.
- Uses IUPAC conventions, underscoring the universal nature of chemistry.

Chemistry Calculations for Beginners

John Obimakinde
Samuel Obimakinde
Ebenezer Obimakinde
Fredrick Akinbolade

CRC Press
Taylor & Francis Group
Boca Raton   London   New York

CRC Press is an imprint of the
Taylor & Francis Group, an **informa** business

First edition published 2025
by CRC Press
2385 NW Executive Center Drive, Suite 320, Boca Raton FL 33431

and by CRC Press
4 Park Square, Milton Park, Abingdon, Oxon, OX14 4RN

CRC Press is an imprint of Taylor & Francis Group, LLC

© 2025 John Obimakinde, Samuel Obimakinde, Ebenezer Obimakinde and Fredrick Akinbolade

Reasonable efforts have been made to publish reliable data and information, but the author and publisher cannot assume responsibility for the validity of all materials or the consequences of their use. The authors and publishers have attempted to trace the copyright holders of all material reproduced in this publication and apologise to copyright holders if permission to publish in this form has not been obtained. If any copyright material has not been acknowledged please write and let us know so we may rectify in any future reprint.

Except as permitted under U.S. Copyright Law, no part of this book may be reprinted, reproduced, transmitted, or utilised in any form by any electronic, mechanical, or other means, now known or hereafter invented, including photocopying, microfilming, and recording, or in any information storage or retrieval system, without written permission from the publishers.

For permission to photocopy or use material electronically from this work, access www.copyright.com or contact the Copyright Clearance Center, Inc. (CCC), 222 Rosewood Drive, Danvers, MA 01923, 978-750-8400. For works that are not available on CCC please contact mpkbookspermissions@tandf.co.uk

Trademark notice: Product or corporate names may be trademarks or registered trademarks and are used only for identification and explanation without intent to infringe.

ISBN: 978-1-032-85415-1 (hbk)
ISBN: 978-1-032-85414-4 (pbk)
ISBN: 978-1-003-51805-1 (ebk)

DOI: 10.1201/9781003518051

Typeset in Times
by SPi Technologies India Pvt Ltd (Straive)

For Mum, Ruth Olufunmilayo Obimakinde, who has since gone to be with the Lord.

Contents

Preface viii
Acknowledgements xii
About the Authors xiii

Chapter 1 Introduction to Chemistry 1

Chapter 2 Approximations and Standard Form 40

Chapter 3 Chemical Formula and Masses 50

Chapter 4 The Mole Concept and Chemical Formulae 59

Chapter 5 Chemical Equations and Stoichiometry 86

Chapter 6 Redox Reactions 113

Chapter 7 Atomic Structure 129

Chapter 8 Gas Laws 137

Chapter 9 Properties of Solutions 164

Chapter 10 Chemical Energetics 200

Chapter 11 Chemical Equilibria 234

Chapter 12 Electrochemistry 264

Chapter 13 Volumetric Analysis 292

Chapter 14 Rates of Chemical Reactions 327

Chapter 15 Nuclear Chemistry 356

Appendix 372
Answers 390
Index 411

Preface

A chemist who does not know mathematics is seriously handicapped.

—Irving Langmuir

Over our decades of combined experience teaching and applying chemistry, we have identified a recurring challenge: students' difficulty with mastering the fundamental concepts and calculations in chemistry. This challenge has significantly contributed to poor performance in chemistry over the years. The idea of this book was conceived to address this issue. Our objective is to convey the fundamental principles and calculations in chemistry to students at all levels. We aim to address perceived gaps in knowledge and provide beginning chemistry students with a strong foundation essential for understanding more advanced topics in the subject.

THE BOOK'S STRUCTURE

This book has been meticulously crafted to make both learning and teaching enjoyable, providing an immersive experience that fosters a passion for chemistry among teachers and students alike. The content is comprehensive, covering all relevant aspects of chemistry to meet the needs of various curricula and educational systems worldwide.

Chapter 1 introduces the basic principles of chemistry, simplifying and clarifying many key concepts that are often misunderstood and difficult to relate to. Chapter 2 discusses standard form (scientific notation) and approximations, guiding users in the correct scientific procedures for reporting the results of mathematical operations and handling large numbers. Subsequent chapters focus on specific aspects of chemistry, each accompanied by related numerical problems.

OUR APPROACH

Our approach in this book blends explanation of theoretical concepts with extensive problem-solving. Each chapter and section begins with a comprehensive yet accessible explanation that provides deep insight into the concept being introduced. This is followed by numerous representative examples of related calculations, often offering alternative pathways to solving the same problem. The examples are followed by practice problems, allowing readers to gauge their understanding and reinforce learning.

The practice problems are structured in increasing levels of difficulty to mimic actual examination questions. At the end of each chapter, more extensive problems, labeled 'further practice problems,' provide additional opportunities for self-assessment. Answers are provided for all practice problems.

We have also employed unit analysis to help students understand how to handle units correctly in calculations.

TO THE STUDENTS

The fact that chemistry cannot be fully appreciated without proficient knowledge in mathematics and its applications to solving the diverse range of problems encountered in the subject cannot be overemphasised. In fact, it is virtually impossible to optimise the learning and practical experience in chemistry without being adept at performing calculations. Consequently, students with weak problem-solving skills may consistently receive low grades, which may lead to serious, inconvenient changes in academic pursuits or career paths.

Preface

Chemistry Calculations for Beginners is a comprehensive textbook designed to help you master chemistry calculations and gain a deep insight into the fundamental concepts and principles required to excel in the subject. Whether you are a student or a professional in the industry, this book will benefit you immensely.

This book is packed with many salient features to help you succeed, some of which are listed below:

Step-by-Step Solutions
Each calculation is broken down into detailed steps to help you understand the process and reasoning behind each one.

Practice Problems
At the end of each chapter, you will find a variety of problems to practise, ranging from basic to advanced levels.

Real-World Applications
Examples and problems are placed within real-world contexts to help you see the practical applications of chemistry calculations.

Visual Aids
Diagrams, tables and charts are used where appropriate to illustrate concepts and support your understanding.

Alternative Methods
Where applicable, alternative ways of solving problems are provided to enhance your understanding and problem-solving flexibility.

IUPAC Conventions
IUPAC conventions are employed throughout the text to ensure uniformity and adherence to standard procedures. Alternative nomenclature versions are explained where necessary to accommodate the requirements of different curricula.

Using Units in Calculations
One of the fundamental aspects of chemistry calculations is the use of units. Units help you keep track of the quantities you are working with and ensure your calculations are accurate. In this book, you will see units included in every step of the calculations. This is done to help you understand how units interact and how to correctly apply them throughout a problem. However, when it comes to exams, the approach to using units can be slightly different:

Given Data and Final Answer: In an exam setting, it is sufficient to include units in the list of given data at the beginning of your problem and in your final answer.

Intermediate Steps: You do not need to write out the units in every intermediate step of your calculations. This can help you save time and keep your work neat and organised.

For example, if you are asked to find the molar concentration or molarity of a solution:

- List your given data with units, such as 'Volume = $1.0 \, dm^3$' and 'Moles of solute = 0.5 mol.'
- Perform your calculations without repeatedly writing the units.
- Ensure your final answer includes the correct units, such as 'Molar concentration = $0.5 \, mol \, dm^{-3}$.'

PREREQUISITE

Although the calculations in this book are presented in an easy-to-follow format, a solid knowledge of basic mathematics, particularly algebra, is essential for you to benefit optimally from it. If your maths is weak, it is advisable to address this deficiency to ensure you can fully grasp the concepts and perform the calculations accurately.

STUDY TIPS

- Practice Regularly: Use the practice problems at the end of each section and chapter to test your understanding and improve your calculation skills.
- Review Mistakes: When you make mistakes, review them to understand where you went wrong and how to correct it.
- Ask for Help: Don't hesitate to ask your instructor or classmates for help if you are struggling with a concept or problem.
- Stay Organised: Keep your notes and practice problems organised. This will make it easier to review and study for exams.

CONFIDENCE IN EXAMS

By practising with this book and following the structured approach to calculations, you will gain confidence in your ability to tackle chemistry problems. Remember to focus on understanding the concepts and processes, and the accuracy of your final answers.

We hope this book serves as a valuable resource in your studies and helps you succeed in your chemistry course. Happy studying!

TO THE INSTRUCTOR

Thank you for considering *Chemistry Calculations for Beginners* for your class. This book is meticulously crafted to enhance your students' understanding and mastery of essential chemistry principles and calculations, a critical skill set for success in the field of chemistry. Here is why this book will be an invaluable asset to your curriculum:

Comprehensive Coverage

The book spans a wide array of topics, from fundamental principles to advanced calculations. Each chapter is designed to build upon the previous one, ensuring a progressive and thorough understanding of chemistry calculations.

Step-by-Step Methodology

One of the unique features of this book is its detailed, step-by-step approach to problem-solving. Each calculation is broken down into manageable steps, making complex concepts easier for students to grasp and apply. This method not only aids comprehension but also boosts students' confidence in their problem-solving abilities.

Engaging Real-World Applications

To make the subject matter more relatable and engaging, real-world examples and applications are integrated throughout the book. This contextual approach helps students see the relevance of what they are learning, thereby increasing their motivation and interest.

Extensive Practice Opportunities

The book offers a wealth of practice problems at the end of each chapter, ranging from basic to challenging levels. These problems are designed to reinforce learning, provide ample practice and prepare students for exams. Additionally, the inclusion of answers helps students learn from their mistakes and understand the correct methodologies.

Visual Learning Aids

Recognising the diverse learning styles of students, this book incorporates various visual aids, including diagrams, tables and charts. These elements support visual learners and help clarify complex concepts, making the material more accessible.

Focus on Efficiency in Exams
A distinctive feature of this book is the guidance provided on handling units in calculations. While the book includes units in every step for educational purposes, it also advises students on how to efficiently handle units in exams by including them in the list of given data and the final answer only. This practical tip helps students save time and stay organised during exams.

Adaptable to Various Teaching Styles
Whether you prefer a lecture-based approach, interactive problem-solving sessions or flipped classroom methods, this book is versatile and adaptable to your teaching style. Its clear structure and comprehensive content make it an excellent resource for diverse educational settings.

By adopting *Chemistry Calculations for Beginners*, you will provide your students with a robust foundation in chemistry, preparing them for advanced studies and professional success. This book is more than just a textbook; it is a tool designed to foster a deep understanding and appreciation of chemistry.

HOW TO USE THIS BOOK

Chapter Overviews: Begin each chapter with a highlight of the key objectives and concepts. This will give students a clear understanding of what they will learn and why it is important.

Interactive Teaching: Encourage students to work through the example problems interactively in class, discussing each step and addressing any questions or misconceptions as they arise.

Assign Practice Problems: Use the practice problems as homework or in-class activities. This will help students apply what they have learned and gain confidence in their calculation skills.

Review and Reinforce: Regularly review key concepts from previous chapters to reinforce learning and ensure retention. Cumulative review sections at the end of major units can be particularly useful.

Supplementary Materials: Utilise any supplementary materials provided, such as answer keys and online resources. These can provide additional support for both you and your students.

SUPPORTING DIVERSE LEARNERS

Recognising that students have diverse learning styles and needs, this book incorporates various instructional strategies:

Worked Examples: Detailed examples cater to students who benefit from seeing fully worked-out solutions.

Practice Problems: A range of problems, from straightforward to challenging, allows students to progress at their own pace.

Visual Elements: Diagrams and charts aid in the comprehension of complex concepts for visual learners.

FEEDBACK AND ADAPTATION

We value your feedback. Please share your experiences using this book, including what works well and any areas for improvement. Your insights will help us refine future editions to better meet the needs of instructors and students alike.

John Obimakinde
Lagos, Nigeria

Acknowledgements

The success of this book owes much to the contributions of many individuals, too numerous to mention. We would like to extend our heartfelt thanks to the entire team at the Taylor and Francis Group for their professionalism, advice and support in transforming our manuscript into this excellent text. Our particular thanks go to the Senior Publisher Fiona Macdonald, Editorial Manager Hilary Lafoe and Production Editor Rachael Panthier for their patience and unwavering support throughout the publishing process. We also extend our gratitude to our editorial assistants, Varalika Kathuria and Aditya Aman, for their invaluable support and contributions. We would also like to thank Thivya Vasudevan of SPi Technologies India Pvt Ltd (Straive) and her entire team for their invaluable contributions to the successful publication of this project.

Similarly, our deep appreciation goes to Dr. Paul Yates, a renowned chemistry teacher and writer, who, despite his busy schedule, found time to review the proposal for this project and made invaluable suggestions. His insights and feedback have greatly enhanced the quality of this work.

In addition, we are grateful to our professional colleagues and friends, particularly Veronica Uzoma and Lanre Ogunjana, who made significant sacrifices to help us get the manuscript ready for transmission to the publisher. Their encouragement and assistance were vital to this project's completion.

We are equally grateful to the owners of the science education website www.psiberg.com for permission to reproduce Figure 15.3.

I would like to express my deepest gratitude to my co-authors, Samuel Obimakinde, Ebenezer Obimakinde and Fredrick Akinbolade. Their expertise, dedication and invaluable contributions have been instrumental in bringing this book to fruition.

Lastly, but by no means least, we convey our heartfelt appreciation to our families for their unwavering support and understanding. Their patience and the concessions they made while we were writing a seemingly endless book have been invaluable.

Thank you all for your contributions and support.

John Obimakinde
Lagos, Nigeria

Authors

John Obimakinde has a degree in chemical and polymer engineering from Lagos State University, Lagos, Nigeria. His academic journey also included studies at the University of Ilorin in Nigeria, Brandenburg University of Technology, Cottbus, Germany and Rhein-Waal University of Applied Sciences, Kleve, Germany. From 2005 to present he has dedicated his career to education, specialising in mathematics, chemistry and physics. Over the years, he has held teaching roles at various institutions. Currently, he serves as a principal lecturer at the Imperial STEM Academy, Lagos.

Samuel Obimakinde holds a first-class degree in industrial chemistry from Bowen University, Iwo, Nigeria, and a Master's degree in environmental chemistry from the University of Ibadan, Nigeria. Additionally, a certificate in data analytics from CoHarvester Technological Institute in Manchester, United Kingdom. His career has been enriched with diverse teaching and research experiences in environmental and analytical chemistry, spanning both school and undergraduate levels. From 2009 to 2014, he served as a lecturer and researcher at Bowen University, teaching courses in environmental, industrial and analytical chemistry. Subsequently, from 2014 to 2019, he continued his academic journey as a Lecturer and researcher at Cape Peninsula University of Technology in Cape Town, South Africa. In 2017, he contributed as a consultant to Damelin and INTEC college in South Africa, proofreading examination papers and marking scripts. Between 2020 and 2022, his focus shifted to tutoring mathematics, computer and science at the School of Hope in Cape Town. Presently, he serves as a mathematics and science tutor at Destin Kids Academy, Cape Town, South Africa.

Ebenezer Obimakinde graduating with a Bachelor of Technology degree in environmental biology from Ladoke Akintola University of Technology, Ogbomoso, Nigeria, furthered his academic journey with Master's and Ph.D. degrees in applied parasitology from the Federal University of Technology, Akure, Nigeria. His extensive teaching and research experience spans both high school and undergraduate levels. He has held positions including science teacher at Vega College, Lagos, Nigeria, from 2013 to 2014, and at Amorokeni Community Senior Grammar School, Ogbia, Nigeria. Presently, he is a Lecturer at the University Advanced Basic Sciences at the Federal University of Technology, Akure, Nigeria. His contributions to academia extend to numerous publications in reputable journals, focusing on topics such as malaria prevalence, parasitic contamination of water sources and efficacy of plant extracts against malaria vectors. Additionally, he engages in peer-review activities for journals such as *Annual Research & Review in Biology* and *International Journal of Tropical Disease & Health*. As a member of the Public Health Society of Nigeria (PPSN), he is dedicated to advancing knowledge in the field.

Fredrick Akinbolade is a dedicated educator with a passion for inspiring young minds. He holds a BSc in Industrial Chemistry from Adekunle Ajasin University, Akungba Akoko, Ondo State, Nigeria. Since 2015, Fredrick has taught chemistry, basic science and mathematics at various schools, including Jolabell Schools, Winsome Model College, Greenland Hall Schools, and currently HOD Sciences Christhill Schools Amunwo Odofin Lagos. His strong academic background is complemented by hands-on experience in chemical manufacturing, including roles in the paint and textile industries. A Google Certified Educator, Fredrick is highly proficient in Google Apps. for Education, leveraging digital tools to enhance learning and foster student engagement. He is committed to making complex scientific concepts accessible, engaging and interactive helping students develop critical thinking and problem-solving skills. Beyond the classroom, he enjoys football and traveling, experiences that enrich his teaching and broaden his perspective.

1 Introduction to Chemistry

1.1 CHEMISTRY: THE CENTRAL SCIENCE

Natural science is a branch of science concerned with the study of the physical or natural world and the phenomena that occur within it. As shown in Figure 1.1, natural science is divided into three main branches: physical sciences, Earth sciences and life sciences. Chemistry is a physical science that focuses on the composition, structure and properties of matter, and with the transformations it undergoes. Chemistry is often called the *Central Science* because its principles provide the foundation for understanding all the other physical sciences.

1.2 A BRIEF HISTORY OF CHEMISTRY

Learning chemistry is as fascinating as it is rewarding. It takes you on an exciting journey spanning millennia of curiosity, exploration, resilience and ground breaking discoveries by some of the most brilliant minds to have ever lived.

Modern chemistry evolved from *alchemy*, a pseudoscience practised in ancient civilisations. Alchemists, using a mix of philosophical speculation, magic and experimentation, attempted to transform base metals into gold. While they ultimately failed in this pursuit, their work led to advances in experimental techniques, an improved understanding of materials and the development of laboratory apparatuses, laying the foundation for modern chemistry.

The Scientific Revolution of the 16th and 17th centuries marked a turning point in science, emphasising the use of the scientific method (systematic experimentation as the only valid means of acquiring scientific knowledge). Scientists designed sophisticated laboratory apparatuses, conducted ingenuous experiments and made significant discoveries. However, the Chemical Revolution of the 18th century is considered the true birth of modern chemistry.

One of the most influential chemists of the Chemical Revolution was Antoine Lavoisier (1743–1794), a French aristocrat and public administrator. His many contributions to chemistry include establishing the law of conservation of mass, discovering the role of oxygen in combustion and respiration, compiling the first extensive list of chemical elements and contributing to the development of the metric system. For his immense contributions, Lavoisier is often called the *Father of Modern Chemistry*. Unfortunately, his role as a tax collector led to his tragic downfall. He was executed by guillotine during the French Revolution.

Other pioneers of modern chemistry include Robert Boyle (1627–1691), who conducted pioneering work on gas behaviour; Joseph Priestley (1733–1804), who discovered oxygen; John Dalton (1766–1844), an English schoolteacher who introduced atomic theory; the Frenchman Joseph Louis Gay-Lussac, known for *Gay-Lussac's law* and his work on the composition of water alongside the German polymath Alexander von Humboldt (1769–1859); and Jöns Jacob Berzelius (1779–1848), a Swedish chemist who made significant contributions to stoichiometry, isomerism, chemical bonding, electrochemistry and allotropy. He also wrote the first chemistry textbook and was awarded the Copley Medal (1836) by the Royal Society of London. Berzelius is fondly known as the *Father of Swedish Chemistry*.

1.3 BRANCHES AND SCOPE OF CHEMISTRY

Chemistry is a vast field of knowledge. As shown in Figure 1.2, it is divided into different branches or areas of specialisation, each focusing on a specific aspect of the subject.

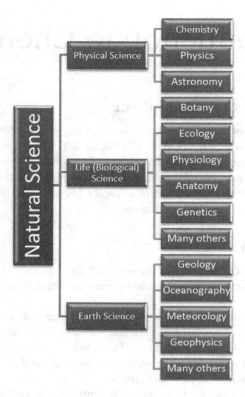

FIGURE 1.1 The branches of natural science.

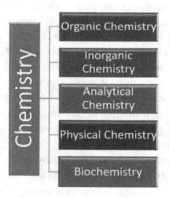

FIGURE 1.2 The major branches of chemistry.

1.4 THE SCIENTIFIC METHOD

The scientific method is a systematic procedure that scientists use to investigate natural phenomena and discover scientific truths. Developed during the Scientific Revolution, it is based on the principle that all ideas and thoughts about matter must be supported by reproducible experiments. As the foundation of scientific work, its ultimate goal is to explain observed natural phenomena.

As shown in Figure 1.3, the scientific method begins with an *observation*. Once a scientist observes the behaviour of matter or a natural phenomenon, they seek to explain it by proposing a

Introduction to Chemistry

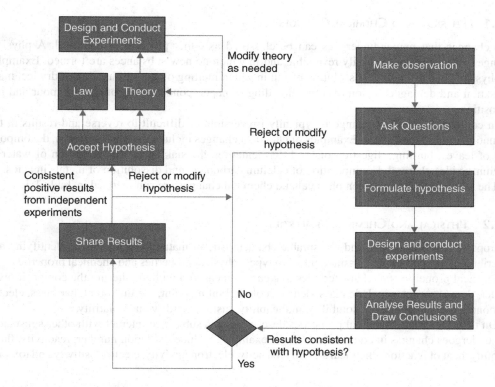

FIGURE 1.3 The scientific method.

hypothesis—a reasonable explanation that accounts for the observation. The next step is to design and conduct experiments to test the validity of the hypothesis. If the experimental results do not support the hypothesis, the scientist may either modify it or formulate a new one, depending on the degree of inconsistency. This process is repeated until the hypothesis is fully supported by experimental evidence.

If experimental results consistently confirm a hypothesis, it is accepted and shared with other scientists in the field for further verification. A hypothesis that is repeatedly validated through experimentation becomes a *theory*. Scientific theories explain natural phenomena based on extensive evidence and are widely accepted within the scientific community.

A verified hypothesis can also be formulated into a law or principle. A scientific law or principle describes a fundamental relationship in nature that has been extensively tested and proven true without exception. Chemistry is governed by numerous laws and principles, including Hess's law, Boyle's law, Graham's law, conservation of mass, Faraday's first and second laws of electrolysis, the law of definite proportions, the law of multiple proportions, the first and second laws of thermodynamics, Heisenberg's uncertainty principle, Avogadro's law, Gay-Lussac's law, ideal gas law, the periodic law, Aufbau principle, Pauli exclusion principle, Dalton's law and the law of radioactive decay. A solid understanding of these fundamental laws and principles is essential for mastering chemistry.

1.5 MATTER

As previously mentioned, chemistry revolves around the study of matter. Matter is anything that has mass and occupies space. Everything in the universe, including yourself, is composed of matter.

1.5.1 Physical and Chemical Changes

The changes that matter undergoes can be classified as either physical or chemical. A physical change is a change that is easily reversible and in which no new substances are formed. Examples of physical changes include dissolution of salt in water, melting of ice, sublimation of dry ice, magnetisation and demagnetisation of iron, shredding of paper, condensation of water vapour and gas deposition.

In contrast, a chemical change is typically irreversible or difficult to reverse and results in the formation of new substances. Examples of chemical changes include the rusting of iron, decomposition of leaves, burning, digestion of food, fermentation, the slaking of lime (addition of water to calcium oxide), thermal decomposition of calcium carbonate, and dissolution of metals in acids.

The key differences between physical and chemical changes are summarised in Table 1.1.

1.5.2 Physical and Chemical Properties

A property is an observable and measurable characteristic of matter, which helps in identifying and describing it. Properties are classified into two types: physical properties and chemical properties.

Physical properties are characteristics that can be measured without altering the composition of matter. Some examples include mass, density, colour, boiling point, melting point, hardness, electrical conductivity, magnetism, solubility, malleability, lustre, ductility and volatility.

On the other hand, chemical properties describe how a substance interact with other substances and undergoes changes in composition. Some examples include oxidation number, reactivity, flammability, heat of reaction, chemical stability, basicity, electronegativity, electropositivity and toxicity.

1.5.3 The Physical States of Matter

Matter exists in three physical states: solid, liquid and gas. The properties of each state matter can be explained using the *particulate theory of matter*, which states that matter consists of tiny, separate particles in constant random motion. Since the particles are always moving, they possess *kinetic energy*, which increases with temperature. The physical state of matter is determined by the separation of its particles and the strength of the forces holding them together. Matter can transition between states by altering the average kinetic energy of its particles through temperature changes.

TABLE 1.1
Differences between Physical and Chemical Changes

Physical Change	Chemical Change
It can be easily reversed through physical means like filtration, evaporation, condensation, etc.	It can only be reversed through a chemical process such as electrolysis
It is not accompanied by significant amounts of heat changes except those involved in change of state	It is accompanied by significant amounts of heat changes
There is no change in composition, i.e., no new substances are formed	There is change in composition, i.e., new substances are formed
There is no change in the mass of a substance undergoing this change	There is a change in the mass of a substance undergoing this change
It affects only the physical properties of a substance	It affects both physical and chemical properties of a substance

1.5.3.1 Solids

As illustrated in Figure 1.4, the particles in solids are tightly packed and held together by strong forces of attraction. Thus, aside from vibrating about fixed positions, the particles of solids cannot move freely or flow. Moreover, solids have a definite shape and fixed volume, and they are difficult to compress. Examples of substances that exist as solids at room temperature include bar soap, metals, glass and plastics. At 0°C, water exists as a crystalline solid called *ice*, while at −78°C, carbon dioxide exists as a white solid known as *dry ice*.

In solids, the conduction of electricity depends on the presence of *free electrons*. For this reason, graphite and all metals are good conductors of heat and electricity. Heat is transferred in solids by conduction.

Solids are classified into two groups: amorphous solids and crystalline solids. Amorphous solids have no definite geometric shape or form. Examples include rubber, powder, glass, tar, starch, cellulose and plastics. In contrast, crystalline solids have a definite geometric shape or form. Examples include sodium chloride, sucrose, the metals, quartz, diamond, solid iodine, metalloids, snowflakes and gypsum. The particles in crystalline solids, or simply crystals, are arranged in a recurring three-dimensional pattern called a *crystal lattice*. The structural and repeating unit of a crystal lattice is called a *unit cell*, and the location of a particle, or part of a particle, in a unit cell is called a *lattice point*. A unit cell is characterised by the length of its edges and the angles α, β and γ between them, as shown in Figure 1.5a. There are 14 kinds of unit cells, which are grouped into seven categories, including cubic, tetragonal, monoclinic, orthorhombic, rhombohedral, hexagonal and triclinic. However, we will consider only the cubic unit cell.

FIGURE 1.4 The particles of a solid are tightly packed.

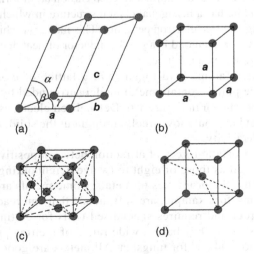

FIGURE 1.5 (a) The properties of a unit cell. (b) A simple cubic structure. (c) A fcc structure. (d) A bcc structure.

A cubic unit cell has all of its edges of equal length and angles of 90°. There are three types of cubic unit cells: simple, or primitive, cubic unit cell, face-centred cubic unit cell (fcc) and body-centred unit cell (bcc). A simple cubic unit cell comprises one-eighth of an atom at each corner of the cube, making a total of one atom per unit cell, as shown in Figure 1.5b. Polonium crystallises in this structure. A face-centred cubic structure has, in addition to one-eighth of an atom at each corner, one-half of an atom at the centre of each of the six faces, making a total of four atoms per unit cell, as shown in Figure 1.5c. Examples of substances that crystallise in the face-centred cubic structure include copper, aluminium, nickel, gamma iron, gold and silver. A body-centred cubic unit cell has, in addition to one-eighth of an atom at the corners, one atom at the centre of the cube making a total of two atoms per unit cell, as shown in Figure 1.5d. Examples of substances with the body-centred cubic structure include vanadium, tungsten, alpha-iron, barium, chromium, lithium, sodium, potassium and niobium.

There are four different types of crystalline solids: ionic solids, molecular solids, covalent solids and metallic solids. An ionic solid consists of alternating positive and negative ions arranged in a regular pattern. Hence, ionic solids contain oppositely charged ions at their lattice points. The ions are held together in the crystal lattice by electrostatic forces of attraction. In the solid state, ionic solids do not conduct electricity because the ions are not free to move. To conduct electricity, the ions must be rendered mobile by dissolving the solid in water or melting it. Ionic solids are also hard and brittle and have high melting points. Examples include sodium chloride, copper(II) sulfate, calcium fluoride, calcium chloride, zinc sulfide and ammonium chloride.

A molecular solid consists of molecules that are arranged in a regular pattern, meaning that the lattice points contain molecules. The molecules are held together by weak intermolecular forces, such as van der Waals forces, hydrogen bond, dipole–dipole interaction and London dispersion forces. Examples of molecular solids include ice, solid iodine, dry ice, naphthalene and solid sulfur dioxide. Unlike ionic crystals, molecular solids are poor conductors of electricity because their electrons are tightly held within molecules and cannot move. Additionally, molecular solids are soft and tend to have low melting points.

Covalent solids, also called covalent network solids, are giant lattices made up of atoms linked together by covalent bonds, meaning that their lattice points contain atoms. The atoms are held together by strong covalent bonds, forming a network of interconnected atoms. Examples include diamond, graphite, silicon dioxide, quartz and fullerenes. Because of their strong covalent bonds, covalent solids are very hard and have high melting points. Diamond, for example, is the hardest known substance, making it ideal for cutting tools. Covalent solids are generally poor conductors of electricity, except graphite. Unlike diamond, where each carbon atom forms four bonds, leaving no free valence electrons, graphite has a hexagonal crystal structure in which each carbon atom bonds to only three others, leaving one free electron per atom. This makes graphite an excellent conductor of heat and electricity. However, diamond is a good conductor of heat due to atomic vibrations that allows heat to travel through its crystal lattice.

Metallic solids consist of atoms held together in a lattice of closely packed spheres by metallic bonds. The lattice points contain metal nuclei surrounded by a cloud of loosely held or delocalised electrons, as shown in Figure 1.6. Delocalised electrons are not associated with a specific atom or covalent bond but move freely throughout the solid, making metals excellent conductors of heat and electricity.

Metallic bonds are electrostatic forces of attraction between positive metal nuclei and their delocalised electrons, as well as those of eight to twelve neighbouring atoms. The strength of these bonds varies, affecting the properties of metals. Some metals are hard, while others are soft. Potassium and sodium, for example, are soft metals that can readily be cut with a knife, whereas iron is much harder and requires specialised tools for cutting. Metals are generally malleable, ductile and lustrous. They have a wide range of melting points, from below room temperature for mercury to 3,442 °C for tungsten. All metals are good conductors of heat and electricity.

Introduction to Chemistry

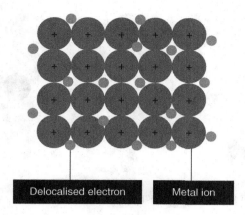

FIGURE 1.6 Metallic bond.

FIGURE 1.7 The arrangement of the particles of liquids

1.5.3.2 Liquids

Liquids are comparatively less compact than solids because they are held together by weaker forces than those in solids (see Figure 1.7). Thus, liquids can flow and undergo translational motion. Like solids, liquids have a fixed volume and are extremely difficult to compress.

Examples of substances that exist as liquids at room temperature include water, ethanol, mercury (a metal), bromine, benzene, silicon tetrachloride, sulfur trioxide, toluene and acetic acid. A liquid has no definite shape; instead, it assumes the shape of its container up to the level it is filled.

Liquid mercury, like all metals, is a good conductor of electricity because it contains free electrons. In the liquid state, all ionic compounds are good conductors of electricity because they contain free-moving ions.

1.5.3.3 Gases

Gases are the least compact of all substances. Their particles are widely separated (see Figure 1.8) with almost negligible forces of attraction. This allows gas particles to move freely in all directions at high speeds unless confined by a container.

Gases have no definite shape or fixed volume; they simply fill their containers completely. Unlike solids and liquids, gases are very easy to compress. Some examples of substances that exist in the gaseous state at room temperature are carbon monoxide, carbon dioxide, methane, hydrogen, nitrogen, helium, argon, oxygen, krypton, chlorine, neon and dinitrogen oxide.

Gases are good conductors of electricity at low pressure and high voltage because of ionisation.

FIGURE 1.8 The particles of gases are widely separated.

1.5.4 CHANGE OF STATE

As mentioned earlier, matter undergoes a change of state when heated or cooled to an appropriate temperature. The various processes involved in state changes are shown in Figure 1.9.

1.5.4.1 Melting and Freezing

When a solid is heated, the average kinetic energy of its particles increases until the forces holding them together weaken, allowing the particles to flow and change into a liquid. This process is known as *melting*, and the temperature at which it occurs is called the *melting point* of the solid.

Conversely, when a liquid is cooled, the average kinetic energy of its particles decreases until the cohesive forces become so strong to prevent movement of particles, causing the substance to solidify. This process is known as *freezing*, and the temperature at which it occurs is the *freezing point* of the substance. The melting and freezing points of a substance are the same.

1.5.4.2 Evaporation (Vaporisation) and Condensation

When the temperature of a liquid is raised, the surface particles can acquire enough kinetic energy to overcome intermolecular forces and escape as gas. This process is known as evaporation or

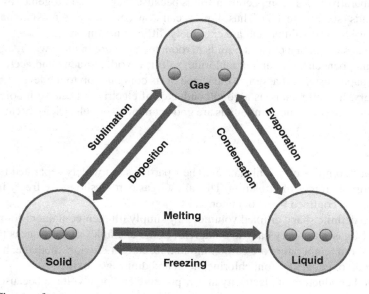

FIGURE 1.9 Change of state.

Introduction to Chemistry

vaporisation, and it occurs at all temperatures. Thus, there is no fixed temperature for evaporation or vaporisation. However, the rate of vaporisation increases with temperature rise.

The rate of evaporation also depends on the type of liquid. Liquids such as petrol, toluene, butane, butyl acetate (ethyl ethanoate), ethanol, benzene, perfumes and acetone, which vaporise easily are called *volatile liquids*. Additionally, an increase in the exposed surface area of a liquid enhances its rate of evaporation.

Conversely, cooling a gas ultimately transforms it back into a liquid in a process known as *condensation*. For example, the water droplets that form around a cold cup results from the condensation of water vapour upon contact with the cup's cool surface.

1.5.4.3 Sublimation and Deposition

Why does a camphor ball shrink over time without melting? The answer lies in a process called *sublimation* — the direct transition of a solid into a gas without first becoming a liquid. Examples of solids that undergo sublimation include iodine, naphthalene, ammonium chloride, camphor, menthol and dry ice.

The reverse of sublimation is called *deposition*, where a gas changes directly into a solid without passing through the liquid state. Examples of deposition in nature include frost formation from water vapour, formation of cirrus clouds, snowflakes formation and soot deposition on chimney walls.

1.6 CLASSIFICATION OF MATTER

As shown in Figure 1.10, matter is classified into three broad categories: elements, mixtures and compounds. Each of these is explained below.

1.6.1 Elements

Elements are the fundamental building blocks of matter. By definition, an element is a substance which cannot be broken down into simpler units by ordinary chemical processes. Alternatively, an element is a substance composed of atoms with the same number of protons.

A commonly encountered but technically inaccurate definition states that an element is a substance that is made up of only one kind of atom. While generally true, this definition does not account for the existence of isotopes, which are variants of atoms of the same element. This important concept is explained in Chapter 7.

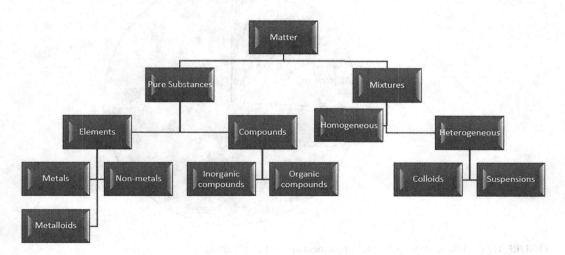

FIGURE 1.10 Classification of matter.

So far, 118 elements have been identified, including hydrogen, helium, sodium, iron, magnesium, copper, uranium, neon, gold, lead, silver, molybdenum, oxygen, nitrogen, mercury, carbon, xenon, iodine, polonium and tin. At room temperature (25°C), 104 elements are solids, including sodium, carbon, potassium, lead, gold, silver, iron, copper and nickel; mercury (a metal) and bromine (a halogen) are liquids, while 11 are gases, including chlorine, oxygen, nitrogen, fluorine, hydrogen, helium, krypton, radon, xenon, neon and argon.

Moreover, 92 elements occur naturally including uranium, aluminium, boron, sulfur, tungsten and all the elements mentioned above, while the rest are artificial. Some artificial elements, such as technetium and promethium, are by-products of nuclear reactors, while others are synthesised in laboratories. In total, 24 elements are synthetic, including americium, copernicium, plutonium, curium, neptunium, californium, einsteinium and lawrencium. Figure 1.11 shows the percent abundance of naturally occurring elements in the Earth's crust.

Elements are classified as *metals*, *non-metals* and *metalloids (also called semi-metals)*. Metals are elements that form ions by electron loss, with the exception of hydrogen, which is a non-metal that ionises by electron loss. Examples include sodium, copper, aluminium, magnesium, gold, mercury, tungsten, platinum, silver, potassium, lithium, cobalt, silver, calcium, iron, etc.

Non-metals, except for hydrogen, form ions by gaining electrons. Examples include hydrogen, carbon, argon, sulfur, oxygen, nitrogen, phosphorus, helium, chlorine, fluorine and iodine. An overview of the differences between the physical and chemical properties of metals and non-metals is given in Tables 1.2 and 1.3.

Metalloids, or semi-metals, exhibit properties intermediate between those of metals and non-metals. Seven elements belong to this category: boron, silicon, arsenic, germanium, antimony, polonium and tellurium. Metalloids are widely used in the electronic industry for producing semiconductors.

Elements are arranged on the Periodic Table (see Figure 1.12) in ascending order of their *atomic numbers*—the number of protons in their atoms. This well-organised structure allows for easy identification of element groups. A bold zigzag line, beginning with boron, separates metals from non-metals: metals are located to the left of the line, non-metals are found on the right, and metalloids along the line. The Periodic Table is the most important Table in chemistry.

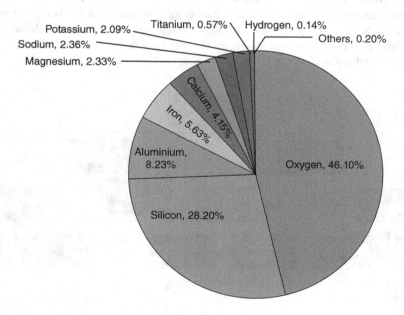

FIGURE 1.11 The percent abundance of elements in the Earth's crust.

TABLE 1.2
Differences between the Physical Properties of Metals and Non-metals

Metals	Non-Metals
With the exception of mercury, they are all solids at room temperature	They may be solids, liquids or gases at room temperature
They have relatively high densities	They generally have lower densities than metals
They are good conductors of heat and electricity	With the exception of graphite, they are poor conductors or non-conductors (insulators)
They are generally malleable, lustrous, ductile and sonorous	They possess none of these properties
They have high tensile strength, i.e., they are hard	They generally have low tensile strength, making them soft or brittle
They generally have high melting and boiling points	They generally have low melting and boiling points

TABLE 1.3
Differences between the Chemical Properties of Metals and Non-metals

Metals	Non-Metals
They generally react with oxygen to form basic oxides	They generally react with oxygen to form acid oxides or acid anhydride
They generally form ionic compounds with non-metals	In addition to forming ionic compounds with metals, they generally combine with one another to form covalent compounds
They form positive ions by losing electrons	They form negative ions by gaining electrons
They are good reducing agents or electron donors	They are good oxidising agents or electron acceptors
Some can displace hydrogen from compounds	None can displace hydrogen from compounds

1.6.1.1 Chemical Symbols

Writing the full names of elements can be inconvenient, especially in chemical equations and formulae. To simplify this, elements are represented using chemical symbols—short forms or abbreviations of their names. The use of chemical symbols dates back to ancient Greece, but the system in use today was developed by the renowned Swedish chemist Jöns Jacob Berzelius in 1813.

Chemical symbols typically consist of one or two letters from the Latin alphabet. If you encounter three-letter symbols, they are placeholders for newly discovered elements before they receive official names from the International Union of Pure and Applied Chemistry (IUPAC), the responsible international body for setting standards for chemical symbols and nomenclature, among other things. For example, tennessine (Ts), element 117 and the most recently discovered element (as of April, 2024), was initially called ununseptium (meaning *one-one-seven* in Latin, a reference to its atomic number) and had the symbol Uus until the IUPAC approved its current name in November 2016. As a result, an up-to-date Periodic Table should contain only one- and two-letter symbols.

On a trivial note, all letters of the Latin alphabet appear on the Periodic Table except J and Q.

As a rule, one-letter symbols are always uppercase (e.g., carbon: C, oxygen: O, nitrogen: N). In two-letter symbols, only the first letter is capitalised (e.g., gold: Au, aluminium: Al, calcium: Ca).

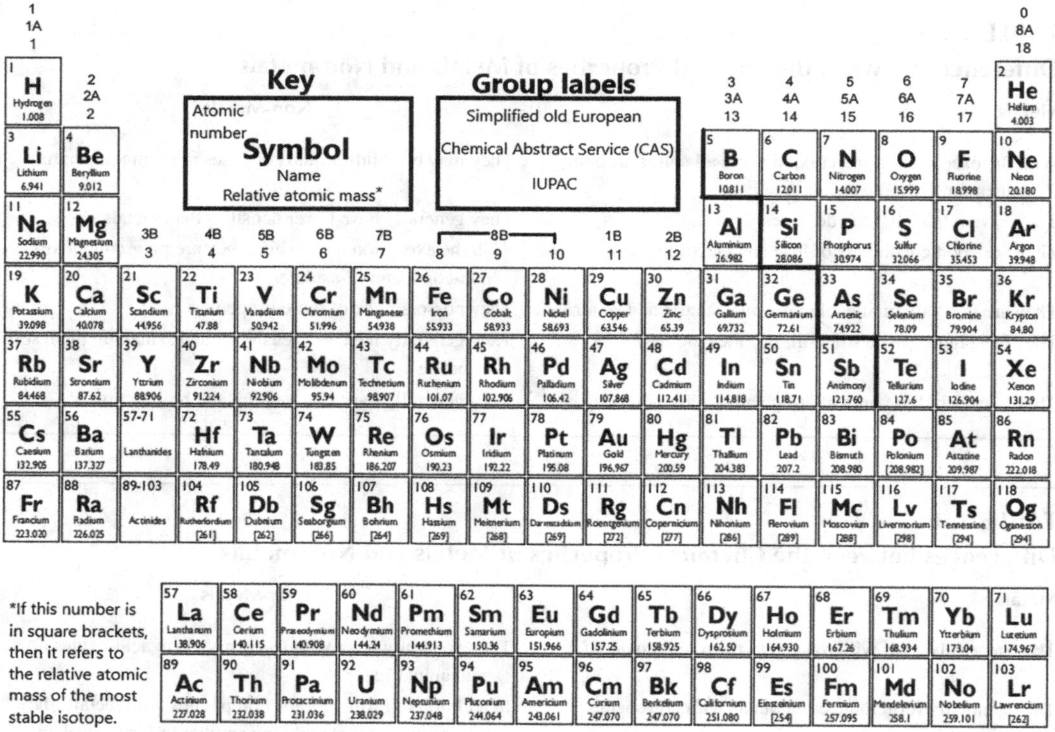

FIGURE 1.12 The modern Periodic Table.

It is essential to learn chemical symbols for proficiency in chemistry. The Periodic Table is an invaluable tool for this purpose. Tables 1.4–1.7 illustrate how chemical symbols are derived, along with many examples.

Notably, tungsten is the only element that does not follow any of the rules outlined in the Tables. Its symbol, W, comes from its German name, *wolfram*, which is how the element is known in most European countries.

1.6.2 COMPOUNDS

A compound is a substance formed as a result of the chemical combination of two or more elements. Unlike mixtures, compounds result from chemical change and have entirely different properties from their constituent elements. For example, when hydrogen and oxygen react, they form water (H_2O):

$$2H_2(g) + O_2(g) \rightarrow 2H_2O(g)$$

Water is a non-flammable liquid that does not support combustion, whereas oxygen and hydrogen are both gases. Oxygen is essential for combustion and hydrogen is highly combustible, yet their combination produces a substance with completely different properties.

Other examples of compounds and their constituents are shown in Table 1.8.

1.6.3 MIXTURES

A mixture forms when two or more substances are combined in any proportion, with each substance retaining its identity. Unlike compounds, the components of a mixture are not chemically bonded.

TABLE 1.4
Some Symbols Based on the First Letter of the English Names of Elements

Element	Symbol
Hydrogen	H
Oxygen	O
Nitrogen	N
Sulfur	S
Carbon	C
Boron	B
Phosphorus	P
Fluorine	F
Uranium	U

TABLE 1.5
Some Symbols Based on the First Two Letters of the English Names of Elements

Element	Symbol
Aluminium	Al
Argon	Ar
Barium	Ba
Beryllium	Be
Bromine	Br
Calcium	Ca
Helium	He
Lithium	Li
Silicon	Si

TABLE 1.6
Some Symbols Based on the First and Some Other Letters—Other than the Second—of the English Names of Elements

Element	Symbol
Chlorine	Cl
Chromium	Cr
Magnesium	Mg
Manganese	Mn
Molybdenum	Md
Zinc	Zn
Platinum	Pt
Strontium	Sr
Radon	Rn

TABLE 1.7
Chemical Symbols Based on Latin Names of Elements

Element	Latin Name	Symbol
Copper	Cuprum	Cu
Gold	Aurum	Au
Iron	Ferrum	Fe
Lead	Plumbum	Pb
Mercury	Hydrargyrum	Hg
Potassium	Kalium	Ka
Silver	Argentum	Ag
Sodium	Natrium	Na
Tin	Stannum	Sn
Antimony	Stibium	Sb

TABLE 1.8
Some Compounds and Their Constituents

Compound	Formula	Constituents
Ammonia	NH_3	Nitrogen and ammonia
Ethanol	C_2H_5OH	Carbon, hydrogen and oxygen
Sodium chloride (table salt)	NaCl	Sodium and chlorine
Glucose	$C_6H_{12}O_6$	Carbon, hydrogen and oxygen
Sodium hydroxide (caustic soda)	NaOH	Sodium, oxygen and hydrogen
Sand	SiO_2	Silicon and oxygen
Sucrose (cane sugar)	$C_{12}H_{22}O_{11}$	Carbon, hydrogen and oxygen
Calcium carbonate (limestone, marble, chalk, pearls, eggshells, etc.)	$CaCO_3$	Calcium, carbon and oxygen
Rust	Fe_2O_3	Iron and oxygen
Sulfuric acid	H_2SO_4	Hydrogen, sulfur and oxygen

For example, dissolving sodium chloride in water produces a saline solution, and air is a mixture of gases.

By volume, air consists of 21% oxygen, 78% nitrogen, 0.03% carbon dioxide, 1% noble gases (mainly argon), variable amounts of water vapour and dust.

Some examples of some mixtures and their constituents are given in Table 1.9. It important to note that hydrogen is not a constituent of air. A common mistake among beginners is to assume its presence, but hydrogen in air would be highly dangerous due to its flammability.

Unlike compounds, the components of a mixture can be elements, compounds or both. In addition, mixture can either be homogeneous or heterogeneous. A homogeneous substance is a mixture that exhibits only one *phase*—a phase is a physically distinct or observable region of matter with a uniform chemical composition. As shown in Figure 1.13a, a salt solution is a homogeneous mixture because it consists of a single phase: there is no way to distinguish the mixture from pure water by mere

Introduction to Chemistry

TABLE 1.9
Mixtures and Their Constituents

Mixture	Constituents
Air	Oxygen, nitrogen, noble gases, carbon dioxide, water vapour and dust
Blood	Water, white blood cells, red blood cells, haemoglobin, hormones, vitamins, mineral salts, sugar and oil
Crude oil	Gas, kerosene, petrol, diesel, bitumen, naphtha, gas oil, etc.
Milk	Sugar, water, vitamins, fat, proteins and mineral salts
Urine	Water, urea and mineral salts
Brass (alloy)	Copper and zinc
Bronze (alloy)	Copper and tin
Seawater	Water, organic matter, mineral salts and bacteria
Orange juice	Water, citric acid, malic acid, ascorbic acid (vitamin c), sugars (sucrose, fructose and glucose), protein, lipids, etc.
Soil	Mineral salts (clay, silt and sand), organic matter, air, water and living organisms
Smoke	Particulate solids, ash, soot (carbon), oils, tar, water vapour, etc.
Coffee	Proteins, mineral salts, lipids, caffeine, soluble fibre, sugars, etc.
Smog	Pollen, dust, sulfur oxides, nitrogen oxides, ammonia gas, ozone, hydrocarbons, etc.
Coca-Cola	Carbonated water, caffeine, caramel colour, phosphoric acid, carbon dioxide, sugar, Coca-Cola concentrate, etc.
Ink	Dyes, solvents, resins, alcohol, pigments, glycerine, lubricants, carbon, etc.
Sweat	Water, sodium chloride, urea, lactic acid, etc.

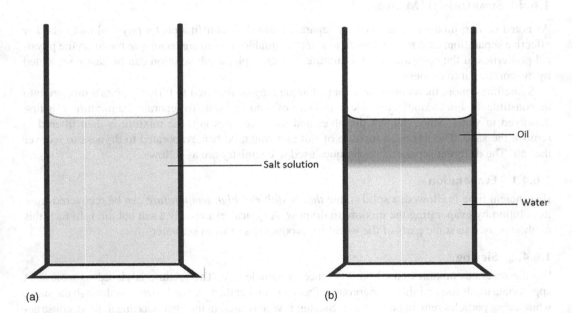

FIGURE 1.13 Liquid mixtures: (a) A salt solution is a homogeneous mixture and (b) an oil–water mixture is a heterogeneous mixture.

observation because it looks identical throughout. In contrast, a heterogeneous mixture has distinct phases. As you can observe in Figure 1.13b, an oil–water mixture separates into two distinct layers, each of which constitutes a different phase.

TABLE 1.10
Differences between Compounds and Mixtures

Compound	Mixture
Can be represented by a chemical formula because constituents are combined in a fixed ratio by mass	Cannot be represented by a chemical formula because constituents can be added in any proportions by mass
Contains elements only	May contain elements, compounds or both
Constituents can only be separated by means of chemical or electrochemical reactions, for example, electrolysis	Constituents can easily be recovered by means of a physical separation technique such as sieving, distillation, etc.
Constituents lose their identity since an entirely new substance is formed	Constituents retain their individual identities, making the properties of a mixture the sum of those of its individual components
It is always homogeneous	It may be homogenous or heterogeneous
It has definite melting and boiling points	It has no definite melting and boiling points
The formation of a compound is accompanied by a large amount heat change	The formation of mixture is accompanied by little to no heat change

Another key property of mixtures is that, unlike compounds with fixed compositions, their constituents can be combined in varying proportions to suit different needs.

Table 1.10 provides an overview of the differences between compounds and mixtures.

1.6.4 SEPARATION OF MIXTURES

As noted earlier, mixtures can be easily separated into their constituents by physical methods. For effective separation, care must be taken to select a suitable *separating technique* based on the physical properties of the constituents of a mixture. For example, a salt solution can be easily separated by evaporating it to dryness.

Sometimes, more than one separation technique may be required to fully separate a mixture into its constituents. For example, consider a mixture of sand and salt. To separate the mixture, it is first dissolved in water. Since the salt dissolves and the sand does not, the mixture is then filtered to remove the sand. The filtrate, a mixture of salt and water, is then evaporated to dryness to recover the salt. The different separation techniques used in chemistry are as follows.

1.6.4.1 Evaporation

If sufficient time is allowed, a solid solute *that is stable at high temperature* can be recovered from its solution by evaporating the mixture to dryness. A typical example is a salt solution. In fact, this method is used in some parts of the world for producing salt from seawater.

1.6.4.2 Sieving

Sieving separates mixtures based on difference in particle size. The mixture is placed on a sieve of appropriate mesh size and shaken vigorously. Particles smaller than the mesh size pass through the sieve, while larger particles remain on the sieve. Sieving is widely used in the pharmaceutical, food, construction and mining industries. In mining, sieving machines separate valuable minerals from impurities.

1.6.4.3 Use of Separating Funnel

A mixture of two immiscible liquids, such as oil and water, is separated using a separating funnel. Since the two liquids form two distinct layers, the lower (denser) layer can be tapped off, leaving the upper layer in the funnel (see Figure 1.14)

Introduction to Chemistry

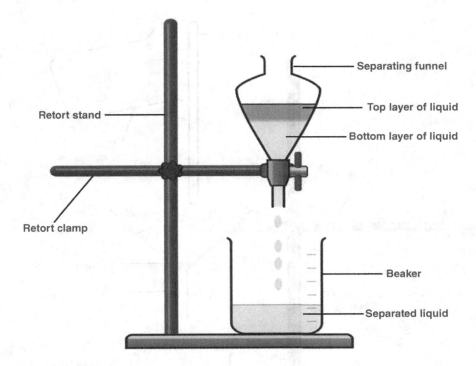

FIGURE 1.14 Separating an oil–water mixture with a separating funnel.

1.6.4.4 Magnetic Separation

A convenient method for separating magnetic substances from a mixture is to use magnetic separation. When such a mixture is brought into a strong magnetic field, the magnetic substances are separated from the non-magnetic ones by getting attracted to the magnet. In the mining industry, this method is used in the concentration of some ores such as cassiterite (SnO_2), magnetite (Fe_3O_4), rutile (TiO_2) and copper pyrite ($CuFeS_2$). Magnetic separation is also used in solid waste treatment to recover magnetic substances from waste.

1.6.4.5 Filtration

Filtration is used to separate insoluble particles, such as chalk or soil particles, from a liquid. The filtration apparatus consists of a filter paper that has been folded into a cone and placed in a funnel, which is in turn placed over a beaker (see Figure 1.15). When the mixture is poured into the funnel, water passes through the pores of the filter paper, leaving the particles behind. The clear water (but impure!) obtained in the beaker is called the *filtrate*, while the solid particles left behind constitute the *residue*.

Filtration is an important step in water treatment. In waterworks, a filter bed is used to separate suspended materials from water before it is treated and distributed to the public. In breweries, filtration is used for removing unwanted particles, such as dead yeast, from beer.

1.6.4.6 Decantation

Decantation is a simple separation technique you have probably used several times. It is used to separate mixtures of a liquid and an insoluble solid (sediment) or of two immiscible liquids. Using a glass rod, the upper layer of liquid can be carefully poured into another beaker, leaving the lower layer of sediment or liquid behind (see Figure 1.16).

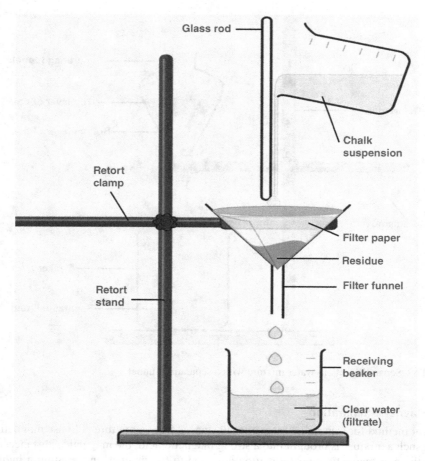

FIGURE 1.15 Filtration.

Some of the applications of decantation include water purification, production of vinegar, wine-making, separation of cream from milk and separation of oil from water, determination of fat in butter and separation of syrups from sugar crystals during sugar production.

1.6.4.7 Sublimation

A sublimable substance can be recovered from a mixture of solids by sublimation. The sublimable substance vaporises (sublimes) when the mixture is heated and can then be recovered.

Examples of solids that can be recovered this way include ammonium chloride, iodine, naphthalene and camphor. The practical applications of sublimation include the purification of sublimable solids, T-shirt printing, freeze-drying in the food industry and forensic science.

1.6.4.8 Flotation

Flotation is a separation technique used to recover solid particles from a liquid. In this process, fine solid particles to be recovered are carried to the surface of the liquid in a foam, called froth, formed with gas bubbles that have been injected into the mixture. The froth is then skimmed or filtered off, producing a *concentrate* of the target particles. The unwanted particles that do not froth, called *tailings*, are discarded as waste.

Flotation is used extensively in the mining industry for the cost-effective extraction of metals from low-grade ores (ores with a low concentration of metals). It is also applied in other areas such as water treatment, wastewater treatment and paper recycling.

Introduction to Chemistry

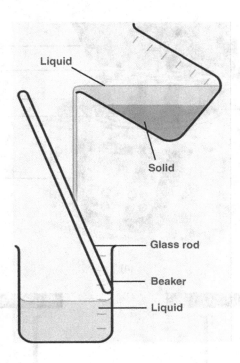

FIGURE 1.16 Decantation.

1.6.4.9 Centrifugation

Centrifugation involves using a high-speed spinning machine, called a centrifuge, to separate suspended particles based on their density. As shown in Figures 1.17a and b, the mixture is placed in test tubes and spun at high speed, causing denser solid particles to settle at the bottom.

In medical laboratories, centrifugation is used to separate blood cells from the plasma (liquid portion of blood). Other applications include the spin-drying of clothes in washing machines, extracting fat from milk to produce skimmed milk, removal of water from moist lettuce in salad spinner, and clarification and stabilisation of wine.

1.6.4.10 Precipitation

Precipitation exploits the difference in solubility of a solute in two miscible liquids to separate it from solution. For example, copper(II) sulfate is soluble in water but not in ethanol. If ethanol is added to a copper(II) sulfate solution, the salt immediately precipitates out as an insoluble solid. The precipitate can then be washed and filtered for recovery. This process of rendering a dissolved solid insoluble by adding a solvent in which it is not soluble is called precipitation.

Important applications include purifying salts and removing dissolved ions from water.

1.6.4.11 Distillation

Distillation is used to recover a solvent from solution, such as obtaining pure water from a salt solution or seawater. As shown in Figure 1.18, the solution is heated, causing the solvent to vaporise and pass through a condenser, where it is cooled and condenses back into liquid form. The recovered solvent is called the *distillate* and is very pure.

Distillation is widely used in water purification, essential oil extraction and ethanol concentration in the alcoholic beverage industry.

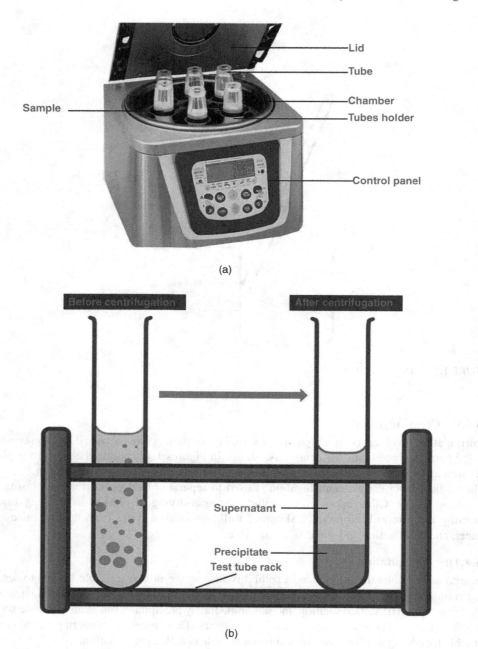

FIGURE 1.17 (a) A centrifuge and (b) Effect of centrifugation on a mixture.

1.6.4.12 Fractional Distillation

Simple distillation is effective for separating a single solvent from a mixture, but to separate multiple miscible liquids with different boiling points, fractional distillation is required. Each component of the mixture is called a *fraction*. The fractional distillation apparatus (see Figure 1.19) is similar to that of simple distillation, but it includes a *fractionating column* between the boiling flask and the condenser. This column maintains a temperature gradient, being hotter at the bottom and cooler at the top.

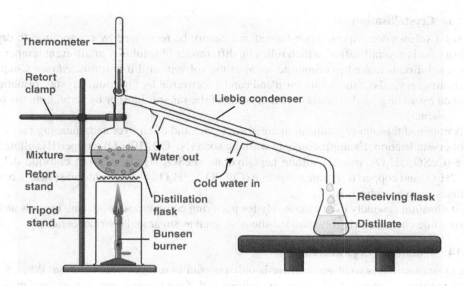

FIGURE 1.18 Distillation apparatus.

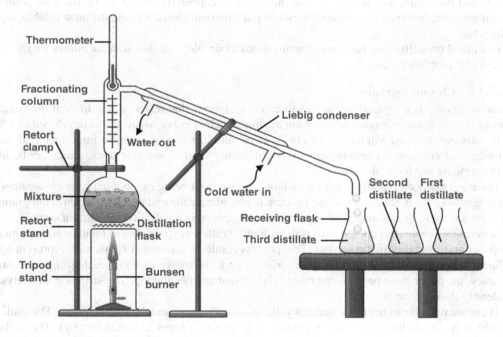

FIGURE 1.19 Fractional distillation apparatus.

Vapours rise through the column, but only those with the lowest boiling point reach the top and enter the condenser first. Higher-boiling-point components condense and return to the boiling flask until their respective boiling points are reached. The process continues until all the fractions are fully recovered.

Fractional distillation is crucial in the oil industry, where it is used in refineries to separate crude oil into fractions such as petrol, gas, diesel, kerosene, natural gases, heavy oil and bitumen.

1.6.4.13 Crystallisation

Some solid solutes decompose when heated and cannot be recovered by evaporation to dryness. An alternative is crystallisation, which relies on differences in solubility at different temperatures. First, the solution is heated to evaporate some of the solvent until it is *saturated* (see Chapter 9). Upon cooling, crystals of the solute form and can be recovered by filtration. Crystal formation can be induced by adding seed crystals (small crystals of the same solute) or by scratching the bottom of the container.

Salts obtained through crystallisation are always pure and often hydrated, meaning they contain water of crystallisation. Examples include washing soda ($Na_2CO_3.10H_2O$), copper(II) sulfate pentahydrate ($CuSO_4.5H_2O$), iron(II) sulfate heptahydrate ($FeSO_4.7H_2O$), calcium chloride dehydrate ($CaCl_2.2H_2O$) and copper(II) nitrate trihydrate ($Cu(NO_3)_2.3H_2O$). Sodium chloride (NaCl), however, crystallises as an anhydrous solid.

Crystallisation is widely used extensively for purifying compounds, producing crystals and manufacturing fine chemicals (highly pure substances), such as sugar and pharmaceuticals.

1.6.4.14 Fractional Crystallisation

Simple crystallisation is used where there is only one salt to recover from solution. When solution contains multiple solutes with different solubilities at different temperatures, fractional crystallisation is used. The process is similar to crystallisation but involves multiple stages to separate each solute in order of increasing solubility. For example, in a solution of sodium chloride, copper(II) sulfate and potassium nitrate. On cooling the solution, copper(II) sulfate, being the least soluble, crystallises first, followed by sodium chloride, while potassium nitrate, being the most soluble, crystallises last.

Fractional crystallisation has similar applications as simple crystallisation but allows for the separation of multiple solutes.

1.6.4.15 Chromatography

Chromatography is a separation and identification technique that relies on the different movement rates of substances over a porous, adsorbent medium, such as gel or paper. It was developed in 1900 by the Russian botanist Mikhail Tsvet (1872–1919) and it is one of the most important analytical techniques in chemistry. Chromatography is commonly used to separate dyes, amino acids, ink, plant pigments and food additives.

There are many chromatographic procedures, the simplest being *ascending paper chromatography* (Figures 1.20). A small spot of the mixture is placed near the end of a strip of chromatographic paper on an insoluble baseline. The paper is then suspended in a solvent in an airtight container. As the solvent moves up the paper, it carries the solutes at different speeds based on their adsorption by the paper and solubility in the solvent. This process results in a series of spots, each representing a different component of the mixture. The series of spots is known as a *chromatogram*. To ensure accuracy, the paper must be removed before the solvent reaches the top. The strip is then analysed to identify the components.

There are two phases in chromatography: the stationary phase and the mobile phase. The stationary phase is the medium that does not move (e.g., paper in paper chromatography). The mobile phase is the solvent that dissolves and carries the solutes (e.g., water and ethanol).

Introduction to Chemistry

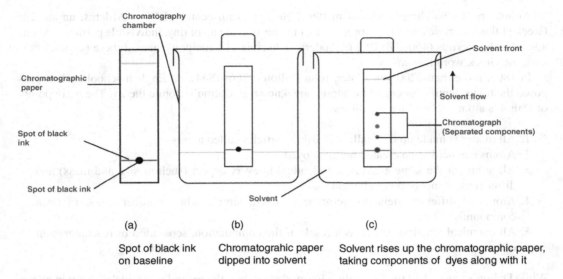

FIGURE 1.20 Ascending paper chromatography

1.6.5 Purification of Substances and Test for Purity

An impure substance is a mixture containing unwanted components. Assessing purity is crucial for safety, accuracy in chemical analysis and maintaining product quality in industries such as food and pharmaceuticals. Impurities in reagents can introduce serious errors in chemical analysis, while contaminants in food can pose health risks and lead to legal consequences.

One way of testing purity is melting or boiling point determination. A pure substance has fixed melting and boiling points, while an impure substance melts at a *lower temperature* and boils at a *higher temperature*. Purity can also be assessed by chromatographic analysis. A pure substance forms a single spot on a paper chromatogram, while an impure substance will form a number of spots.

Impure substances can be purified using appropriate separation techniques, as discussed in previous sections.

1.7 THE PARTICLES OF MATTER

Earlier, we introduced the particulate theory—the idea that matter is made up of tiny, discrete particles. Here, we will take a closer look at these particles. Matter consists of three main types of particles: *atoms*, *molecules* and *ions*. The interactions and behaviour of these particles at the microscopic level determine the macroscopic (observable) properties of matter.

1.7.1 Atom

An atom is defined as the smallest particle of an element that can take part in a chemical reaction. Alternatively, an atom can be defined as the smallest particle of an element that can exist while still retaining the chemical identity of that element.

Atoms are the building blocks of matter. Leucippus (5th century BC) of Miletus, an ancient Greek philosopher, was the first to propose that matter is made up of tiny, indivisible particles. About 430 BC, Democritus (460–370 BC) of Abdera, a disciple of Leucippus, named these particles *atomos*, the Greek word for *indivisible*.

In 1803, more than 2200 years later, John Dalton (1766–1844), an English schoolteacher, proposed the first scientific theory of the atom, now known as Dalton's atomic theory. The main points of Dalton's atomic theory are as follows:

1. All matter is made up of small, indivisible particles called atoms.
2. Atoms can neither be created nor destroyed.
3. All atoms of the same element are identical in every aspect (such as size and mass) and differ from atoms of other elements.
4. Atoms of different elements combine in fixed, simple whole-number ratios to form compounds.
5. All chemical reactions occur as a result of the combination, separation or rearrangement of atoms.

While Dalton's theory laid the foundation for modern atomic theory and was widely accepted by the scientific community, later discoveries revealed that some aspects of it were incorrect. These modifications are outlined below:

1. *All matter is made up of small, indivisible particles called atoms.* This is partly wrong. While atoms are indeed the fundamental units of matter, they are not indivisible. Atoms contain smaller particles, the main ones being protons, neutrons and electrons. These particles are called *subatomic particles*.
2. *Atoms can neither be created nor destroyed.* This statement, based on the law of conservation of mass, is true for ordinary chemical reactions. As we shall see in Chapter 12, atoms undergo transformation in nuclear reactions.
3. *All atoms of the same element are exactly alike in every aspect and differ from atoms of other elements.* While atoms of different elements are indeed different, the discovery of isotopes has shown that atoms of the same element can have different physical properties. Hydrogen, for example, has three forms of atoms (or isotopes): protium, deuterium and tritium with mass numbers 1, 2 and 3, respectively.
4. *Atoms of different elements combine in fixed, simple whole-number ratios to form compounds.* This applies mainly to inorganic compounds. In some organic compounds, atoms do no always combine in simple whole-number ratios. A typical example is sucrose ($C_{12}H_{22}O_{11}$).

The atom can be depicted as a sphere, containing a nucleus, which houses protons and neutrons, surrounded by electrons in shells (see Figure 1.21). Most of the atom's mass is concentrated in the nucleus. The proton is positively charged, while the neutron has no charge, making the nucleus overall positively charged. In contrast, the electron is negatively charged. Since an atom contains an equal number of protons and electrons it is electrically neutral. The characteristics of the subatomic particles are shown in Table 1.11.

1.7.1.1 Atomic Mass

The absolute (actual) mass of the atom is too small, with the heaviest atoms having masses on the order of 4.88×10^{-22} g. To avoid dealing with such inconveniently small numbers, chemists use a non-SI unit called the unified atomic mass unit (amu or U), using carbon-12, the stable and most abundant isotope of carbon, as the standard or reference.

Introduction to Chemistry

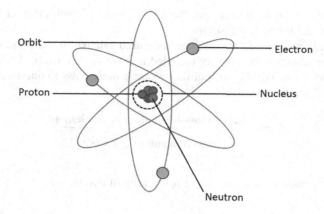

FIGURE 1.21 A simplified structure of the atom.

TABLE 1.11
The Sub-particles of the Atom

Sub-particle	Location	Relative Charge	Relative Mass
Proton	Nucleus	+	1
Electron	Nucleus	−	≈ 0.0005
Neutron	Shell or orbital	Zero	1

An atomic mass unit is defined as one-twelfth the mass of an atom of carbon-12, which is exactly $1.660538921 \times 10^{-27}$ kg. By this definition, the carbon-12 atom has a mass of exactly 12 amu.

As we shall see in Chapter 7, the sum of the protons and neutrons in an atom is approximately equal to its atomic mass.

1.7.1.2 Isotopes and Relative Atomic Mass

Atoms of the same element have the same number of protons; hence, they have the same chemical properties. The number of protons in an atom of an element is called its atomic number, Z. Elements are arranged on the Periodic Table in ascending order of their atomic number. As we pointed out earlier, atoms of the same elements are not exactly alike—they exhibit variations in their masses, leading to differences in their physical properties. Atoms of the same elements with different atomic masses are called isotopes, and the phenomenon is known as isotopy.

The proportion or percentage of each isotope in any sample of an element is called the *relative abundance* of that particular isotope. For example, chlorine has two isotopes: chlorine-35 and chlorine-37, with relative abundances of 75% and 25%, respectively. Isotopy is caused by variations in the number of neutrons in atoms of the same element.

The atomic masses of elements are reported as an average of the masses of their different isotopes, although there are some exceptions. Some elements, including beryllium, fluorine, sodium, aluminium, phosphorus, scandium, manganese, cobalt, arsenic, yttrium, niobium, rhodium, iodine, caesium, praseodymium, terbium, holmium, thulium and gold, have only one naturally occurring isotope on whose mass their atomic masses are based. These elements are said to be *monoisotopic*.

Similarly, some elements, including technetium (atomic number 43) and elements from bismuth (atomic number 83) onward have no stable isotopes. In other words, all isotopes of the elements are radioactive. With the exception of uranium, the atomic mass of each element in this category is based on the mass of its longest-lived isotope.

The average atomic masses of elements are measured relative to the mass of carbon-12. The value of atomic mass obtained in this way is called *relative atomic mass*, A_r. The relative atomic mass of an element is defined as the ratio of its average atomic mass to one-twelfth of the mass of an atom of carbon-12. In other words:

$$A_r = \frac{\text{Average mass of an atom of an element}}{\frac{1}{12} \times \text{mass of an atom of carbon-12}}$$

The relative atomic mass has no units since it is a ratio of two masses.

TABLE 1.12
Approximate Relative Atomic Masses of the First 30 Elements

Element	Relative Atomic Mass
Hydrogen	1
Helium	4
Lithium	7
Beryllium	9
Boron	10.8
Carbon	12
Nitrogen	14
Oxygen	16
Fluorine	19
Neon	20
Sodium	23
Magnesium	24
Aluminium	27
Silicon	28
Phosphorus	31
Sulfur	32
Chlorine	35.5
Argon	40
Potassium	39
Calcium	40
Scandium	45
Titanium	48
Vanadium	51
Chromium	52
Manganese	55
Iron	55.8
Cobalt	59
Nickel	58
Copper	63.5
Zinc	65.4

Introduction to Chemistry

Relative atomic masses are very important in chemistry calculations. However, there is no need to memorise them all, as the values will be provided when required. The approximate relative atomic masses of the first 30 elements are given in Table 1.12.

We will learn how to calculate the relative atomic masses of elements from their isotopic masses in Chapter 7.

It is important to note that, for convenience, the relative atomic masses used in calculations are mostly rounded values. As you will notice on a standard Periodic Table, relative atomic masses are usually not whole numbers. While some whole-number values may appear, these are rounded values.

1.7.2 Molecule

Most non-metal atoms do not exist independently; instead, they bond with other atoms to form molecules. A molecule can be either an element or a compound. A molecular element consists of atoms of the same elements, whereas a molecular compound consists of atoms of different elements.

Thus, a molecule is defined as the smallest particle of a substance (i.e., element or compound) which can exist independently while retaining all the properties of that substance.

Examples of elements that exist as molecules include hydrogen, sulfur, chlorine, phosphorus and oxygen. The number of atoms in a molecule of an element is called its *atomicity*. *Monatomic* molecules (one atom per molecule) include helium, neon, argon, krypton, xenon and radon. *Diatomic* molecules (two atoms per molecule) include oxygen, iodine, chlorine, hydrogen, nitrogen, bromine and fluorine. Polyatomic molecules (three or more atoms per molecule) include sulfur (eight atoms), ozone (three atoms) and phosphorus (four atoms).

Examples of molecular compounds are provided in Section 1.9.

1.7.3 Ion

An ion is an atom or group of atoms that carries an electric charge. There are two types of ions: cations (positively charged ions) and anions (negatively charged ions). Cations are formed when an atom loses one or more electrons. Examples include silver ion Ag^+, calcium ion Ca^{2+}, cobalt(II) ion Co^{2+}, sodium ion Na^+, iron(III) iron Fe^{3+} and ammonium NH_4^+.

On the other hand, anions are formed when an atom gains one or more electrons. Examples include chloride ion Cl^-, oxygen ion O^{2-}, sulfide ion S^{2+}, carbonate CO_3^{2-}, hydrogen carbonate (bicarbonate) HCO_3^-, hydroxide OH^-, nitrate NO_3^- and sulfate SO_4^{2-}.

An ion consisting of multiple atoms is called a *polyatomic* ion. Some examples are bicarbonate, ammonium, nitrate, sulfate and hydroxide. Polyatomic ions are sometimes referred to as *radical* in older literature. A polyatomic ion behaves as a single unit, meaning that its charge applies to the entire group of atoms, not to any individual atom within it.

As we shall see below, ionic compounds are composed of ions, not molecules. An ionic compound consists of equal number of cations and anions, making it electrically neutral.

1.8 VALENCY

Valency is the combining capacity or power of an element defined by the number of electrons an atom can donate, accept or share to form a chemical bond. For example, sodium defined readily donates its single valence electron to form a bond, giving it a valency of 1. Oxygen typically accepts two electrons from most metallic elements to form a bond, meaning that it has a valency of 2. If it bonds with another non-metal, it shares two electrons instead. The valencies of some elements are given in Table 1.13.

Some elements exhibit variable valency, meaning they can form multiple compounds with different oxidation states. Iron can have a valency of 2 or 3 forming iron(II) chloride ($FeCl_2$) and iron(III) chloride ($FeCl_3$) with chlorine, for example. Sulfur can exhibit valencies of 2, 4 or 6 depending on the compound it forms.

TABLE 1.13
The Valencies of Some Elements

Element	Valency
Hydrogen	1
Lithium	1
Sodium	1
Potassium	1
Calcium	2
Aluminium	3
Gold	1
Tin	2 or 4
Lead	2 or 4
Carbon	2 or 4
Copper	1 or 2
Iron	2 or 3
Mercury	1 or 2
Chlorine	1
Sulfur	2, 4 or 6
Nitrogen	3 or 5
Cobalt	2 or 3
Zinc	2
Argon	0
Helium	0
Neon	0
Phosphorus	3 or 5
Oxygen	2
Iodine	1
Manganese	2, 4 or 7
Vanadium	4 or 5
Magnesium	2

1.9 CHEMICAL BONDS AND BONDING

The force of attraction between the atoms of the constituent elements of a compound is called a chemical bond, while the process leading to the formation of a chemical bond is called chemical bonding. There are two main types of chemical bonding: *ionic (electrovalent)* bonding and *covalent* bonding. Covalent bonding is further classified into two variants: *ordinary covalent* bonding and *coordinate (dative) covalent* bonding. The ability of an element to react or bond with other substances depends on its electronic configuration, i.e., the arrangement of electrons around the nucleus, with the valence electrons (the electrons in the outermost shell) being of particular interest as these electrons participate in chemical bonding.

There are two ways of writing electronic configurations: the KLMN notation, which shows how electrons are distributed within energy levels or shells, and a more elaborate form that depicts how subshells or orbitals within shells fill with electrons. For now, we will focus on the former, while the latter will be discussed in Chapter 7. Beginning with the closest shell to the nucleus of an atom—the lowest energy level or ground state—shells are designated with the principal quantum number n = 1, 2, 3, and so on up to infinity. These shells are named with the letters K, L, M, N, etc., respectively.

Introduction to Chemistry

For example, the second and third shells from the nucleus, i.e., the L and M shells, have the principal quantum numbers 2 and 3, respectively. Note that the principal quantum number can only have integral values, and that electrons at the same energy level are found in the same shell.

How do the shells fill with electrons? As a rule, a shell can accommodate a maximum possible number of $2n^2$ electrons, where n is the principal quantum number. For example, the K (n = 1), L (n = 2) and M (n = 3) can accommodate a maximum of 2, 8 and 18 electrons, respectively. The electronic configuration of an element is written by listing the number of electrons in each shell, separated by commas, starting from the lowest energy level or quantum number. For example, sodium, with atomic number of 11, has two, eight and one electrons in the K, L and M shells, respectively. Thus, its electronic configuration is 2, 8, 1.

There is an issue we need to address, though. The electronic configuration of potassium should be 2, 8, 9 (since the M shell can hold a maximum of 18 electrons), but it is actually 2, 8, 8, 1, as shown in Figure 1.22. To explain this, we must consider that, as we shall see in Chapter 7, electrons occupy subshells or sublevels (each with a maximum number of electrons) within shells in order of increasing energy. In some cases, the last subshell of a shell is at higher energy level than the first subshell of the next shell, meaning that electrons fill the latter first. If any electrons remain, they return to the former. However, if there are not enough electrons to completely fill the former, the shell hosting it will have fewer electrons than expected.

For potassium, the shells and their subshells are as follows:

K: = {1s}
L: = {2s, 2p}
M: = {3s, 3p, 3d}
N: = {4s, 4p, 4d, 4f}

The s, p, d and f subshells can hold a maximum of 2, 6, 10 and 14 electrons, respectively. The 3d subshell is at a higher energy level than the 4s subshell, and since potassium has 19 electrons, they fill the subshells in order of increasing energy. This gives potassium the electronic configuration 2, 8, 8, 1.

The electronic configurations of the noble gases—helium, neon, argon, krypton, xenon and radon—are shown in Table 1.14. Apart from helium, which have two valence electrons, all noble gases have eight electrons each in their valence shells. Helium is said to have a *duplet* (from the

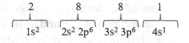

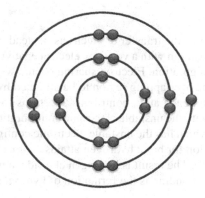

FIGURE 1.22 The electronic configuration of potassium.

TABLE 1.14
The Electronic Configurations of the Noble Gases

Noble Gas	Atomic Number	Electronic Configuration				
		K	L	M	N	O	P
Helium	2	2					
Neon	10	2	8				
Argon	18	2	8	8			
Krypton	36	2	8	18	8		
Xenon	54	2	8	18	18	8	
Radon	86	2	8	18	32	18	8

Latin for 2) electronic structure configuration, while the other noble gases have an *octet* (from the Latin for 8) electronic configuration. Since noble gases are chemically inert under ordinary conditions, duplet and octet electronic configurations are associated with stability. Other elements undergo chemical reaction at different rates to attain the stable electronic configuration of the noble gases.

To attain stability, an element, via its valence shell, has one of three options: donate electrons to other atoms, accept electrons from other atoms or share electrons with other atoms. This is known as the *duplet and octet rules*. A few elements, including hydrogen, lithium and beryllium follow the duplet rule, while most others follow the octet rule.

1.9.1 ELECTROVALENT BONDING

In electrovalent bonding, each metallic atom donates electrons to a non-metallic atom, resulting in the formation of ions—electrically charged atoms—and a stable electronic configuration. The metallic atom forms a positively charged ion (cation) due to a surplus of protons, while the non-metallic atom forms a negatively charged ion (anion) due to excess electrons. The resulting compound consists of these two types of ions held together by strong electrostatic forces of attraction and is represented by a formula unit.

The formation of sodium chloride, NaCl, from sodium and chlorine atoms is illustrated in Figure 1.23. A sodium atom, with has unstable electronic configuration of 2, 8, 1, transfers its valence electron to a chorine atom, which has the unstable electronic configuration of 2, 8, 7. This results in a sodium ion with the stable electronic configuration of 2, 8 and a chloride ion with the stable electronic configuration of 2, 8, 8. These ions are held strongly together by strong electrostatic forces of attraction.

1.9.2 COVALENT BONDING

Covalent bonding does not involve the transfer of electrons. Instead, two non-metallic atoms—or a metallic atom and a non-metallic atom with a very low electronegativity difference—share electrons to attain a stable electronic configuration. Electron sharing occurs when the orbitals of atoms overlap, allowing the electrons in the overlapping region to be attracted by the nuclei of both atoms.

In ordinary covalent bonding, each atom contributes electrons to the shared pair. As shown in Figure 1.24, a hydrogen atom, with the unstable electronic configuration of 1, shares its only valence electron with a chlorine atom, which has the unstable electronic configuration of 2, 8, 7. This results in a stable electronic configuration for both: hydrogen attains a duplet structure (2) and the chlorine attains an octet structure (2, 8, 8). The result is hydrogen chloride, a molecular compound.

Another example of covalent bonding is the formation of hydrogen molecule, where two hydrogen atoms share one electron each.

Introduction to Chemistry

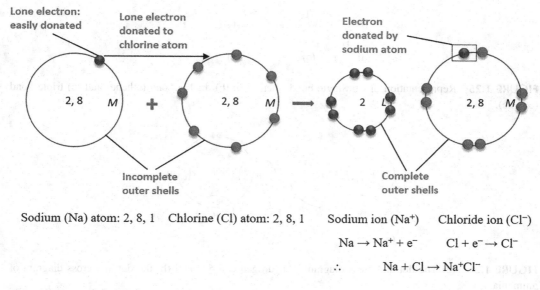

FIGURE 1.23 Electrovalent bonding: formation of sodium chloride.

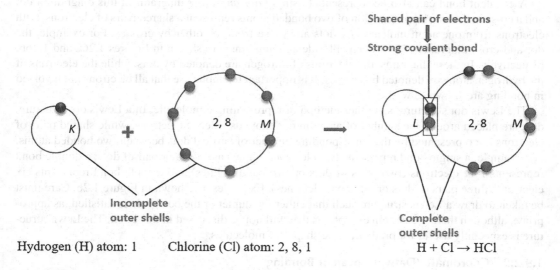

FIGURE 1.24 Covalent bonding: formation of hydrogen chloride.

Covalent bonds can be single, double or triple, depending on the number of shared electron pairs. In a single covalent bond, one pair of electrons is shared; in a double bond, two pairs are shared; and in triple bond, three pairs are shared. For example, hydrogen and hydrogen chloride molecules contain single bonds, while an oxygen molecule contains a double bond, as each oxygen atom contributes two electrons. A typical example of a molecule containing triple bond is carbon monoxide.

1.9.2.1 Representation of Covalent Bonds

Conventionally, a single bond is represented by a line (–) between two bonded atoms, a double bond by two parallel lines (=), and a triple bond by three parallel lines (≡). This is illustrated in Figure 1.25a–c.

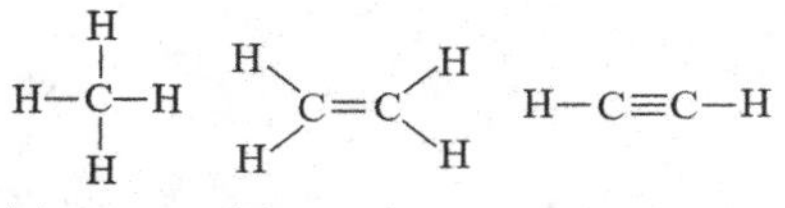

$$H-\underset{\underset{H}{|}}{\overset{\overset{H}{|}}{C}}-H \qquad \overset{H}{\underset{H}{\diagdown}}C=C\overset{\diagup H}{\diagdown H} \qquad H-C\equiv C-H$$

(a)   (b)   (c)

FIGURE 1.25 Representation of (a) Single bond (methane); (b) double bond (ethene) and (c) triple bond (ethyne).

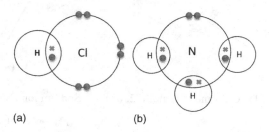

(a)   (b)

FIGURE 1.26 (a) The dot and cross diagram of hydrogen chloride and (b) the dot and cross diagram of ammonia.

A covalent bond can also be represented using a *dot-and-cross* diagram. In this diagram, a dot and a cross in the overlapping region of two bonded atoms represent a shared pairs of electrons, with electrons from one atom indicated by dots and those from the other by crosses. For example, the dot-and-cross diagrams of hydrogen chloride and ammonia are shown in Figures 1.26a and 1.26b, respectively. In these diagrams, the electrons of nitrogen are denoted by dots, while the electrons of the hydrogen atom are denoted by crosses. It is important to emphasise that all electrons not involved in bonding are shown as well.

The Lewis dot structure is another method of representing a molecule. In a Lewis dot structure, dots are placed around the symbol of an atom to denote its valence electrons, while shared pairs of electrons are represented by the corresponding number of vertical dots between two bonded atoms. For example, a single bond represents two electrons, shown as a dash or pair of dots; a double bond represents four electrons (two pairs of dots or two parallel lines) and a triple bond represents six electrons (three pairs of dots or three parallel lines). Examples are shown in Figure 1.26. Care must be taken to draw a Lewis structure such that either the duplet or the octet rule is satisfied, as appropriate, although there are certain exceptions that will not be discussed in this book. The Lewis structure is especially useful for predicting the shapes of molecules.

1.9.2.2 Coordinate (Dative) Covalent Bonding

Unlike ordinary covalent bonding in which each atom contributes one or more electrons to form the bond, a coordinate covalent bond is formed when only one of the two bonded atoms provides the shared pair of electrons. Thus, to form a coordinate covalent bond, one of the participating atoms must have a *lone pair* of electrons. By definition, a lone pair is a pair of electrons on an atom that is not involved in chemical bonding with other atoms. As shown in Figures 1.28a–1.28c, the nitrogen atom in ammonia possesses a lone pair of electrons, while the oxygen atom in a water molecule possesses two lone pairs. Consequently, these molecules can undergo coordinate covalent bonding. A hydrogen ion (H^+, a proton), possibly derived from hydrochloric acid, can attain electronic stability by bonding with a lone pair of electrons in either molecules to form a cation. In the case of ammonia, ammonium ion NH_4^+, is formed; whereas with water the oxonium ion H_3O^+, is produced.

Introduction to Chemistry

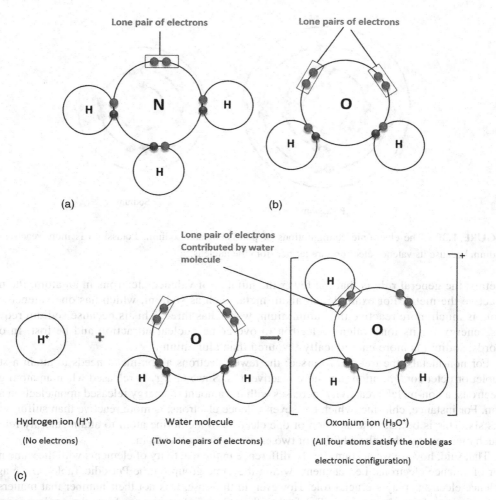

FIGURE 1.27 The Lewis dot structures of (a) methane, (b) methanal and (c) ethyne

FIGURE 1.28 (a) Ammonia molecule showing a lone pair of electrons, (b) water molecule showing two lone pairs of electrons and (c) formation of oxonium ion.

Other examples of substances containing coordinate covalent bonds include the silver diammine ion $[Ag(NH_3)_2]^+$, tetraammine copper(II) ion $[Cu(NH_3)_4]^+$, hydrated copper(II) ion $[Cu(H_2O)_4]^+$, ammonia boron trifluoride BF_3NH_3 and the dimer aluminium chloride Al_2Cl_6.

In a structural formula, a coordinate covalent bond is represented by an arrow pointing from the donor atom to the acceptor atom, as shown in Figure 1.29.

1.9.3 Why Are Some Elements More Reactive than Others?

How do we explain why some elements are more reactive than others? This ultimately comes down to their electronic configuration, particularly the arrangement of valence electrons. For

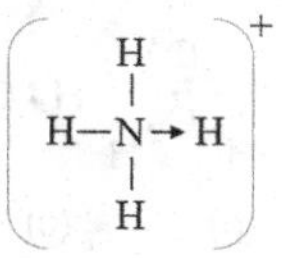

$$\left[\begin{array}{c} H \\ | \\ H-N\rightarrow H \\ | \\ H \end{array} \right]^+$$

FIGURE 1.29 The structural formula of ammonium, NH_4^+, showing the coordinate covalent bond.

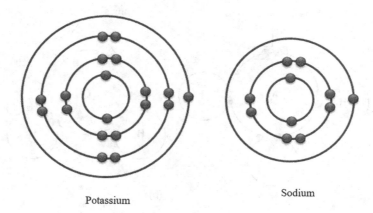

FIGURE 1.30 The electronic configurations of potassium and sodium. Potassium is more reactive than sodium because its valence electrons are farther from the nucleus.

metals, the general rule is that the fewer the number of valence electrons in an atom, the more reactive the metal. For example, an alkali metal such as sodium, which has one valence electron, is much more reactive than aluminium, which has three. This is because sodium requires less energy for its lone valence electron to overcome nuclear attraction and be lost. In other words, sodium is more energetically favoured than aluminium.

For non-metals, the trend is reversed: the fewer electrons a non-metal needs to attain a stable duplet or octet configuration, the more reactive it is. Since energy is released when an atom gains electrons, a non-metal's reactivity increases with the amount of energy released upon electron addition. For instance, chlorine, which has seven valence electrons, is more reactive than sulfur, which has six. This is because the addition of one electron to a chlorine atom to attain stability releases much more energy than the addition of two electrons to a sulfur atom.

That said, how do we account for the difference in the reactivity of elements with the same number of valence electrons, i.e., elements within the same group on the Periodic Table? Once again, valence electrons play a crucial role. However, in this case, it is not their number that matters but their distance from the nucleus. The farther the valence electrons are from the nucleus, the less energy is required for the atom to achieve the stable electronic configuration and, leading to higher reactivity. This means that metals can lose electrons more easily, while non-metallic atoms can attract electrons more easily.

For example, as shown in Figure 1.30, potassium (electronic configuration 2, 8, 8, 1) is comparatively more reactive than sodium (electronic configuration 2, 8, 1) because its valence electron is farther from the nucleus.

1.9.4 ELECTROVALENT (IONIC) AND COVALENT COMPOUNDS

The two kinds of bond formation discussed earlier result in the formation of two classes of compounds: covalent and ionic (electrovalent) compounds.

Introduction to Chemistry

1.9.4.1 Covalent Compounds

Covalent bonding leads to the formation of covalent compounds. A covalent compound consists of electrically neutral, discrete tiny particles called molecules. Thus, covalent compounds are also referred to as molecules. They are primarily formed between non-metals. Examples include carbon dioxide, water, ammonia, methane, benzene, toluene, nitrogen dioxide, glucose, fructose and phosphorus trichloride.

A few metals can also form covalent compounds with non-metals, particularly when the electronegativity difference between the elements is small. Typical examples include beryllium chloride and aluminium chloride, formed when beryllium and aluminium combine with chlorine, respectively.

1.9.4.2 Electrovalent Compounds

Electrovalent or ionic compounds are compounds formed through electrovalent bonding. Unlike covalent compounds, ionic compounds do not exist as molecules. Instead, they consist of oppositely charged ions held together by strong electrostatic forces of attraction. Ionic compounds are typically formed between metals and non-metals with a significant electronegativity difference. Examples include sodium chloride (table salt), sodium hydroxide, barium chloride, magnesium oxide, copper(II) sulfate and silver nitrate.

Ionic compounds are represented by formula units, indicate their constituents ions. However, for simplicity, the charges on the ions are usually omitted. For instance, the formula units of sodium chloride and copper(II) sulfate are NaCl and $CuSO_4$, respectively.

1.9.4.3 Differences between Ionic and Covalent Compounds

Ionic compounds generally have high melting and boiling points due to the strong electrostatic forces holding their ions together. As a result, they are mostly solids at room temperature. For example, calcium chloride has a melting point of 772°C and a boiling point of 1,935°C.

TABLE 1.15
The Melting and Boiling Points of Some Compounds

Compound	Physical State (At Room Temperature)	Nature of Bonding	Melting Point (°C)	Boiling Point (°C)
Calcium chloride	Solid	Ionic	772	1935
Sodium chloride	Solid	Ionic	800	1430
Calcium oxide	Solid	Ionic	2572	2850
Siver chloride	Solid	Ionic	455	1550
Potassium iodide	Solid	Ionic	681	1330
Calcium fluoride	Solid	Ionic	1418	2500
Lithium chloride	Solid	Ionic	605	1382
Copper(II) phosphate	Solid	Ionic	300	158
Magnesium sulfate	Solid	Ionic	1124	330
Ammonium carbonate	Solid	Ionic	58	333.6
Water	Liquid	Covalent	0	100
Ammonia	Gas	Covalent	−77.7	−33.3
Aluminium chloride	Solid	Covalent	192.4	180
Carbon dioxide	Gas	Covalent	−56.6	−78.5
Methane	Gas	Covalent	−182.5	−161.5
Benzene	Liquid	Covalent	5.5	80.1
Ethanol	Liquid	Covalent	−114.1	78.4
Ethene	Gas	Covalent	−169.2	−103.7

TABLE 1.16
Differences between Ionic and Covalent Compounds

Ionic Compounds	Covalent Compounds
They are mostly solids at room temperature	They can be solids, liquids or gases at room temperature
They generally have high melting and boiling points	They generally have low melting and boiling points
They are mostly soluble in polar solvents like water	Most do not dissolve in water but are soluble in non-polar solvents like toluene and benzene. However, some covalent compounds, such as ammonia, ethanol and acetic acid are soluble in water due to their ability to form hydrogen bonding with the solvent
They are electrolytes	They are either non-electrolytes or weak electrolytes
Their representative units are called molecules	Their representative units are called formula units
They are hard, brittle crystalline solids	They can either form amorphous or crystalline solids

On the other hand, covalent compounds typically have low meting and boiling points, which increase with increasing molecular mass. This is due to the weak intermolecular forces between their molecules. For instance, water has a melting point of 0°C and a boiling point of 100°C. Despite its low molecular mass of 18 g mol^{-1}, water has a relatively high boiling point due to the presence of *hydrogen bonds*. A hydrogen bond is a weak intermolecular force of attraction that arises when hydrogen is covalently bonded to highly electronegative elements like oxygen or nitrogen. The melting and boiling points of selected compounds are listed in Table 1.15.

Additionally, ionic compounds in solution or in molten form are good conductors of electricity and are therefore classified as *electrolytes*. In contrast, covalent compounds are generally *non-electrolytes*, meaning their solutions or molten forms do not conduct electricity. However, certain covalent compounds, such as hydrogen chloride, conduct electricity effectively when dissolved in water. Others, like ammonia and acetic acid (ethanoic acid), conduct electricity to a limited extent in solution, making them *weak electrolyte*s. Table 1.16 gives the main differences between ionic and covalent compounds.

PRACTICE PROBLEMS

1. (a) What is matter?
   (b) State the three states of matter and describe the characteristics of each.

2. (a) Differentiate between physical and chemical properties.
   (b) Classify each of the following as either physical or chemical property.
   i. Ionisation energy, ii. Atomic mass, iii. Melting and boiling points, iv. Reaction with air, v. Ability to form complex ions, vi. Toxicity, vii. Solubility, viii. Flammability, ix. Volatility, x. Ability to act as a catalyst.

3. Classify each of the following properties of carbon dioxide as either a physical property or a chemical property.
   i. It is a colourless, odourless gas with a sharp, refreshing taste ii. It dissolves in water to form carbonic acid. iii. It reacts with alkalis to form carbonates. iv. It is about 1.5 times denser than air. v. It changes red-hot carbon to carbon monoxide. vi. It reacts with burning magnesium to produce carbon and magnesium oxide. vii. It changes damp blue litmus paper to pink.

Introduction to Chemistry 37

4. (a) Define physical change and chemical change.
   (b) State the differences between physical change and chemical change.

5. Classify the following into either a physical change or chemical change.
   (i) Rusting, (ii) Decomposition, (iii) Sublimation of dry ice, (vi) Freezing of water, (v) The hardening of cement, (vi) The slaking of lime, (vii) The melting of candle wax, (viii) The dissolution of iron filings in concentrated hydrochloric acid, (ix) The dissolution of copper(II) nitrate in water, (x) The crystallisation of sodium chloride from solution, (xi) Boiling an egg.

6. Explain the following:
   (a) Molecular crystals
   (b) Covalent solids
   (c) Ionic solids
   (d) Metallic solids
   (e) Crystal lattice

7. (a) What is a unit cell?
   (b) What is a cubic crystal structure? State and explain the three types of cubic structures.

8. (a) Define the following terms.
      i. Element, ii. Compound, iii. Mixture
   (b) State the differences between a mixture and a compound.

9. Write the names and symbols for two elements whose names are derived from the following:
   (a) The names of scientists
   (b) The names of countries
   (c) States or Cities in the United States
   (d) Countries in Europe
   (e) Names of ores
   (f) Names of continents

10. Write the name of each of the following elements.
    (a) Po
    (b) Zr
    (c) Pb
    (d) Hg
    (e) Ge
    (f) Te
    (g) Lr
    (h) Na
    (i) W
    (j) Ba

11. Write the symbol of each of the following elements
    (a) Hydrogen
    (b) Tin
    (c) Antimony
    (d) Iron
    (e) Cadmium
    (f) Uranium
    (g) Radon
    (h) Francium
    (i) Titanium
    (j) Iodine

12. Classify the following substances as an element, compound or mixture.
    (i) Solder, (ii) Stainless steel, (iii) Ammonia, (iv) Vanadium, (v) Scandium, (vi) Water, (vii) Hydrogen, (viii) Aniline, (ix) Hydrogen peroxide, (x) Milk, (xi) Molybdenum, (xii) Sea water, (xiii) Cement, (xiv) Soap, (xv) Curium

13. (a) State the three sates of matter and enumerate their characteristic.
    (b) Explain the mechanism of change of state.
    (c) Explain the following:
       i. Sublimation, ii. Evaporation, iii. Freezing, iv. Melting, v. Deposition, vi. Condensation, vii. Melting point.

14. (a) State the difference between a homogeneous mixture and a heterogeneous mixture.
    (b) Classify each of the following as either a homogeneous mixture or a heterogeneous mixture.
       i. Urine, ii. Air, iii. Chalk suspension, iv. Carbonated beverage, v. Kerosene–water mixture, vi. Smog, vii. Oil-vinegar mixture, viii. Sweat, ix. Sugar solution, x. Salad.

15. Explain how you would separate each of the following mixtures.
    (a) Sand, sodium chloride and iron filings
    (b) Petrol, ethanol and water
    (c) Crude oil
    (d) Ammonium chloride, sodium chloride and chalk particles
    (e) Calcium carbonate and sodium carbonate
    (f) A mixture of three soluble solids with different solublities at different temperatures
    (g) Carotenoids—the natural pigments present in fruits and vegetables

16. State the main physical property on which each of the following separation techniques is based.
    (a) Sieving
    (b) Fractional distillation
    (c) Fractional crystallisation
    (d) Chromatography
    (e) Centrifugation
    (f) Separation funnel method
    (g) Magnetic separation
    (h) Sublimation method
    (ix) Precipitation
    (x) Decantation

17. How would you test for the purity of the following:
    (a) Ice cubes
    (b) Food colouring
    (c) Ethanol
    (d) Sodium chloride
    (e) Potassium

18. (a) State the basic postulates of Dalton's atomic theory.
    (b) Explain how Dalton's atomic theory has been modified.

19. Define the following terms.
    (a) Atoms
    (b) Molecules
    (c) Ions
    (d) Isotopes
    (e) Relative atomic mass

Introduction to Chemistry

20. (a) Explain why relative atomic masses are not usually whole numbers.
    (b) What are subatomic particles? State three examples and their characteristics.

21. Explain the following terms.
    (a) Chemical bond
    (b) Electrovalent bonding
    (c) Covalent bonding
    (d) Dative or coordinate covalent bonding
    (e) Hydrogen bonding

22. Draw the Lewis structure and the dot-and-cross diagram of each of the following molecules.
    (a) Water
    (b) Hydrogen chloride
    (c) Sulfur trioxide
    (d) Methane
    (e) Carbon dioxide

23. (a) Explain the formation of potassium chlorate by potassium and chlorine.
    (b) Explain the formation of water by hydrogen and oxygen.
    (c) Explain the formation of oxonium ion by hydrogen ion and ammonia molecule.

24. (a) State three examples each of electrovalent and covalent compounds.
    (b) Sate the differences between electrovalent and covalent compounds.
    (c) Why do you think electrovalent compounds are called electrolytes?

25. (a) What are valence electrons? What roles do valence electrons play in chemical bonding?
    (b) Why are the noble gases unreactive under normal conditions?
    (c) Using chlorine and fluorine as examples, explain why elements in the same group of the Periodic Table have different chemical reactivities.

2 Approximations and Standard Form

2.1 APPROXIMATIONS

In chemistry, you will often need to round off the results of your calculations to a specific number of decimal places or significant figures. Therefore, mastering calculations alone is not enough—you must also be skilled in approximations to avoid losing precious marks.

2.1.1 DECIMAL NUMBERS

A fraction is a number that represents a part of a whole. Examples include 3/4, 1/2, 5/2 and 10/3. A fraction consists of two parts: the number above the line, known as the *numerator*, and the number below the line, known as the *denominator*. In the first fraction, for example, the numerator is 3, while the denominator is 4. When the numerator of a fraction is smaller than the denominator, the fraction is called a *proper fraction*. A proper fraction is greater than zero but less than 1. On the other hand, an *improper fraction* results when the numerator of a fraction is bigger than the denominator. For example, 5/2 is an improper fraction.

The *decimal number system* is an alternative way of denoting fractions. A decimal number is obtained from a fraction by dividing the denominator into the numerator. For example, the decimal equivalents of the fractions 3/4 and 5/2 are 0.75 and 2.5, respectively. As shown in Figure 2.1, the point in a decimal number is known as a *decimal point*. In the decimal number equivalent of a proper fraction, the number to the left of the decimal point is zero, while it is a whole number in an improper fraction. The number to the right of the decimal point is called a *decimal fraction*.

As shown in Figure 2.2, the place value of a digit in a decimal fraction depends on its position, starting from tenths, followed by hundredths, thousandths and so on. The digits in the whole-number part have their usual place values of units, tens, hundreds, thousands and so on, starting from the closest digit to the decimal point. The number of digits in the decimal part of a decimal number is known as its *number of decimal places*. For example, the numbers 2.5 and 0.75 are expressed correct to one and two decimal places, respectively.

Decimal places are used to show the degree of accuracy in a measured or calculated value. The greater the number of decimal places to which a value is reported, the higher the degree of accuracy it reflects. For example, 0.75 shows a degree of accuracy to the hundredth place. Therefore, it reflects a higher degree of accuracy than 2.5, which provides degree of accuracy to the tenth place.

2.1.1.1 Rounding Off Decimal Numbers

Measured and calculated values must be reported to the correct number of decimal places to reflect the degree of accuracy of a measuring instrument. For example, a calliper can measure accurately to the tenth of a millimetre, i.e., 0.1 mm. A micrometre screw gauge is even more accurate, measuring to the hundredth of a millimetre, i.e., 0.01 mm.

The number of decimal places to which a calculated value must be rounded depends on the type of calculation. In calculations involving money, for example, results must be rounded to two decimal places to show that the currency is divided into hundred units. For instance, an amount of $30.20 is expressed to two decimal places to reflect the cents. In other contexts, the number of decimal places is determined by a set of rules.

Approximations and Standard Form

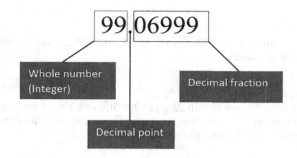

FIGURE 2.1 Parts of a decimal number.

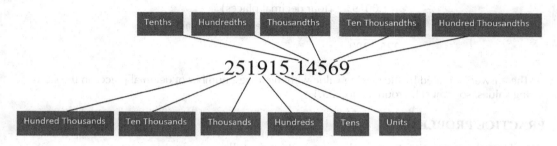

FIGURE 2.2 Place values in a decimal number.

In addition and subtraction operations involving numbers with different number of decimal places, the results must be rounded to the same number of decimal places as the number of decimal places in the starting value with the least number of decimal places.

In multiplication and division problems, the result must be expressed with same number of significant figures as the starting value with the fewest number of significant figures, as discussed in the next section.

To round a number to a specific number of decimal places, we count the required number of digits in its decimal fraction and check the first digit to right of the last counted digit. If the digit is 5 or greater, we round up by adding 1 to the last retained digit and dropping all digits to the right. If it is less than 5, we round down by keeping the last retained digit unchanged and dropping all digits to the right. The following examples illustrate this process.

Example 1

Determine the number of decimal places in each of the following numbers.

(a) 35.5
(b) 55.85
(c) 125.0007
(d) 50.000
(e) 0.70800

Solution

As mentioned earlier, the number of decimal places in a decimal number is obtained by counting the number of digits in its decimal fraction, including all zeros. So we have one decimal place in (a), two in (b), four in (c), three in (d) and five in (e).

Note that zeros in a number, after which no non-zero digit follows, are called *trailing zeros*. As illustrated in this example, trailing zeros contribute to the number of decimal places in a decimal number. Adding trailing zeros does not change the value of a decimal number, but it does increase the number of decimal places. This is particularly useful in avoiding confusion when adding or subtracting numbers with different number of decimal places without a calculator.

Example 2

Add the relative atomic masses 1.0079, 6.941, 26.98 and 190.2, and round the answer to the correct number of decimal places.

Solution

Using an electronic calculator, we obtain 225.1289. Without a calculator, we need to set up the numbers as shown below. Note that trailing zeros (in bold) have been added to the relevant values to ensure the same number of decimal places throughout.

$$
\begin{array}{ll}
190.2\mathbf{000} & \text{(originally one decimal place)} \\
26.98\mathbf{00} & \text{(originally two decimal places)} \\
1.0079 & \text{(four decimal places)} \\
+\ 6.941\mathbf{0} & \text{(originally three decimal places)} \\
\hline
225.1289 &
\end{array}
$$

The answer is limited by the starting value with the fewest number of decimal places in the starting values, so it must be rounded to 225.1.

PRACTICE PROBLEMS

1. State the number of decimal places in each of the following numbers.
   (a) 10.002008
   (b) 15000
   (c) 12.0019
   (d) 2.050000
   (e) 0.01
2. Find the sum of 25.00 dm^3, 4.7 dm^3 and 50.01 dm^3.
3. Evaluate 106.413 g + 39.10 g − 23.901 g.

2.1.2 SIGNIFICANT FIGURES

Every measurement is associated with some degree of error or uncertainty. Scientists use the concept of significant figures to account for uncertainty in measurements. In a measurement, significant figures consist of all digits that are known with certainty plus one uncertain or estimated digit. The last digit in a measurement is usually considered to have an uncertainty of ±1. Consider, for example, a length measurement taken with a vernier calliper and reported as 11.5 mm. In this value, only the first two digits are known with certainty, while the 5 is an estimate which could range from 4 to 6. This implies that the length could be as high as 11.6 mm or as low as 11.4 mm.

A measured value should include as many decimal places—and, by extension, significant figures—as necessary to reflect the precision or degree of accuracy of the instrument used. For instance, it would make no sense to include a hundredth digit in a measurement taken with a vernier calliper and reported in millimetres. If greater accuracy is required, an instrument with a higher degree of accuracy or precision, such as a micrometre screw gauge, should be used.

2.1.2.1 Counting Significant Figures

It is important to to know how determine the number of significant figures in a reported value. The following rules will help with this.

1. All non-zero digits in a number are significant. For example, 3,551 and 22.98977 have four and seven significant figures, respectively.

Approximations and Standard Form

2. All trailing zeros in a decimal number are significant. For example, 22.9897700 and 50.00 have nine and four significant figures, respectively.
3. Trailing zeros in a number without a decimal fraction are insignificant. For example, 5,000 and 15,900 have one and three significant figures, respectively.
4. All zeros located between non-zero digits are significant. For example, 1.0079 and 50.0024 have five and six significant figures, respectively.
5. All zeros that appear before the first non-zero digit of a number—often called leading zeros—are insignificant. For example, 0.001 and 0.000845 have one and three significant figures, respectively.

PRACTICE PROBLEMS

State the number of significant figures in the following numbers.
1. 15.994915
2. 12.000000
3. 14.003242
4. 70,500
5. 0.0000105

2.1.2.2 Significant Figures in Arithmetic Operations

The result of an arithmetic operation must be rounded off to the correct number of significant figures based on the following rules.

1. Addition and subtraction: an answer should have as many number of decimal places as the starting value with the fewest number of decimal places. This has already been illustrated in the previous section.
2. Multiplication and addition: the number of significant figures in an answer is limited to the number of significant figures in the starting value with the fewest number of significant figures.
3. Combined operations: the number of significant figures in the final answer of a combined addition/subtraction and multiplication/division operations is determined by the final operation. Avoid rounding intermediate results, as this can introduce errors in the final answer.
4. Numbers should be rounded up or down, based on the value of the first non-significant figure. If the first non-significant figure is 5 or higher, add 1 to the least significant figure. If the first non-significant figure is less than 5, retain the least significant figure. Conclude by replacing all non-significant figures with zeros in a whole number and dropping them in a decimal fraction.

These guidelines are illustrated with the following examples.

Example 1

The radius of the Earth is 6,378 km at the equator. Express this value to two significant figures.

Solution

All digits in the given value are significant. The first two digits are the required significant figures. Since the first non-significant figure is greater than 5, then, by Rule 4, the answer is 6,400 km.

Example 2

Express the atomic mass of helium-3, 3.016029 u, to three significant figures.

Solution

All digits in the given value are significant. The first three digits are the required significant figures. The first non-significant digit is greater than 5, then, by Rule 4, the answer is 3.02 u.

Example 3

Express the atomic mass of boron-10, 10.012937 u, to three significant figures.

Solution

All digits are significant. The first three digits are the required number of significant figures. Since the first non-significant figure is less than 5, then, by Rule 4, the answer is 10.0 u.

Example 4

Express 0.0024 dm^3 to one significant figure.

The only significant figures are the last two digits. The first non-zero digit is the only required significant figure. Since the first non-significant digit is less than 5, then, by Rule 4, the answer is 0.002 dm^3.

Example 5

Divide 45.98 by 3.

Solution

Dividing 45.98 by 3 gives15.326666666…, a recurring decimal—a decimal number that has a digit or group of digits that repeats forever; hence, infinite number of significant figures. By Rule 2, the answer is limited to only four significant figures because there are four significant figures in 45.98. Thus, the answer is 15.33. It is important to emphasise that the concept of significant figure does not apply to 3 since it is not a measured value.

Example 6

Evaluate (4.50123 ×2.56) ÷ 0.80.

Solution

The operation gives 14.403936, which is limited to two significant figures by the number of significant figures in 0.80—the starting value with the fewest number of significant figures. Hence, the answer is 14.

Example 7

Evaluate (5.145001 + 4.0011) × 0.123.

Solution

Applying the principle of BODMAS, we first deal with the operation in the brackets to obtain 9.146101. By Rule 3, we now carry over all significant figures in the intermediate result to the last step of the calculation to obtain 9.146101 × 0.123 = 1.124970423. The final answer must now be rounded to 1.12 according to Rule 2.

Approximations and Standard Form

PRACTICE PROBLEMS

1. The atomic mass of the only naturally occurring isotope of fluorine is 18.998403 u. Express this mass to:
   (a) One significant figure.
   (b) Two significant figures.
   (c) Three significant figures.
   (d) Four significant figures.
   (e) Five significant figures.
2. Express 0.00905 to one significant figure.
3. Evaluate $\dfrac{6.865 + 65600}{1200}$.

2.2 STANDARD FORM (SCIENTIFIC NOTATION)

Some numbers are rather too large or too small to be conveniently written and handled in the ordinary form. Consider, for example, Avogadro's number and absolute charge of an electron: 602,000,000,000,000,000,000,000 mol⁻¹ and 0.00000000000000000016 C, respectively. Clearly, these numbers are cumbersome in their current form. The *standard form* or *scientific notation* provides a more efficient way to write and manipulate such extremely large or small values. A number is said to be in the standard form when it is written as $A \times 10^n$ (pronounced A exponential n), where $1 \leq A < 10$ and n is an integer. The indicated range of values of A implies that the corresponding power of 10 can be multiplied either by a single-digit whole number or a decimal number with a single non-zero digit before the decimal point. In a numbers greater than 1, n is positive, whereas in number less than 1, n is negative. For example, the seemingly intractable numbers given above, for example, are expressed in the standard form as 6.02×10^{23} mol⁻¹ and 1.6×10^{-19} C, respectively. This should make the difference between the standard form and the ordinary form of writing very large and very small numbers crystal clear.

All the digits in a number expressed in the standard form are significant. Thus, the standard form is particularly very useful in expressing numbers to a specific number of significant figures without any ambiguity. For example, 6.02×10^{23} mol⁻¹ and 1.6×10^{-19} C have three and two significant figures, respectively. Should we, for any reason, decide to express the charge on the electron to three significant figures, it will be written as 1.60×10^{-19} C.

2.2.1 Expressing Numbers in Standard Form

The following rules are involved in the conversion of ordinary numbers to the standard form.

1. For numbers greater than 1: Divide the number by the appropriate power of 10 to obtain a value with a single non-zero digit before the decimal point. If the number is a multiple of 10, ensure the result is a single-digit whole-number. Drop all trailing zeros, then multiply the resulting number by the corresponding power of 10, expressed in index form with a positive power or exponent.
2. For numbers less than 1: Multiply the number by the appropriate power of 10 to obtain a value with a single non-zero digit before the decimal point. Then, multiply the resulting number by the corresponding power of 10, expressed in index form with a negative power or exponent.

These rules are illustrated in the following examples.

Example 1

Express 1905000 in standard form.

Solution

There are seven digits in the number 1905000. To obtain a value with a unit digit before the decimal point, we have to divide the number by the power of 10 with six zeros, i.e., 1000000.

So

$$\frac{1905000}{1000000} = 1.905000$$

Finally, we now drop the trailing zeros and multiply the result by 10^6 (the index form of 1000000) to obtain the standard form 1.905×10^6.

Example 2

Express 200000000 in standard form.

Solution

Here we have a multiple of 10 with nine digits; hence, we have to divide the number by the power of 10 with eight zeros, i.e., 100000000.

$$\frac{200000000}{100000000} = 2$$

We now conclude by multiplying the result by 10^8 (the index form of 100000000) to obtain the standard form 2×10^8.

Example 3

Express 0.0000005050 in standard form.

Solution

The given number is less than 1. There are six zeros to the right of the decimal point before the first non-zero digit. To obtain a number with a single non-zero digit before the decimal point, we multiply by a power of 10 with seven zeros, i.e., 10000000.
So

$$0.0000005050 \times 10000000 = 5.050$$

Finally, we multiply this result by 10^7 (the index form of 10000000), but since the original number is less than 1, we assign a negative exponent to obtain the standard form 5.050×10^{-7}.

Note: In each of these examples, the standard form retains the same number of significant figures as the original number.

PRACTICE PROBLEMS

Express the following in standard form.
1. 1.25
2. 0.109
3. 0.00508
4. 102500001
5. 0.00000007650

2.2.2 MULTIPLICATION AND DIVISION IN STANDARD FORM

The easiest way to multiply and divide numbers in standard form is by using an electronic calculator. However, it is quite straightforward to perform these operations without a calculator, especially when the digits are easy to multiply and divide manually.

Approximations and Standard Form

To multiply two numbers in standard form without a calculator, first multiply their digits together. Then, multiply the result by 10 raised to the sum of the exponents or powers. In other words,

$$(A \times 10^n) \times (B \times 10^m) = (A \times B) \times (10^n \times 10^m) = (A \times B) \times 10^{n+m}$$

To divide a number in standard form by another in the same form, divide the digits of the dividend by those of the divisor. Then, multiply the result by 10 raised to the difference of the exponents or powers. In other words,

$$(A \times 10^n) \div (B \times 10^m) = (A \div B) \times (10^n \div 10^m) = (A \div B) \times 10^{n-m}$$

Example 1

Evaluate $3 \times 10^8 \times 2 \times 10^6$.

Solution

$$3 \times 10^8 \times 2 \times 10^6 = (3 \times 2) \times 10^{8+6}$$
$$= 6 \times 10^{14}$$

To solve the problem with a calculator, press:

Example 2

Evaluate $2.5 \times 10^{10} \times 5 \times 10^{-8}$.

Solution

$$2.5 \times 10^{10} \times 5 \times 10^{-8} = (2.5 \times 5) \times 10^{10+(-8)}$$
$$= 12.5 \times 10^2$$

Rewriting in proper standard form:

$$12.5 \times 10^2 = 1.25 \times 10^1 \times 10^2 = 1.25 \times 10^3$$

To solve the problem with a calculator, press:

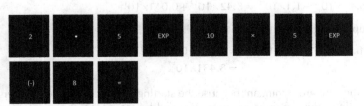

Example 3

Evaluate $\dfrac{4.25 \times 10^{-5}}{1.5 \times 10^{-8}}$.

Solution

$$\dfrac{4.25 \times 10^{-5}}{1.5 \times 10^{-8}} = \left(\dfrac{4.25}{1.5}\right) \times 10^{-5-(-8)}$$
$$= 2.8 \times 10^{3}$$

To solve the problem with a calculator, press:

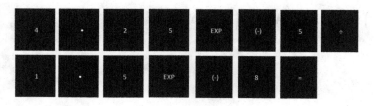

PRACTICE PROBLEM

1. Solve, without using a calculator, the following problems in standard form.
   (a) $8.42 \times 10^{12} \times 1.1 \times 10^{-10}$
   (b) $1.32 \times 10^{-6} \div 1.2 \times 10^{8}$
2. Solve, using a calculator, the following problems in standard form.
   (a) $6.02 \times 10^{23} \times 1.92 \times 10^{-10}$
   (b) $9.32 \times 10^{6} \div 3.21 \times 10^{8}$

2.2.3 ADDITION AND SUBTRACTION IN STANDARD FORM

Numbers, no matter how small or large, can be conveniently added or subtracted in the standard form without a calculator. The first step is to express the number with the lower power of 10 in a form that matches the power of 10 of the other number. The final step is to multiply the sum or difference of the digits by 10, raised to the common exponent or power. In other words,

$$(A \times 10^{n}) \pm (B \times 10^{n}) = (A \pm B) \times 10^{n}$$

Example 1

What is the sum of 8.42×10^{12} and 1.1×10^{10}.

Solution

We have to start by rendering the powers of 10 in the two numbers equal. The number 1.1×10^{10} has the lower power of 10, which can be set to 12 by moving the decimal point backwards twice (that is, dividing 1.1 by 100) and then adding 2 to 10.

$$8.42 \times 10^{12} + 1.1 \times 10^{10} = 8.42 \times 10^{12} + 0.011 \times 10^{12}$$

We should now sum up the digits and multiply the result by 10^{12}.

$$8.42 \times 10^{12} + 0.011 \times 10^{12} = (8.42 + 0.011) \times 10^{12}$$
$$= 8.431 \times 10^{12}$$

Note that all the digits in the sum are significant because the starting values are whole numbers. If we were to write the starting values in ordinary form, we would discover that they contain no decimal points.

Approximations and Standard Form

Example 2

Evaluate $9.84 \times 10^7 - 1.92 \times 10^9$.

Solution

We will proceed as we did above.

$$9.84 \times 10^7 - 1.92 \times 10^8 = 0.984 \times 10^8 - 1.92 \times 10^8$$
$$= (0.984 - 1.92) \times 10^8$$
$$= -0.936 \times 10^8$$
$$= -9.36 \times 10^7$$

PRACTICE PROBLEMS

1. Evaluate $1.32 \times 10^{11} + 1.92 \times 10^7$.
2. Subtract 8.64×10^8 from 1.64×10^{11}.

FURTHER PRACTICE PROBLEMS

1. State the number of decimal places in each of the following.
   (a) 1.32
   (b) 0.0110
   (c) 11.804
   (d) 0.001340
   (e) 0.100000
2. Solve each of the following problems.
   (a) $1.1 + 10.0025 + 1.0089 + 2.23$
   (b) $4.5059 - 4.550 + 3.142 - 2.2202$
3. How many significant figures are there in each of the following?
   (a) 8750000
   (b) 50401000
   (c) 4001.001
   (d) 0.00101000
   (e) 5.000
4. Evaluate the following:
   (a) 66.008×1.11
   (b) 0.001115×0.911
   (c) $23.983 \div 5.00$
   (d) $\sqrt[3]{8.01 \times 10^{-9}}$
   (e) $\sqrt{\dfrac{1.910 \times 10^5}{2.25 \times 10^3}}$
5. Evaluate the following:
   (a) $2.89 \times 10^7 + 2.89 \times 10^{11}$
   (b) $1.04 \times 10^8 - 5.64 \times 10^5$
   (c) $4.14 \times 10^{11} + 1.05 \times 10^9 - 1.999 \times 10^{10}$
   (d) $3.315 \times 10^{13} - 1.05 \times 10^{10} \times 9.1 \times 10^{-31}$
   (e) $(6.150 \times 10^{-25} + 7.050 \times 10^{-23}) \div (9.1 \times 10^{-31} - 6.626 \times 10^{-34})$

3 Chemical Formula and Masses

3.1 CHEMICAL FORMULAE OF IONIC COMPOUNDS

A chemical formula is a concise representation of a chemical substance using the chemical symbols of its constituent elements. For ionic compounds, a chemical formula represents the simplest ratio of ions and is referred to as a *formula unit*. For instance, the chemical formulae of nitric acid and water are HNO_3 and H_2O, respectively. Chemical formulae play a crucial role in chemistry, serving as essential tools for calculating masses and elemental compositions of chemical substances.

Since a compound must be electrically neutral, a correctly written chemical formula must possess no net electric charge. For ionic compounds, this means that the number of cations and anions must be such that their total positive and negative charges cancel out. This is achieved by assigning a subscript corresponding to the valency of one ion to the other. If a polyatomic ion is involved, it must be enclosed in parentheses before adding a subscript. A subscript is only included when the number of atoms or ions is greater than 1. The final formula is obtained by dropping the charges on the ions.

For example, consider an ionic compound containing the ions A^{a+} and B^{b-}. The chemical formula of the compound is written as

$A_b B_a$, where a and b are the valencies of the elements A and B, respectively.

As illustrated above, the valencies of elements and polyatomic ions are obtained from their ionic charges by simply dropping the positive (+) and negative (−) signs.

Example 1

Write the chemical formula sodium chloride.

Solution

We begin by writing the constituent ions of the compound. These are sodium ions Na^+, and chloride ions Cl^-. Since sodium and chlorine both have a valency of 1, then, according to the above guideline, the formula of the compound is NaCl. Note that as indicated earlier, subscripts of 1 are not written but assumed.

Example 2

Deduce the chemical formula of calcium chloride.

Solution

The constituent ions of calcium chloride are calcium ions, Ca^{2+}, and chloride ions, Cl^-. Since the valencies of calcium and chlorine are 2 and 1, respectively, it follows that the formula of calcium chloride is $CaCl_2$.

Example 3

Write the formula of aluminium sulfate.

Solution

The constituent ions are aluminium ions, Al^{3+}, and sulfate ions, SO_4^{2-}, a polyatomic ion. It is known that the valencies of aluminium and the polyatomic ion SO_4^{2-} are 3 and 2, respectively; hence, the formula of aluminium sulfate is $Al_2(SO_4)_3$. Note that, as mentioned earlier, polyatomic ions should be enclosed in parentheses before adding subscripts.

Chemical Formula and Masses

Example 4

Write the chemical formula of iron(II) hydroxide.

Solution

Iron is a typical example of a multivalent metal, i.e., a metal that forms multiple ions. The Roman numeral II in parentheses indicates that the iron ion in the compound carries a charge of 2+, i.e., Fe^{2+}. The anion is the hydroxide ion, OH^-. Since the valencies of iron and the hydroxide ion are 2 and 1, respectively, it follows that the formula of iron(II) hydroxide is $Fe(OH)_2$.

PRACTICE PROBLEMS

Write the formulae of the following compounds.
1. Silver chloride
2. Calcium hydroxide
3. Ammonium sulfate
4. Iron(II) chloride
5. Magnesium nitrate

3.2 RELATIVE FORMULA AND FORMULA MASSES

The relative formula mass, M_r, of a substance is the number of times the average mass of one mole or molecule of that substance is as heavy as one-twelfth the mass of an atom of carbon-12, i.e.,

$$M_r = \frac{\text{Average mass of a mole or molecule of a substance}}{\frac{1}{12} \times \text{mass of an atom of carbon-12}}$$

The relative formula mass of a covalent compound or molecule is also called the *relative molecular mass*. The relative formula mass has no units since it is a ratio of two masses.

The relative formula mass of a substance is obtained from its chemical formula by summing the relative atomic masses of its constituent elements in the numbers shown in the formula.

The formula mass or molar mass, M, of a substance is the mass of one mole of the substance. This corresponds to the relative formula mass of the substance, expressed in grams per mole (g mol^{-1}).

Example 1

Determine the relative molecular mass of oxygen molecule.

$$(O = 16)$$

Solution

The oxygen molecule is represented by the formula O_2.

$$O_2 = O + O$$
$$\therefore M_r = 16 + 16 = 32$$

Example 2

Calculate the molar mass of the chlorine atom.

$$(Cl = 35.5)$$

Solution

The molar mass of an atom of an element is its relative atomic mass expressed in g mol^{-1}.

$$\therefore M = 35.5 \text{ g mol}^{-1}.$$

Example 3

Calculate the molar mass of calcium oxide, CaO.

$$(O = 16, \ Ca = 40)$$

Solution

$$CaO = Ca + O$$

$$\therefore M = (40 + 16) \text{ g mol}^{-1} = 56 \text{ g mol}^{-1}$$

Example 4

Calculate the relative molecular mass of water, H$_2$O.

$$(H = 1, \ O = 16)$$

Solution

$$H_2O = 2H + O$$

$$\therefore M_r = 1 + 1 + 16 = (2 \times 1) + 16 = 18$$

Example 5

Determine the molar mass of zinc hydroxide, Zn(OH)$_2$.

$$(H = 1, \ O = 16, \ Zn = 65)$$

Solution

$$Zn(OH)_2 = Zn + 2(O + H)$$

$$\therefore M = \left(65 + 2\{16 + 1\}\right) \text{ g mol}^{-1} = 99 \text{ g mol}^{-1}$$

Chemical Formula and Masses

Example 6

Determine the molar mass of iron(III) sulfate, $Fe_2(SO_4)_3$.

$$(O = 16, \ S = 32, \ Fe = 56)$$

Solution

$$Fe_2(SO_4)_3 = 2Fe + 3S + 12O$$

$$\therefore M = (\{2 \times 56\} + \{3 \times 32\} + \{12 \times 16\}) \ \text{g mol}^{-1} = 400 \ \text{g mol}^{-1}$$

Example 7

Determine the relative formula mass of cobalt(II) chloride hexahydrate, $CoCl_2 \cdot 6H_2O$.

$$(H = 1, \ O = 16, \ Cl = 35.5, \ Co = 59)$$

Solution

$$CoCl_2 \cdot 6H_2O = Co + 2Cl + 12H + 6O$$

$$\therefore M_r = 59 + (2 \times 35.5) + (12 \times 1) + (6 \times 16) = 238$$

Example 8

Determine the molar mass of an alkane that is represented as C_xH_{20}.

$$(H = 1, \ C = 12)$$

Solution

We begin by determining the molecular formula of the given alkane. To do this, we have to compare the subscripts in C_xH_{20} and the general formula of alkanes, C_nH_{2n+2}.

$$n = x \qquad (3.1)$$

$$2n + 2 = 20 \qquad (3.2)$$

Solving for n in Equation (3.2), we have

$$2n = 18$$

$$\therefore n = \frac{18}{2} = 9$$

From Equation (3.1), we know that $n = x = 9$. Inserting this into C_xH_{20} gives C_9H_{20}, the molecular formula of nonane. Now, we calculate the molar mass M:

$$C_9H_{20} = 9C + 20H$$

$$\therefore M = (\{9 \times 12\} + \{20 \times 1\}) \ \text{g mol}^{-1} = 128 \text{g mol}^{-1}$$

PRACTICE PROBLEMS

Determine the molar mass of each of the following compounds.
1. CO
2. CaO
3. NH_3
4. N_2O
5. $CaCl_2$
6. $CaCO_3$
7. H_2CO_3
8. $Ca(OH)_2$
9. NH_4NO_3
10. $FeSO_4.7H_2O$

$(H = 1, C = 12, N = 14, O = 16, S = 32, Cl = 35.5, Ca = 40, Fe = 56)$

3.3 MASS AND PERCENT COMPOSITIONS

The mass composition of a substance represents the contribution of each of its constituent elements towards its overall mass. The mass, m, of an element in a substance is given by the relation:

$$m = \frac{\text{Number of atoms of element} \times \text{relative atomic mass of element}}{\text{Relative formula mass of substance}} \times \text{mass of sample of substance}$$

Mass composition is often expressed in percentage, called *mass percent composition*. The percentage by mass, $\%m$, of an element in a substance is given by the relation:

$$\%m = \frac{\text{Number of atoms of element} \times \text{relative atomic mass of element}}{\text{Relative formula mass of the substance}} \times 100\%$$

With the exception of rounding errors, the mass percentages of all the elements in a substance must sum to 100%. The mass percent composition of a substance remains constant for any pure sample of the substance, regardless of quantity.

Example 1

Calculate the mass of each element in 5.500 g of C_2H_4.

$$(H = 1, C = 12)$$

Solution

We will apply the relation:

$$m = \frac{\text{Number of atoms of element} \times \text{relative atomic mass of element}}{\text{Relative formula mass of } C_2H_4} \times \text{mass of sample of substance}$$

$$C_2H_4 = 2C + 4H$$

$$\therefore M_r = (2 \times 12) + (4 \times 12) = 28$$

Mass of sample = 5.5 g

C = ?

H = ?

Chemical Formula and Masses

Substituting, we have

$$C = \frac{2 \times 12}{28} \times 5.500\,g = 4.7\,g$$

$$H = \frac{4}{28} \times 5.500\,g = 0.79\,g$$

Alternatively, we can determine the mass of hydrogen by subtraction:

$$H = (5.500 - 4.7)\,g = 0.8\,g$$

Note: The slight difference arises due to rounding.

Example 2

The combustion of a hydrocarbon sample produces 2.2 g of carbon dioxide and 1.8 g of water. Determine the masses of carbon and hydrogen in the original sample.

$$(H = 1,\ C = 12,\ O = 16)$$

Solution

As usual, we will apply the relation:

$$m = \frac{\text{Number of atoms of element} \times \text{relative atomic mass of element}}{\text{Relative formula mass of substance}} \times \text{mass of sample of substance}$$

The mass of carbon in CO_2 is equivalent to its mass in the original hydrocarbon sample.

$$CO_2 = C + 2O$$

$$\therefore M_r = 12 + (2 \times 16) = 44$$

$$\text{Mass of } CO_2 = 2.2\,g$$

$$C = ?$$

Substituting, we have

$$C = \frac{12}{44} \times 2.2\,g = 0.60\,g$$

Similarly, the mass of hydrogen in H_2O is equivalent to its mass in the original hydrocarbon sample.

$$H_2O = 2H + O$$

$$\therefore M_r = (2 \times 1) + 16 = 18$$

$$\text{Mass of } H_2O = 1.8\,g$$

$$H = ?$$

Substituting, we have

$$H = \frac{1 \times 2}{18} \times 1.8\,g = 0.20\,g$$

Example 3

The combustion of 1.2 g of an organic compound containing carbon, hydrogen and oxygen yields 1.3 g of carbon dioxide and 0.54 g of water. Determine the mass composition of the sample.

$$(H = 1, C = 12, O = 16)$$

Solution

$$m = \frac{\text{Number of atoms of element} \times \text{relative atomic mass of element}}{\text{Relative formula mass of substance}} \times \text{mass of sample of substance}$$

As usual, the mass of carbon in CO_2 is equivalent to its mass in the original hydrocarbon sample.

$$CO_2 = C + 2O$$

$$\therefore M_r = 12 + (2 \times 16) = 44$$

$$\text{Mass of } CO_2 = 1.3 \text{g}$$

$$C = ?$$

We now substitute to obtain

$$C = \frac{1 \times 12}{44} \times 1.3 \text{g} = 0.35 \text{g}$$

Similarly, the mass of hydrogen in H_2O is equivalent to its mass in the original hydrocarbon sample.

$$H_2O = 2H + O$$

$$\therefore M_r = (2 \times 1) + 16 = 18$$

$$\text{Mass of } H_2O = 0.54 \text{g}$$

$$H = ?$$

Substituting, we have

$$H = \frac{2}{18} \times 1.8 \text{g} = 0.060 \text{g}$$

The mass of the third element, oxygen, is obtained by deduction. Since the mass of the sample is 1.2 g, it then follows that

$$O = 1.2 \text{ g} - (0.35 + 0.060) \text{ g} = 0.8 \text{g}$$

Example 4

Determine the percent composition of copper(II) oxide, CuO.

Solution

$$(O = 16, Cu = 64)$$

Chemical Formula and Masses

We will apply the relation:

$$\%m = \frac{\text{Number of atoms of element} \times \text{relative atomic mass of element}}{\text{Relative formula mass of CuO}} \times 100\%$$

$$CuO = Cu + O$$

$$\therefore M_r = 64 + 16 = 80$$

$$Cu = ?$$

$$O = ?$$

Substituting, we have

$$Cu = \frac{64}{80} \times 100\% = 80\%$$

$$O = \frac{16}{80} \times 100\% = 20\%$$

The percent composition should sum to 100%: $(80 + 20)\% = 100\%$.

Example 5

Calculate the percent composition of calcium nitrate, $Ca(NO_3)_2$.

$$(N = 14, \ O = 16, \ Ca = 40)$$

Solution

$$\%m = \frac{\text{Number of atoms of element} \times \text{relative atomic mass of element}}{\text{Relative formula mass of } Ca(NO_3)_2} \times 100\%$$

$$Ca(NO_3)_2 = Ca + 2N + 6O$$

$$\therefore M_r = 40 + (2 \times 14) + (6 \times 16) = 164$$

$$Ca = ?$$

$$N = ?$$

$$O = ?$$

Substituting we have

$$Ca = \frac{40}{164} \times 100\% = 24.4\%$$

$$N = \frac{2 \times 14}{164} \times 100\% = 17.1\%$$

$$O = \frac{6 \times 16}{164} \times 100\% = 58.5\%$$

As usual, the values must sum up to 100%: $(24.4 + 17.1 + 58.5)\% = 100\%$.

PRACTICE PROBLEMS

1. Determine the mass of each element in 1.5 g of carbon monoxide, CO.

2. A hydrocarbon sample was burnt completely to produce 15.7 g and 6.4 g of carbon dioxide and water, respectively. Determine the masses of carbon and hydrogen in the original sample.

3. A 5.0-g sample of a compound containing carbon, hydrogen and oxygen was burnt completely to produce 9.57 g and 5.85 g of carbon dioxide and water, respectively. Determine the mass composition of the original sample.

4. Determine the percent compositions of the following compounds.
   (a) NH_3
   (b) CuO
   (c) Cu_2O
   (d) $CaCO_3$
   (e) H_2SO_4
   (f) $Ca(OH)_2$
   (g) $(NH_4)_2SO_4$
   $\left(H = 1, C = 12, N = 14N = 14, O = 16, S = 32, Ca = 40, Cu = 63.5\right)$

FURTHER PRACTICE PROBLEMS

1. Determine the chemical formulae of the following compounds.
   (a) Calcium hydrogen phosphate
   (b) Potassium dichromate
   (c) Iron(III) sulfate
   (d) Perchloric acid
   (e) Manganese(II) nitrate
   (f) Ammonium carbonate

2. Calculate the molar masses of the following compounds.
   (a) $C_{18}H_{24}N_2O_6$
   (b) $Al_2(SO_4)_3$
   (c) $CuSO_4.5H_2O$
   (d) $Na_2CO_3.10H_2O$
   (e) $KAl(SO_4)_2.12H_2O$
   $\left(H = 1, C = 12, N = 14, O = 16, Na = 23, Al = 27, S = 32, K = 39, Cu = 63.5\right)$

3. Determine the percent compositions of the following compounds.
   (a) $C_6H_{12}O_6$
   (b) $C_{32}H_{39}NO_4$
   (c) $Na_2CO_3.10H_2O$
   $\left(H = 1, C = 12, N = 14, O = 16, Na = 23\right)$

4 The Mole Concept and Chemical Formulae

4.1 THE MOLE

The mole—abbreviated as mol and denoted by n—is the SI unit for measuring the amount of a chemical substance. It is one of the most fundamental concepts in chemistry. One mole is defined as the amount of a substance that contains the same number of discrete particles as 12 g of carbon-12. In simpler terms, one mole corresponds to precisely 12 g of carbon-12. A particle could be an atom, ion, molecule, electron or any other specified entity. As we shall see below, one mole contains approximately 6.02×10^{23} particles.

As shown in Figure 4.1, the number of moles of any sample of a substance can be calculated from four different quantities, depending on its physical form. These include mass, number of particles, volume of solution and volume of gas at STP. We shall now consider each of these relationships in turn.

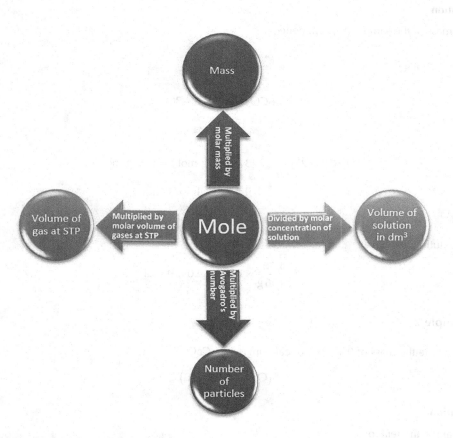

FIGURE 4.1 The mole and other quantities.

DOI: 10.1201/9781003518051-4

4.1.1 THE MOLE AND MASS

The molar mass of a substance is the mass of 1 mol of the substance, expressed in g mol^{-1}. A 58.5 g sample of sodium chloride, for example, contains one mole of the compound. Thus, to obtain the number of moles in a given solid sample, we use the following relation:

$$n = \frac{m}{M}$$

where

m = mass of substance
M = molar mass of substance

Example 1

Determine the number of moles in 4.50 g of sodium carbonate, Na_2CO_3.

$$(C = 12, O = 16, Na = 23)$$

Solution

The mass of the sample is given; hence,

$$n = \frac{m}{M}$$

$$Na_2CO_3 = 2Na + C + 3O$$

So

$$M = (\{2 \times 23\} + 12 + \{3 \times 16\}) \text{ g mol}^{-1} = 106 \text{ g mol}^{-1}$$

$$m = 4.50 \text{ g}$$

$$n = ?$$

Substituting, we have

$$n = \frac{4.50 \text{ g}}{106 \text{ g}} \times 1 \text{ mol} = 0.0420 \text{ mol}$$

Example 2

Determine the mass of 0.15 mol of calcium oxide, CaO.

$$(O = 16, Ca = 40)$$

Solution

We apply the relation:

$$n = \frac{m}{M}$$

The Mole Concept and Chemical Formulae

$$\therefore m = n \times M$$
$$CaO = Ca + O$$

So

$$M = (40 + 16) \text{ g mol}^{-1} = 56 \text{ g mol}^{-1}$$
$$n = 0.15 \text{ mol}$$
$$m = ?$$

Substituting, we have

$$m = 0.15 \text{ mol} \times \frac{56 \text{ g}}{1 \text{ mol}} = 8.4 \text{ g}$$

Example 3

Determine the molar mass of a salt if 0.0150 mol of the salt weighs 0.802 g.

Solution

We use the relation:

$$n = \frac{m}{M}$$
$$\therefore M = \frac{m}{n}$$
$$n = 0.015 \text{ mol}$$
$$m = 0.802 \text{ g}$$
$$M = ?$$

Substituting, we have

$$M = \frac{0.802 \text{ g}}{0.0150 \text{ mol}} = 53.5 \text{ g mol}^{-1}$$

Example 4

Determine the value of x in the compound $Na_2CO_3 \cdot xH_2O$ if 0.019 mol of the compound weighs 5.5 g.

$$(H = 1, C = 12, O = 16, Na = 23)$$

Solution

The first step is to determine the molar mass of the compound using the following relation:

$$n = \frac{m}{M}$$

$$\therefore M = \frac{m}{n}$$

$$n = 0.019 \text{ mol}$$

$$m = 5.5 \text{ g}$$

$$M = ?$$

Now, we substitute to obtain

$$M = \frac{5.01 \text{ g}}{0.0175 \text{ mol}} = 286 \text{ g mol}^{-1}$$

We can now determine the value of x by equating this value to the relative formula mass of $Na_2CO_3.xH_2O$.

$$Na_2CO_3.xH_2O = 2Na + C + 3O + x(2H + O)$$

So

$$(2 \times 23) + 12 + (3 \times 16) + x(\{2 \times 1\} + 16) = 286$$

$$106 + 18x = 286$$

$$18x = 180$$

$$\therefore x = \frac{180}{18} = 10$$

Hence, the full formula of the compound is $Na_2CO_3.10H_2O$.

Example 5

0.0411 mol of a compound XCl weighs 1.50 g. Identify the element X by matching its relative atomic mass with one of the following values given in parentheses below.

$$(H = 1, Na = 23, Cl = 35.5, Ag = 108)$$

Solution

The first step is to determine the molar mass of the compound using the relation:

$$n = \frac{m}{M}$$

$$\therefore M = \frac{m}{n}$$

$$n = 0.0411 \text{ mol}$$

$$m = 1.50 \text{ g}$$

$$M = ?$$

Substituting, we have

$$M = \frac{1.50 \text{ g}}{0.0411 \text{ mol}} = 36.5 \text{ g mol}^{-1}$$

The Mole Concept and Chemical Formulae

We can now determine the relative atomic mass of the element X by equating 36.5 to the relative formula mass of XCl.

$$XCl = 36.5$$

So

$$X + 35.5 = 36.5$$
$$\therefore X = 36.5 - 35.5 = 1$$

This corresponds to the relative atomic mass of hydrogen; hence, the compound in question is HCl (hydrogen chloride).

PRACTICE PROBLEMS

1. Determine the number of moles in each of the following:
   (a) A 1.0 g sample of calcium carbide, CaC_2.
   (b) A 0.12 g sample of silver nitrate, $AgNO_3$.
   (c) A 1.15 g sample of calcium carbonate ($CaCO_3$).
2. Determine the mass of each of the following:
   (a) 0.15 mol of hydrochloric acid, HCl.
   (b) 0.014 mol of zinc oxide, ZnO.
   (c) 2.5 mol of ammonium chloride, NH_4Cl.
3. Determine the molar mass in the following cases.
   (a) A substance containing 0.014 mol of the substance in 1.04 g.
   (b) A substance containing 0.11 mol of the substance in 0.20 g.
   (c) A substance containing 2.50 mol of the substance in 91.3 g.
4. Determine the full formula of $Na_2CO_3.xH_2O$ if 0.0121 mol of the compound weighs 1.5 g.

$$\left(H = 1,\ C = 12,\ N = 14,\ O = 16,\ Na = 23,\ Cl = 35.5,\ Ca = 40,\ Zn = 65,\ Ag = 108\right)$$

4.1.2 THE MOLE AND NUMBER OF PARTICLES

A mole of any substance contains approximately 6.02×10^{23} particles. This value is known as Avogadro's number or Avogadro's constant, N_A. Thus, the number of moles in a sample of any substance, given the number of particles, N, is determined using the following relation:

$$n = \frac{N}{N_A}$$

Example 1

Determine the number of moles of in a sample of chlorine containing 5.98×10^{22} atoms.

$$\left(N_A = 6.02 \times 10^{23}\,mol^{-1}\right)$$

Solution

The number of particles is known; hence, we apply the relation:

$$n = \frac{N}{N_A}$$

$$N = 5.98 \times 10^{22}$$

$$N_A = 6.02 \times 10^{23} \text{ mol}^{-1}$$

$$n = ?$$

Substituting, we have

$$n = \frac{5.98 \times 10^{22}}{6.02 \times 10^{23}} \times 1 \text{ mol} = 0.993 \text{ mol}$$

Example 2

A sample of hydrogen chloride contains 3.5×10^{23} molecules. Determine the number of moles in the sample.

$$\left(N_A = 6.02 \times 10^{23} \text{ mol}^{-1}\right)$$

Solution

As usual, we apply the relation:

$$n = \frac{N}{N_A}$$

$$N = 3.5 \times 10^{23}$$

$$N_A = 6.02 \times 10^{23} \text{ mol}^{-1}$$

$$n = ?$$

We now substitute to obtain

$$n = \frac{3.5 \times 10^{23}}{6.02 \times 10^{23}} \times 1 \text{ mol} = 0.58 \text{ mol}$$

Example 3

A solution contains 0.058 mol of hydrogen ions. Calculate the number of hydrogen ions in the solution.

$$\left(N_A = 6.02 \times 10^{23} \text{ mol}^{-1}\right)$$

Solution

We will apply the relation:

$$n = \frac{N}{N_A}$$

$$\therefore N = n \times N_A$$

$$N_A = 6.02 \times 10^{23} \text{ mol}^{-1}$$

The Mole Concept and Chemical Formulae

$$n = 0.058 \text{ mol}$$
$$N = ?$$

Substituting, we have

$$N = 0.058 \text{ mol} \times \frac{6.02 \times 10^{23}}{1 \text{ mol}} = 3.5 \times 10^{22}$$

Example 4

A sample of glucose, $C_6H_{12}O_6$, contains 6.05×10^{21} molecules. Determine:

(a) the number of moles of the sample;
(b) the mass of the sample.

$$\left(H = 1,\ C = 12,\ O = 16;\ N_A = 6.02 \times 10^{23}\ \text{mol}^{-1}\right)$$

Solution

(a) As usual, we have to apply the relation:

$$n = \frac{N}{N_A}$$

$$N = 6.05 \times 10^{21}$$

$$N_A = 6.02 \times 10^{23}\ \text{mol}^{-1}$$

$$n = ?$$

Substituting, we have

$$n = \frac{6.05 \times 10^{21}}{6.02 \times 10^{23}} \times 1 \text{ mol} = 0.01 \text{ mol}$$

(b) The mass is determined using the relation:

$$n = \frac{m}{M}$$

$$\therefore m = n \times M$$

$$C_6H_{12}O_6 = 6C + 12H + 6O$$

So

$$M = \left(\{6 \times 12\} + \{12 \times 1\} + \{6 \times 16\}\right)\ \text{g mol}^{-1} = 180\ \text{g mol}^{-1}$$

$$n = 0.01 \text{ mol}$$

$$m = ?$$

Finally, we now substitute to obtain

$$m = 0.01 \text{ mol} \times \frac{180 \text{ g}}{1 \text{ mol}} = 1.8 \text{ g}$$

Example 5

Determine the number of molecules in 5.0 g of calcium carbonate, $CaCO_3$.

$$(C = 12, O = 16, Ca = 40; N_A = 6.02 \times 10^{23} \text{mol}^{-1})$$

Solution

The first step is to determine the number of moles in the sample using the relation:

$$n = \frac{m}{M}$$

$$CaCO_3 = Ca + C + 3O$$

$$\therefore M = (40 + 12 + \{3 \times 16\}) \text{ g mol}^{-1} = 100 \text{ g mol}^{-1}$$

$$m = 5.0 \text{ g}$$

$$n = ?$$

Substituting, we have

$$n = \frac{5.0 \text{ g}}{100 \text{ g}} \times 1 \text{ mol} = 0.05 \text{ mol}$$

Finally, we can now determine the number of molecules in the sample as follows.

$$n = \frac{N}{N_A}$$

$$\therefore N = n \times N_A$$

$$N_A = 6.02 \times 10^{23} \text{ mol}^{-1}$$

$$n = 0.05 \text{ mol}$$

$$N = ?$$

We now substitute to obtain

$$N = 0.05 \text{ mol} \times \frac{6.02 \times 10^{23}}{1 \text{ mol}} = 3.01 \times 10^{22}$$

Example 6

A 0.51 g sample of a metal contains 1.33×10^{22} atoms. Determine:

(a) the number of moles of the sample;
(b) the molar mass of the metal.

$$(N_A = 6.02 \times 10^{23} \text{ mol}^{-1})$$

The Mole Concept and Chemical Formulae

Solution

(a) Since the number of particles is known, we apply the relation:

$$n = \frac{N}{N_A}$$

$$N = 1.33 \times 10^{22}$$

$$N_A = 6.02 \times 10^{23} \text{ mol}^{-1}$$

$$n = ?$$

Substituting, we have

$$n = \frac{1.33 \times 10^{22}}{6.02 \times 10^{23}} \times 1 \text{ mol} = 0.0221 \text{ mol}$$

(b) We will apply the relation:

$$n = \frac{m}{M}$$

$$\therefore M = \frac{m}{n}$$

$$m = 0.51 \text{ g}$$

$$n = 0.0221 \text{ mol}$$

$$M = ?$$

Finally, we now substitute to obtain

$$M = \frac{0.51 \text{ g}}{0.0221 \text{ mol}} = 23 \text{ g mol}^{-1}$$

PRACTICE PROBLEMS

1. Determine the number of moles of the following:
   (a) 4.55×10^{24} atoms of lead.
   (b) 3.00×10^{20} molecules of oxygen gas.
   (c) 9.08×10^{23} molecules of methane.
2. Determine the number of particles in each of the following:
   (a) 2.5 mol of uranium.
   (b) 0.11 mol of calcium carbide.
   (c) 5.0 mol of calcium ions.
3. Determine the mass in each of the following:
   (a) 5.50×10^{23} atoms of helium gas, He.
   (b) 1.06×10^{23} molecules of carbon monoxide, CO.
   (c) 8.01×10^{23} molecules of hydrogen sulfide, H_2S.
4. Determine the molar mass of the following:
   (a) A gas containing 1.35×10^{24} atoms in 9.0 g of the gas.
   (b) A metal containing 6.02×10^{22} atoms in a 6.5-g sample of the metal.
   (c) A gas containing 1.51×10^{23} molecules in 4.0 g of the gas.

$$\left(H = 1,\ He = 4,\ C = 12,\ O = 16,\ S = 32;\ N_A = 6.02 \times 10^{23} \text{ mol}^{-1}\right)$$

4.1.3 THE MOLE AND VOLUME

The number of moles a substance in solution is calculated using the following relation:

$$n = C \times V$$

where

C = concentration of solution in mole per cubic decimetre (mol dm^{-3})
V = volume of solution in cubic decimetre (dm^3)

Note: A cubic decimetre and a litre are equivalent units. A cubic decimetre is divided into 1000 cubic centimetres (cm^3), i.e., 1000 cm^3 = 1000 mL = 1 L = 1 dm^3.

Example 1

A beaker contains 50.00 cm^3 of 0.15 mol dm^{-3} sodium hydroxide solution. Calculate the number of moles of sodium hydroxide in the beaker.

Solution

Since we are dealing with a solution, we use the relation:

$$n = C \times V$$

$$C = 0.15 \text{ mol dm}^{-3}$$

$$V = 50.00 \text{ cm}^3 = 0.050 \text{ dm}^3$$

$$n = ?$$

Substituting, we have

$$n = \frac{0.15 \text{ mol}}{1 \text{ dm}^3} \times 0.050 \text{ dm}^3 = 0.0075 \text{ mol}$$

Example 2

A sodium carbonate solution has a concentration of 0.50 mol dm^{-3}. Determine the volume of the solution that would contain 0.25 mol of the compound.

Solution

We apply the relation:

$$n = C \times V$$

$$\therefore V = \frac{n}{C}$$

$$n = 0.25 \text{ mol}$$

$$C = 0.15 \text{ mol dm}^{-3}$$

$$V = ?$$

Substituting, we have

$$V = \frac{0.25 \text{ mol}}{0.15 \text{ mol}} \times 1 \text{ dm}^3 = 1.6 \text{ dm}^3$$

The Mole Concept and Chemical Formulae

Example 3

50.00 cm³ of a hydrochloric acid solution contains 0.010 mol of the acid. Determine the concentration of the solution.

Solution

We will apply the relation:

$$n = C \times V$$

$$\therefore C = \frac{n}{V}$$

$$n = 0.010 \text{ mol}$$

$$V = 50.00 \text{ cm}^3 = 0.050 \text{ dm}^3$$

$$C = ?$$

Substituting, we have

$$C = \frac{0.010 \text{ mol}}{0.050 \text{ dm}^3} = 0.20 \text{ mol dm}^{-3}$$

Example 4

A conical flask contains 25.00 cm³ of 0.15 mol dm⁻³ sodium hydroxide, NaOH, solution. Calculate:

(a) the number of moles of sodium hydroxide in the compound;
(b) the mass of sodium hydroxide in the solution.

$$(H = 1, O = 16, Na = 23)$$

Solution

(a) As usual, we apply the relation:

$$n = C \times V$$

$$C = 0.15 \text{ mol dm}^{-3}$$

$$V = 25 \text{ cm}^3 = 0.025 \text{ dm}^3$$

$$n = ?$$

Substituting, we have

$$n = \frac{0.15 \text{ mol}}{1 \text{ dm}^3} \times 0.025 \text{ dm}^3 = 0.0038 \text{ mol}$$

(b) The mass is determined using the relation:

$$n = \frac{m}{M}$$

$$\therefore m = n \times M$$

$$NaOH = Na + O + H$$

So

$$M = \{(23+16+1)\} \text{ g mol}^{-1} = 40 \text{ g mol}^{-1}$$

$$n = 0.00375 \text{ mol}$$

$$m = ?$$

Substituting, we have

$$m = 0.0038 \text{ mol} \times \frac{40 \text{ g}}{1 \text{ mol}} = 0.2 \text{ g}$$

Example 5

Determine the number of hydrogen ions in 250 cm³ solution if the concentration of the solution is 0.50 mol dm⁻³.

$$\left(N_A = 6.02 \times 10^{23} \text{mol}^{-1}\right)$$

Solution

We begin by calculating the number of moles in the ions using the relation:

$$n = C \times V$$

$$C = 0.50 \text{ mol dm}^{-3}$$

$$V = 250 \text{ cm}^3 = 0.25 \text{ dm}^3$$

$$n = ?$$

Substituting, we have

$$n = \frac{0.50 \text{ mol}}{1 \text{ dm}^3} \times 0.25 \text{ dm}^3 = 0.125 \text{ mol}$$

We can now determine the number of hydrogen ions using the relation:

$$n = \frac{N}{N_A}$$

$$\therefore N = n \times N_A$$

$$n = 0.125 \text{ mol}$$

$$N_A = 6.02 \times 10^{23} \text{ mol}^{-1}$$

$$N = ?$$

Finally, we now substitute to obtain

$$N = 0.125 \text{ mol} \times \frac{6.02 \times 10^{23}}{1 \text{ mol}} = 7.5 \times 10^{22}$$

The Mole Concept and Chemical Formulae

Example 6

25 cm³ of a 0.60 mol dm⁻³ solution contains 0.84 g of a base. Determine:

(a) the number of moles of the base in the solution;
(b) the molar mass of the base.

Solution

(a) We have to apply the relation:

$$n = C \times V$$

$$C = 0.60 \text{ mol dm}^{-3}$$

$$V = 25 \text{ cm}^3 = 0.025 \text{ dm}^3$$

$$n = ?$$

Substituting, we have

$$n = \frac{0.60 \text{ mol}}{1 \text{ dm}^3} \times 0.025 \text{ dm}^3 = 0.015 \text{ mol}$$

(b) The molar mass is determined using the relation:

$$n = \frac{m}{M}$$

$$\therefore M = \frac{m}{n}$$

$$m = 0.84 \text{ g}$$

$$n = 0.015 \text{ mol}$$

$$M = ?$$

We now substitute to obtain

$$M = \frac{0.84 \text{ g}}{0.015 \text{ mol}} = 56 \text{ g mol}^{-1}$$

PRACTICE PROBLEMS

1. Determine the mass and number of moles in each in the specified substances.
   (a) 25.00 cm³ of a 0.011 mol dm⁻³ solution of potassium hydroxide.
   (b) 150 cm³ of a 0.25 mol dm⁻³ solution of silver nitrate.
   (c) 20.00 cm³ of 0.10 mol dm⁻³ solution of sodium hydroxide.
2. Determine the volume in each of the following cases.
   (a) 2.5 mol of a 0.22 mol dm⁻³ solution of sodium nitrate.
   (b) 0.0050 mol of a 0.15 mol dm⁻³ solution of nitric acid.
   (c) 0.75 mol of a 1.15 mol dm⁻³ solution of copper(II) sulfate.
3. Determine the concentration in each of the following cases.
   (a) 50.00 cm³ of a solution containing 0.050 mol of acetic acid.
   (b) 250 cm³ of a solution containing 1.5 mol of ethanol.
   (c) 150 cm³ of a solution containing 0.10 mol of hydrochloric acid.

4. Determine the number of particles in each of the following:
   (a) 100.00 cm³ of a 1.5 mol dm⁻³ solution of chloride ions.
   (b) 150 cm³ of a 0.55 mol dm⁻³ solution of carbonate ions.
   (c) 250 cm³ of 0.10 mol dm⁻³ of a solution of sucrose.
5. Determine the molar mass in each of the following:
   (a) 500.00 cm³ of a 1.00 mol dm⁻³ solution containing 53 g of a substance.
   (b) 150 cm³ of a 0.66 mol dm⁻³ solution containing 4.0 g of a substance.
   (c) 1.5 dm³ of a 0.10 mol dm⁻³ of a solution containing 14.7 g of a substance.
   (H = 1, N = 14, O = 16, Na = 23, K = 39, Ag = 108; N_A = 6.02 × 10²³ mol⁻¹)

4.1.4 THE MOLE AND MOLAR VOLUME OF GASES AT STP

The volume of a gas changes with temperature and pressure. For this reason, scientists have adopted standard conditions of temperature and pressure for measuring gas volumes. The volume occupied by one mole of a gas at any temperature and pressure is called *molar volume*, V_m. The two most widely used standards at this level are the room temperature and pressure (RTP) and standard temperature and pressure (STP).

Room temperature and pressure correspond to 20°C and 1 atm, respectively. The molar volume at RTP is 24 dm³ mol⁻¹. In other words, one mole of a gas occupies a volume of 24 dm³ at RTP. The relationship between the number of moles of a gas at RTP and its molar volume at RTP is as follows:

$$n = \frac{V}{V_m} = \frac{V}{24\,\text{dm}^3\,\text{mol}^{-1}}$$

where

V = volume of a gas in dm³ at RTP
V_m = molar volume of gases at RTP = 24 dm³ mol⁻ᵐ

Standard temperature and pressure correspond to 273 K (0°C) and 760 mmHg (1 atm), respectively. The molar volume at STP is 22.4 dm³ mol⁻¹. In other words, one mole of a gas occupies a volume of 22.4 dm³ at STP. The relationship between the number of moles of a gas at STP and its molar volume at STP is as follows:

$$n = \frac{V}{V_m} = \frac{V}{22.4\,\text{dm}^3\,\text{mol}^{-1}}$$

where

V = volume of a gas in dm³ at STP
V_m = molar volume of gases at STP = 22.4 dm³ mol

Example 1

Calculate the number of moles in 250 cm³ of oxygen at RTP.

$$\left(V_m = 24\,\text{dm}^3\,\text{mol}^{-1}\right)$$

Solution

The applicable relation is:

$$n = \frac{V}{24\,\text{dm}^3\,\text{mol}^{-1}}$$

The Mole Concept and Chemical Formulae

$$V = 250 \text{ cm}^3 = 0.25 \text{ dm}^3$$

$$n = ?$$

We now substitute to obtain

$$n = \frac{0.25 \text{ dm}^3}{24 \text{ dm}^3} \times 1 \text{ mol} = 0.010 \text{ mol}$$

Example 2

Determine the volume of 0.25 mol of carbon dioxide at STP.

$$\left(V_m = 22.4 \text{ dm}^3 \text{ mol}^{-1}\right)$$

Solution

We will apply the relation:

$$n = \frac{V}{22.4 \text{ dm}^3 \text{ mol}^{-1}}$$

$$\therefore V = n \times V_m$$

$$n = 0.25 \text{ mol}$$

$$V = ?$$

Substituting, we have

$$V = 0.25 \text{ mol} \times \frac{22.4 \text{ dm}^3}{1 \text{ mol}} = 5.6 \text{ dm}^3$$

Example 3

Determine the number of molecules in 550 cm³ of carbon dioxide at STP.

$$\left(V_m = 22.4 \text{ dm}^3 \text{ mol}^{-1}, \; N_A = 6.02 \times 10^{23} \text{ mol}^{-1}\right)$$

Solution

The first step is to calculate the number of moles of the gas at STP using the relation:

$$n = \frac{V}{22.4 \text{ dm}^3 \text{ mol}^{-1}}$$

$$V = 550 \text{ cm}^3 = 0.55 \text{ dm}^3$$

$$n = ?$$

We now substitute to obtain

$$n = \frac{0.55 \text{ dm}^3}{22.4 \text{ dm}^3} \times 1 \text{ mol} = 0.02455 \text{ mol}$$

Finally, we can now calculate the number of molecules of the gas at STP using the relation:

$$n = \frac{N}{N_A}$$

$$\therefore N = n \times N_A$$

$$n = 0.02455 \text{ mol}$$

$$N_A = 6.02 \times 10^{23} \text{ mol}^{-1}$$

$$N = ?$$

Substituting, we have

$$N = 0.02455 \text{ mol} \times \frac{6.02 \times 10^{23}}{1 \text{ mol}} = 1.5 \times 10^{22}$$

Example 4

Determine the mass of 150 cm³ of hydrogen sulfide, H_2S, at RTP.

$$\left(H = 1, \ S = 32; \ V_m = 24 \text{ dm}^3 \text{ mol}^{-1}\right)$$

Solution

The first step is to determine the number of moles of the gas at STP using the relation:

$$n = \frac{V}{24 \text{ dm}^3 \text{ mol}^{-1}}$$

$$V = 150 \text{ cm}^3 = 0.15 \text{ dm}^3$$

$$n = ?$$

Substituting the value, we have

$$n = \frac{0.15 \text{ dm}^3}{24 \text{ dm}^3} \times 1 \text{ mol} = 0.00625 \text{ mol}$$

Finally, we can obtain the mass of the gas at RTP.

$$n = \frac{m}{M}$$

$$\therefore m = n \times M$$

$$H_2S = 2H + S$$

So

$$M = \left(\{2 \times 1\} + 32\right) \text{ g mol}^{-1} = 34 \text{ g mol}^{-1}$$

$$n = 0.00625 \text{ mol}$$

$$m = ?$$

We now substitute to obtain

$$m = 0.00625 \text{ mol} \times \frac{34 \text{ g}}{1 \text{ mol}} = 0.21 \text{ g}$$

The Mole Concept and Chemical Formulae

Example 5

5.0 g of a gas occupies a volume of 1400 cm³ at STP. Calculate the molar mass of the gas.

$$(V_m = 22.4 \text{ dm}^3 \text{ mol}^{-1})$$

Solution

The very first step is to calculate the number of moles of the gas at STP.

$$n = \frac{V}{22.4 \text{ dm}^3 \text{ mol}^{-1}}$$

$$V = 1400 \text{ cm}^3 = 1.4 \text{ dm}^3$$

$$n = ?$$

We now substitute to obtain

$$n = \frac{1.4 \text{ dm}^3}{22.4 \text{ dm}^3} \times 1 \text{ mol} = 0.0625 \text{ mol}$$

Finally, we can determine the molar mass of the gas using the relation:

$$n = \frac{m}{M}$$

$$\therefore M = \frac{m}{n}$$

$$m = 5.0 \text{ g}$$

$$n = 0.0625 \text{ mol}$$

$$M = ?$$

Substituting, we have

$$M = \frac{5.0 \text{ g}}{0.0625 \text{ mol}} = 80 \text{ g mol}^{-1}$$

PRACTICE PROBLEMS

1. Determine the number of moles at RTP in the following:
   (a) 550 cm³ of neon.
   (b) 250 cm³ of sulfur dioxide.
   (c) 1500 cm³ of methane.
2. Determine the volume at STP of the following:
   (a) 2.5 mol of hydrogen gas.
   (b) 0.15 mol of nitrogen gas.
   (c) 1.25 mol of benzene gas.
3. Determine the number of molecules at RTP in the following:
   (a) 5500 cm³ of nitrogen dioxide.
   (b) 150 cm³ of carbon monoxide.
   (c) 250 cm³ of hydrogen sulfide.

4. Determine the mass at STP of the following:
   (a) 250 cm³ of sulfur trioxide, SO_3.
   (b) 2500 cm³ of nitrogen gas, N_2.
   (c) 510 cm³ of hydrogen chloride, HCl.
5. Determine the molar mass of the gas in the following:
   (a) 7.0 g occupying a volume of 6.0 dm³ at RTP.
   (b) 4.5 g occupying a volume of 3.18 dm³ at RTP.
   (c) 2.0 g occupying a volume of 3.0 dm³ at RTP.
   (H = 1, N = 14, O = 16, S = 32, Cl = 35.5; N_A = 6.02 × 10²³ mol⁻¹; V_m (STP) = 22.4 dm³ mol⁻¹ V_m(RTP) = 24 dm³ mol⁻¹)

4.2 CHEMICAL FORMULAE

A chemical formula is the symbolic representation of the composition or structure of a substance. The most common types of chemical formulae include the empirical formula, molecular formula, structural formula and projection formula. This section focuses on the calculation of empirical and molecular formulae.

4.2.1 Empirical Formula

The empirical formula is the simplest formula of a substance which shows its constituent elements and the whole-number ratio (indicated by subscripts of the elements) in which their atoms combine. For example, the empirical formula of a water molecule, H_2O, shows that water is made up of hydrogen and oxygen, whose atoms are combined in a 2:1 ratio. The chemical formulae of ionic compounds, which we learned how to write in Chapter 3, are also examples of empirical formula.

Ionic compounds have unique empirical formulae, which are the formulae with which they are represented. In other words, the empirical formula of an ionic compound corresponds to its formula unit. In contrast, for molecules, empirical formulae can either be unique or shared by two or more compounds. Examples of molecules with a unique empirical formula include water, H_2O; ammonia, NH_3; methane, CH_4; carbon dioxide, CO_2; and ethanol, C_2H_6O. Examples of molecules that share the same empirical formula are benzene and ethyne, both of which have the empirical formula CH; methanal and glucose, which share the empirical formula CH_2O; and sulfur monoxide and disulfur dioxide, which both have the empirical formula SO.

Empirical formulae are determined from the mass or percent composition of a compound. The procedure is as follows:

1. Determine the number of moles of atoms of each element in the compound.
2. Determine the simplest whole-number mole ratio of atoms by dividing each number of moles by the least value.
3. Round values to the nearest whole number if the difference is 5% or less. For clarity, round decimals from 0.9 and above up to the nearest whole number and 0.1 and below down to the nearest whole number. Other values should be converted to fractions and multiplied by the lowest common multiple (LCM) of the values to obtain the whole-number mole ratio of atoms. Some examples of decimal fractions encountered in empirical formula calculations, along with their equivalent fractions, are given in Table 4.1.
4. Write the empirical formula of the compound by adding the numbers obtained in Step 2 or 3 as subscripts of the corresponding elements.

The Mole Concept and Chemical Formulae

TABLE 4.1
Some Fractions Encountered in Empirical Formula Calculations

Decimal	Equivalent Fraction	Example
0.500	1/2	2.500 = 2 + 1/2 = 5/2
0.330	1/3	2.330 = 2 + 1/3 = 7/3
0.670	2/3	2.670 = 2 + 2/3 = 8/3
0.250	1/4	2.250 = 2 + 1/4 = 9/4
0.750	3/4	2.750 = 2 + 3/4 = 11/4
0.400	2/5	2.400 = 2 + 2/5 = 12/5
0.200	1/5	2.200 = 2 + 1/5 = 11/5
0.167	1/6	2.167 = 2 + 1/6 = 13/6
0.600	3/5	2.350 = 2 + 3/5 = 13/5
0.800	4/5	2.800 = 2 + 4/5 = 14/5

Example 1

Determine the empirical formula of sodium carbonate, Na_2CO_3.

Solution

As stated earlier, the empirical formula of an ionic compound is its formula unit; hence, the empirical formula of sodium carbonate is Na_2CO_3.

Example 2

On analysis, a sample of a hydrocarbon yields 18.5 g of carbon and 1.5 g of hydrogen. Determine the empirical formula of the compound.

$$(H = 1,\ C = 12)$$

Solution

We begin by calculating the number of moles of atoms of each using the relation:

$$n = \frac{m}{M}$$

For carbon:

$$m = 18.5\ g$$

$$M = 12\ g\ mol^{-1}$$

$$n = ?$$

Substituting, we have

$$n = \frac{18.5\ g}{12\ g} \times 1\ mol = 1.5\ mol$$

For hydrogen:

$$m = 1.5 \text{ g}$$
$$M = 1 \text{ g mol}^{-1}$$
$$n = ?$$

Substituting we obtain

$$n = \frac{1.5 \text{ g}}{1 \text{ g}} \times 1 \text{ mol} = 1.5 \text{ mol}$$

Note: It is important to remember that the number of moles of atoms are calculated using their molar masses—the relative atomic values expressed in g mol^{-1} rather than the molar masses of the element.

Next, we must determine the whole-number ratio of atoms by dividing the number of moles by 1.5 mol.

$$C = \frac{1.5 \text{ mol}}{1.5 \text{ mol}} = 1$$

$$H = \frac{1.5 \text{ mol}}{1.5 \text{ mol}} = 1$$

Finally, we can write the empirical formula of the compound as CH, indicating that the ratio of carbon to hydrogen atoms in the compound is 1:1.

Example 3

A hydrated salt has the following composition by mass.

Sodium = 16.1%
Carbon = 4.20%
Oxygen = 16.8%
Water of crystallisation = 62.9%

Determine the empirical formula of the compound.

$$(H = 1, C = 12, O = 16, Na = 23)$$

Solution

As usual, we begin by calculating the number of moles of each using the relation:

$$n = \frac{m}{M}$$

Given the percent composition of a compound, the mass composition of any of its samples is can be determined by assuming a sample mass of 100 g. Thus, for example, an element whose mass constitutes 16.1% of a compound's mass will have a mass of 16.1 g. For sodium atoms:

$$m = 16.1 \text{ g}$$
$$M = 23 \text{ g mol}^{-1}$$
$$n = ?$$

Substituting, we have

$$n = \frac{16.1\text{ g}}{12\text{ g}} \times 1\text{ mol} = 0.70\text{ mol}$$

For carbon atoms:

$$m = 4.20\text{ g}$$
$$M = 12\text{ g mol}^{-1}$$
$$n = ?$$

Substituting, we have

$$n = \frac{4.20\text{ g}}{12\text{ g}} \times 1\text{ mol} = 0.35\text{ mol}$$

For oxygen atoms:

$$m = 16.8\text{ g}$$
$$M = 16\text{ g mol}^{-1}$$
$$n = ?$$

Substituting, we have

$$n = \frac{16.8\text{ g}}{16\text{ g}} \times 1\text{ mol} = 1.05\text{ mol}$$

For water of crystallisation, H_2O:

$$m = 62.9\text{ g}$$
$$H_2O = 2H + O$$
$$\therefore M = (\{2 \times 1\} + 16)\text{ g mol}^{-1} = 18\text{ g mol}^{-1}$$
$$n = ?$$

Substituting, we have

$$n = \frac{6.29\text{ g}}{12\text{ g}} \times 1\text{ mol} = 3.49\text{ mol}$$

The next step is to divide each value by the least number of moles.

$$Na = \frac{0.70\text{ mol}}{0.35\text{ mol}} = 2$$

$$C = \frac{0.35\text{ mol}}{0.35\text{ mol}} = 1$$

$$O = \frac{1.05\text{ mol}}{0.35\text{ mol}} = 3$$

$$H_2O = \frac{3.49\text{ mol}}{0.35\text{ mol}} = 10$$

Finally, we can write the empirical formula of the compound as $Na_2CO_3.10H_2O$.

Example 4

The percentage by mass of hydrogen in a hydrocarbon is 11.1%. Calculate the empirical formula of the compound.

$$(H = 1, C = 12)$$

Solution

The first step is to calculate the number of moles of each atom using the relation:

$$n = \frac{m}{M}$$

As usual, we will assume a 100-g sample of the compound.
For hydrogen atoms:

$$m = 11.1 \text{ g}$$

$$M = 1 \text{ g mol}^{-1}$$

$$n = ?$$

Substituting, we have

$$n = \frac{11.1 \text{ g}}{1 \text{ g}} \times 1 \text{ mol} = 11.1 \text{ mol}$$

A hydrocarbon consists only of carbon and hydrogen; hence, for carbon:

$$m = (100 - 11.1) \text{ g} = 88.9 \text{ g}$$

$$M = 12 \text{ g mol}^{-1}$$

$$n = ?$$

Substituting, we have

$$n = \frac{88.9 \text{ g}}{12 \text{ g}} \times 1 \text{ mol} = 7.4 \text{ mol}$$

We can now determine the whole-number mole ratio of atoms as follows.

$$H = \frac{11.1 \text{ mol}}{7.4 \text{ mol}} = 1.5 = \frac{3}{2}$$

$$C = \frac{7.4 \text{ mol}}{7.4 \text{ mol}} = 1$$

Here we encounter a fraction. To obtain the required whole-number mole ratio of atoms, we need to multiply both values by 2, the LCM of 1 and 2, resulting in:

$$H = 3$$

$$C = 2$$

Finally, we can now write the empirical formula of the compound as C_2H_3.
  It is important to reiterate here that a value should only be rounded if the difference between it and the nearest whole number is not more than 5%. Thus, we cannot simply round 1.5 to 2, as this would result in the incorrect empirical formula CH_2.

The Mole Concept and Chemical Formulae

PRACTICE PROBLEMS

1. Determine the empirical formulae of the following compounds.
   (a) C_5H_8
   (b) C_4H_{10}
   (c) $CaCO_3$
   (d) C_2H_5COOH
   (e) $C_8H_{10}N_4O_2$
2. Determine the empirical formulae of the compounds with the following compositions.
   (a) C = 80%, H = 20%
   (b) C = 27.3%, O = 72.7%
   (c) C = 52.2%, H = 13.0%, O = 34.8%
   (d) Cu = 25.4%, S = 12.8%, O = 25.7%, H_2O = 36.1%
   (e) C = 73%, H = 5.4%, O = 21.6%

4.2.2 Molecular Formula

A molecular formula is a type of chemical formula that shows the constituent elements of a molecule as well the actual number of their atoms. Unlike an empirical formula, which can refer to both molecular or ionic compounds, the term molecular formula is used exclusively for molecules. Molecular formulae are unique and represent molecular compounds. For instance, the molecular formulae of benzene and ethyne are C_6H_6 and C_2H_2, respectively.

An empirical formula is derived from the molecular formula by reducing the subscripts of atoms in the empirical formula by their greatest common factor to obtain the simplest whole-number ratio. This factor corresponds to the number of empirical formula units in the molecular formula. For instance, the empirical formula of benzene—CH—is obtained from its molecular formula—C_6H_6—by dividing the subscripts of atoms by 6; hence, there are six empirical formula units in one molecular formula of benzene.

In many cases, the molecular and empirical formulae of a compound are identical. This occurs when the subscripts of atoms in a molecular formula do not have a common factor by which the molecular formula can be further reduced. Examples of compounds that have the same molecular formulae as their empirical formulae include water, H_2O; ammonia, NH_3; carbon dioxide, CO_2; methane, CH_4; ethanol, C_2H_6O; and methanal, CH_2O.

The first step in calculating the molecular formula is to determine the number of empirical formula units in the molecular formula using the following relation:

$$(\text{Empirical formula})n = M_r$$

where

M_r = relative molecular mass
n = number of empirical formula units

Conversely, the molecular formula is obtained by multiplying the subscripts of atoms in the empirical formula by the number of empirical formula units, n.

Example 1

Determine the empirical formula of ethene if its molecular formula is C_2H_4.

Solution

The solution is straightforward. The subscripts of atoms in the molecular formula C_2H_4 have a common factor of 2. This means that, there are two empirical formula units in ethene's molecular formula. Dividing the subscripts by 2 gives the empirical formula CH_2.

Example 2

Determine the empirical formula of ethanol if its molecular formula is C_2H_5OH.

Solution

Rewriting the molecular formula as C_2H_6O shows that the subscripts have no common factor. Therefore, the empirical formula is the same as the molecular formula C_2H_6O.

Example 3

The empirical formula of a compound is CH. Determine its molecular formula, given that its relative molecular mass is 78.

$$(H = 1, C = 12)$$

Solution

The first step is to determine the number of empirical formula units, n, in the molecular formula as follows.

$$(\text{Empirical formula})n = M_r$$
$$\text{Empirical formula} = CH$$
$$M_r = 78$$
$$n = ?$$

Substituting, we have

$$(CH)n = 78$$

So

$$(12+1)n = 78$$
$$13n = 78$$
$$\therefore n = \frac{78}{13} = 6$$

Finally, we should now multiply the number of atoms in the empirical formula by 6 to obtain C_6H_6, which is the molecular formula of benzene.

Example 4

The empirical formula of a compound is CH_2O. Determine its molecular formula if its molar mass is 180 g mol^{-1}.

$$(H = 1, C = 12, O = 16)$$

Solution

We follow the procedure as in the previous example.

$$(\text{Empirical formula})n = M_r$$

The Mole Concept and Chemical Formulae

$$\text{Empirical formula} = CH_2O$$

$$M_r = 180$$

$$n = ?$$

Substituting, we have

$$(CH_2O)n = 180$$

So

$$(12 + \{2 \times 1\} + 16)n = 180$$

$$30n = 180$$

$$\therefore n = \frac{180}{30} = 6$$

Thus, the required molecular formula is $C_6H_{12}O_6$, which is the empirical formula of glucose—or its isomer, fructose.

Example 5

The empirical formula of a hydrocarbon is C_7H_8. Determine its molecular formula if 0.25 mol of the compound weighs 23 g.

$$(H = 1, \ C = 12, \ O = 16)$$

Solution

We will apply the same method used in the previous examples.

$$(\text{Empirical formula})n = M_r$$

The relative molecular mass of the compound is not stated explicitly, so, we need to calculate it as follows.

$$n = \frac{m}{M}$$

$$\therefore M = \frac{m}{n}$$

$$m = 23 \text{ g}$$

$$n = 0.25 \text{ mol}$$

$$M = ?$$

So

$$M = \frac{23 \text{ g}}{0.25 \text{ mol}} = 92 \text{ g mol}^{-1}$$

$$\therefore M_r = 92$$

$$\text{Empirical formula} = C_7H_8$$

$$n = ?$$

We can now substitute to obtain

$$(C_7H_8)n = 92$$

$$(\{7 \times 12\} + \{8 \times 1\})n = 92$$

$$92n = 92$$

$$n = \frac{92}{92} = 1$$

Since there is only one empirical formula unit in the molecular formula, the molecular formula is the same as the empirical formula, C_7H_8.

PRACTICE PROBLEMS

Determine the molecular formulae of the compounds whose empirical formulae and relative molecular masses are given below.
1. $C_{10}H_{22}$, 142
2. CHO_2, 90
3. H_2CO_2, 46
4. C_3H_6O, 58
5. CH_2O, 180

$$(H = 1, \ C = 12, \ O = 16, \ S = 32, \ Cu = 63.5)$$

FURTHER PRACTICE PROBLEMS

1. The vapour density of a gas is 16. Determine:
   (a) the number of moles in 2.5 g of the gas;
   (b) the number of molecules in 16 g of the gas.

$$(N_A = 6.02 \times 10^{23} \, \text{mol}^{-1})$$

2. There are approximately 8.50×10^{23} atoms in a sample of gas. What mass of calcium chloride, $CaCl_2$, must be weighed to obtain the number of moles in the gas sample?

$$(Cl = 35.5, \ Ca = 40; \ N_A = 6.02 \times 10^{23} \, \text{mol}^{-1})$$

3. 1.55 dm³ of a gas weighs 4.13 g at RTP. How many moles are there in 1.50 g of the gas at RTP?

$$(V_m = 24 \, \text{dm}^3 \, \text{mol}^{-1})$$

4. A 0.15 mol dm⁻³ solution contains the same number of sodium ions as 5.0 g of sodium chloride. Determine the volume of the solution.

$$(Na = 23, \ Cl = 35.5)$$

The Mole Concept and Chemical Formulae

5. 0.50 mol of a mono-alkanoic acid weighs 44 g. Determine its molecular formula and name.

$$(H = 1, C = 12, O = 16)$$

6. The analysis of a hydrocarbon sample yields 87% carbon. Given that 0.25 mol of the compound weighs 69 g, determine:
   (a) the molar mass of the compound;
   (b) the empirical formula of the compound;
   (c) the molecular formula of the compound.

$$(H = 1, C = 12)$$

7. The combustion of a 1.5-g sample of lycopene, the compound responsible for the colour of ripe tomatoes, yields 4.93 g of carbon dioxide. Given that 0.0225 mol of the compound weighs 12.06 g, determine:
   (a) the molar mass of the compound;
   (b) the mass percentage composition of lycopene;
   (c) the empirical formula of lycopene;
   (d) the molecular formula of lycopene.

$$(H = 1, C = 12)$$

8. A 2.50-g sample of a compound containing carbon, hydrogen and oxygen produces 4.58 g and 1.25 g of carbon dioxide and water, respectively, upon complete combustion. If 0.15 mol of the compound weighs 10.8 g, determine:
   (a) the molar mass of the compound;
   (b) the mass composition of the sample;
   (c) the empirical formula of the compound;
   (d) the molecular formula of the compound.

$$(H = 1, C = 12, O = 16)$$

5 Chemical Equations and Stoichiometry

5.1 CHEMICAL EQUATIONS

A chemical equation is a concise representation of a chemical reaction using only the chemical symbols and formulae of reactants and products. In a chemical equation, an arrow (→) points from the reactants—on the left-hand side (LHS)—to the products—on the right-hand side (RHS). A reversible reaction is indicated by two arrows with single barbs pointing in opposite directions (⇌). Some examples of chemical equations are given below:

1. The equation for the decomposition of calcium carbonate into calcium oxide and carbon dioxide: $CaCO_3(s) \rightarrow CaO(s) + CO_2(g)$.
2. The equation for the reaction between hydrogen and oxygen gases to produces team: $2H_2(g) + O_2(g) \rightarrow 2H_2O(g)$.
3. The reversible reaction between nitrogen and hydrogen gases to produce ammonia gas: $N_2(g) + 3H_2(g) \rightleftharpoons 2NH_3(g)$

The number that precedes a chemical formula or symbol in a chemical equation is called a *stoichiometric coefficient*—or simply the *coefficient*. The contribution of a product or reactant to the total number of atoms of an element in a chemical equation is determined by multiplying its coefficient by the subscript of the atom.

5.1.1 BALANCING CHEMICAL EQUATIONS

Chemical reactions involve only the reorganisation of atoms to form new substances; no atoms are created or destroyed. Hence, mass is conserved in ordinary chemical reactions. This principle is known as *the law of conservation of matter*. Therefore, it is an inviolable rule that the number of each type of atom on the LHS of an equation must be equal to the number on the RHS. A chemical equation written in this way is called a balanced equation. For example, consider the following equation:

$$2H_2(g) + O_2(g) \rightarrow 2H_2O(g)$$

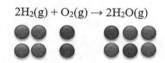

There are four hydrogen atoms and two oxygen atoms on both sides of the equation; hence, the equation is balanced as written.

To balance an unbalanced chemical equation, appropriate coefficients—preferably whole numbers—must be added to the reactants and products. There are two common methods of determining the correct coefficients: inspection method and algebraic method. The inspection method is a trial-and-error approach that involves adjusting the coefficients of reactants and products until the equation is balanced. The algebraic method requires a good understanding of algebra. It involves representing unknown coefficients with letters, usually placed in front of reactants and products with a coefficient of 1. The lowest possible whole-number values of these unknowns are then determined by solving algebraic equations derived from the unbalanced equation.

Both methods will be illustrated with the fellowing example.

Chemical Equations and Stoichiometry

Example

Determine which of the following equations are unbalanced, and balance them accordingly.

1. $2CO(g) + O_2(g) \rightarrow 2CO_2(g)$
2. $H_2(g) + I_2(g) \rightarrow HI(g)$
3. $CH_4(g) + Cl_2(g) \rightarrow CH_3Cl(g) + H_2(g)$
4. $2HCl(aq) + Zn(s) \rightarrow ZnCl_2(aq) + H_2(g)$
5. $C_6H_{12}O_6(s) + O_2(g) \rightarrow CO_2(g) + 6H_2O(l)$

Solution

We must ensure that the total number of atoms of each element is the same on both sides of the equation.

1. The atom count in $2CO(g) + O_2(g) \rightarrow 2CO_2(g)$ is as follows.

Atom	LHS	RHS
C	$2 \times 1 = 2$	$2 \times 1 = 2$
O	$(2 \times 1) + (1 \times 2) = 4$	$2 \times 2 = 4$

As shown in the table, the number of atoms of each element is the same on both sides of the equation; hence, the equation is balanced as written.

2. The atom count in $H_2(g) + I_2(g) \rightarrow HI(g)$ is as follows.

Atom	LHS	RHS
H	$1 \times 2 = 2$	$1 \times 1 = 1$
I	$1 \times 2 = 2$	$1 \times 1 = 1$

The atom count shows that the equation is not balanced. To balance the equation by the inspection method, we can add a 2 in front of HI, resulting in $H_2(g) + I_2(g) \rightarrow 2HI(g)$. This ensures that the number of atoms of each element is equal on both sides of the equation. Thus, the balanced equation is $H_2(g) + I_2(g) \rightarrow 2HI(g)$.

Alternatively, with the algebraic method, we begin by adding letter coefficients to the reactants and products.

$$xH_2(g) + yI_2(g) \rightarrow zHI(g)$$

Since the number of each type of atom must be equal on both sides of the equation we can set up the following algebraic equations:

$$H: \quad 2x = z \tag{5.1}$$

$$I: \quad 2y = z \tag{5.2}$$

In Equation (5.1), we need to choose a value of x that results in the smallest possible whole-number value of z. This is only possible when $x = 1$, which gives $z = 2$. Substituting this value into Equation (5.2) yields $y = 1$. Thus, we have $x = 1$, $y = 1$ and $z = 2$. Finally, we can substitute the values into the chemical equation to obtain the balanced equation: $H_2(g) + I_2(g) \rightarrow 2HI(g)$.

It should be emphasised that a chemical equation must not be balanced by adding subscripts, as this would alter the identity of the substance involved or result in a non-existent substance. For instance, it is incorrect to balance the above equation as follows:

$$H_2(g) + I_2(g) \rightarrow H_2I_2(g)$$

3. We begin by counting the atoms of each element in $CH_4(g) + Cl_2(g) \rightarrow CH_3Cl(g) + H_2(g)$.

Atom	LHS	RHS
C	$1 \times 1 = 1$	$1 \times 1 = 1$
H	$1 \times 4 = 4$	$(1 \times 3) + (1 \times 2) = 5$
Cl	$1 \times 2 = 2$	$2 \times 1 = 2$

We can see from the table that the equation is not balanced. Using the inspection method, we will find that the equation becomes balanced if we place a 2 in front of CH_4 and CH_3Cl. Thus, the balanced equation is: $2CH_4(g) + Cl_2(g) \rightarrow 2CH_3Cl(g) + H_2(g)$.

Alternatively, we can use the algebraic method as follows.

$$xCH_4(g) + wCl_2(g) \rightarrow yCH_3Cl(g) + zH_2(g)$$

So

$$C: \quad x = y \quad (5.3)$$

$$H: \quad 4x = 3y + 2z \quad (5.4)$$

$$Cl: \quad 2w = y \quad (5.5)$$

We see from Equation (5.5) that the lowest possible whole-number value of y is 2, which is obtained when $w = 1$. hence, from Equation (5.3), $x = 2$. Substituting x, w and y into Equation (5.4), we have

$$8 = 6 + 2z$$

So

$$2z = 2$$

$$\therefore z = \frac{2}{2} = 1$$

Thus, $x = 2$, $w = 1$, $y = 2$ and $z = 1$. Substituting these values into the equation, we obtain $2CH_4(g) + Cl_2(g) \rightarrow 2CH_3Cl(g) + H_2(g)$.

4. As before, we begin by counting each type of atom in $2HCl(aq) + Zn(s) \rightarrow ZnCl_2(aq) + H_2(g)$.

Atom	LHS	RHS
H	$2 \times 1 = 2$	$1 \times 2 = 2$
Cl	$2 \times 1 = 2$	$1 \times 2 = 2$
Zn	$1 \times 1 = 1$	$1 \times 1 = 1$

The atom count shows that the equation is balanced as written.

5. The atom count in $C_6H_{12}O_6(s) + O_2(g) \rightarrow CO_2(g) + 6H_2O(l)$ is as follows.

Atom	LHS	RHS
C	$1 \times 6 = 6$	$1 \times 1 = 1$
H	$1 \times 12 = 12$	$6 \times 2 = 12$
O	$(1 \times 6) + (1 \times 2) = 8$	$(1 \times 2) + (6 \times 1) = 8$

Chemical Equations and Stoichiometry

It is now clear that the equation is unbalanced. Using the inspection method, we can balance it by placing a 6 in front of O_2 and CO_2, resulting in the balanced equation: $C_6H_{12}O_6(s) + 6O_2(g) \rightarrow 6CO_2(g) + 6H_2O(l)$.

Alternatively, we can apply the algebraic method as follows.

$$xC_6H_{12}O_6(s) + yO_2(g) \rightarrow zCO_2(g) + 6H_2O(l)$$

So

C: $\quad 6x = z \quad$ (5.6)

H: $\quad 12x = 12 \quad$ (5.7)

O: $\quad 6x + 2y = 2z + 6 \quad$ (5.8)

In Equation (5.6), we set the value of x to 1 to obtain $z = 6$, the lowest possible whole-number value of z. The value of x can also be determined from the second equation. Substituting the values of x and z into Equation (5.8), we obtain

$$6 + 2y = 12 + 6$$

So

$$2y = 12$$

$$\therefore y = \frac{12}{2} = 6$$

Thus, $x = 1$, $y = 6$ and $z = 6$. Substituting into the equation, we have

$$C_6H_{12}O_6(s) + 6O_2(g) \rightarrow 6CO_2(g) + 6H_2O(l)$$

PRACTICE PROBLEMS

Carefully check the following equations and balance any that are unbalanced.

1. $2H_2O_2(aq) \rightarrow H_2O(l) + O_2(g)$
2. $CH_4(g) + O_2(g) \rightarrow CO_2(g) + H_2O(g)$
3. $CaCO_3(s) + 2HCl(aq) \rightarrow CaCl_2(aq) + CO_2(g) + H_2O(l)$
4. $Na_2CO_3(aq) + H_2SO_4(aq) \rightarrow Na_2SO_4(aq) + CO_2(g) + H_2O(l)$
5. $2Na(s) + 2H_2O(l) \rightarrow NaOH(aq) + H_2(g)$
6. $C_5H_{12}(g) + O_2(g) \rightarrow CO_2(g) + H_2O(g)$
7. $Ca(OH)_2(aq) + H_3PO_4(aq) \rightarrow Ca_3(PO_4)_2(aq) + H_2O(l)$
8. $2AgI(aq) + Na_2S(s) \rightarrow Ag_2S(s) + NaI(aq)$
9. $3CaCl_3(aq) + 2Na_3PO_4(aq) \rightarrow Ca_3(PO_4)_2(aq) + 6NaCl(aq)$
10. $4FeS(s) + 7O_2(g) \rightarrow 2Fe_2O_3(s) + 4SO_2(g)$

5.1.1.1 Information from Chemical Equations

A balanced chemical equation provides the following information:

1. There reactants and products, represented by their symbols and formulae.
2. The direction of the action.
3. The physical states of substances, indicated in parentheses next to their symbols or formulae. A gas is denoted as (g), a liquid as (l), a solid as (s) and an aqueous solution of a substance (a substance dissolved in water) as (aq).

4. The mole ratio of reactants to products. This is based on the stoichiometric coefficients of the substances in a chemical equation, as illustrated below.

$$2H_2(g) + O_2(g) \rightarrow 2H_2O(g)$$

Moles	2	1	2
Mole ratio	2	: 1	: 2

5. Extra information about a reaction, such as the conditions required for the reaction to occur and the heat absorbed or released, can also be included. For instance, the reaction below shows that the formation of sulfur trioxide from sulfur dioxide and oxygen requires vanadium(V) oxide, V_2O_5, as a catalyst and releases 99.1 kJ of heat per mole of sulfur dioxide reacted:

$$2SO_2(g) + O_2(g) \xrightarrow{V_2O_5} 2SO_3(g) \qquad \Delta H = -99.1 \text{ kJ mol}^{-1}$$

5.2 STOICHIOMETRY

Stoichiometry is the branch chemistry that deals with the quantitative relationships between reactants and products in a chemical reaction. As mentioned earlier, the mole ratio of reactants to products is a key piece of information derived from a balanced chemical equation. Additionally, a defining characteristic of a chemical reaction is that it occurs in fixed proportions. Stoichiometry applies this principle to determine the amount of any substance consumed or produced in a chemical reaction.

Example 1

What mass of oxygen would be produced from the decomposition of 5.0 mol of hydrogen peroxide? The equation of the reaction is $2H_2O_2(aq) \rightarrow 2H_2O(l) + O_2(g)$.

$$(H = 1, O = 16)$$

Solution

The first step is to calculate the number of moles of oxygen that would be produced:

$$2H_2O_2(aq) \rightarrow 2H_2O(l) + O_2(g)$$

Moles	2	2	2
Mole ratio	2	: 2	: 1

So, mole ratio of O_2 to $H_2O_2 = \dfrac{1}{2}$.

Since the amount of H_2O_2 supplied is 5.0 mol, it then follows that the number of moles, n, of O_2 produced is:

$$n = \frac{1}{2} \times 5.0 \text{ mol} = 2.5 \text{ mol}$$

Chemical Equations and Stoichiometry

A similar way of presenting this calculation is as follows.
Let n be the amount of O_2 produced from 5.0 mol of H_2O_2. Since, according to the equation,

$$\frac{n}{5 \text{ mol of } H_2O_2} = \frac{1 \text{ mol of } O_2}{2 \text{ mol of } H_2O_2}$$

So

$$n \times 2 \text{ mol} = 5.0 \times 1 \text{ mol}$$

$$\therefore n = \frac{5.0 \text{ mol} \times 1 \text{ mol}}{2 \text{ mol}} = 2.5 \text{ mol}$$

The last step is to convert this amount of oxygen to mass using the following relation:

$$n = \frac{m}{M}$$

$$\therefore m = n \times M$$

$$n = 2.5 \text{ mol}$$

$$M = (2 \times 16) \text{ g mol}^{-1} = 32 \text{ g mol}^{-1}$$

$$m = ?$$

Substituting, we have

$$m = 2.5 \text{ mol} \times \frac{32 \text{ g}}{1 \text{ mol}} = 80 \text{ g}$$

Alternatively, we can first convert the given amount of H_2O_2 to mass and then multiply it by the mass ratio of H_2O_2 to O_2.

$$n = \frac{m}{M}$$

$$\therefore m = n \times M$$

$$n = 5.0 \text{ mol}$$

$$M = (\{2 \times 1\} + \{2 \times 16\}) \text{ g mol}^{-1} = 34 \text{ g mol}^{-1}$$

$$m = ?$$

We now substitute to obtain

$$m = 5.0 \text{ mol} \times \frac{34 \text{ g}}{1 \text{ mol}} = 170 \text{ g}$$

	$2H_2O_2(aq)$	$\rightarrow$	$2H_2O(l)$	$+$	$O_2(g)$
Moles	2		2		1
Mass	(2×34) g		(2×18) g		32 g
Mass ratio	68	:	36	:	32

So, the mass ratio of O_2 to $H_2O_2 = \frac{32}{68}$.

Since the mass of H_2O_2 supplied is 170 g, it follows that the mass, m, of oxygen that would be produced is:

$$m = \frac{32}{68} \times 170 \text{ g} = 80 \text{ g}$$

We can also set up the calculation as follows.

Let m be the mass of O_2 produced from 170 g of H_2O_2. According to the equation, 68 g of H_2O_2 produces 32 g of O_2; it follows that 170 g of H_2O_2 would produce m, i.e.,

$$\frac{68 \text{ g of } H_2O_2}{32 \text{ g of } O_2} = \frac{170 \text{ g of } H_2O_2}{m}$$

So

$$m \times 68 \text{ g} = 170 \text{ g} \times 32 \text{ g}$$

$$\therefore m = \frac{32 \text{ g} \times 170 \text{ g}}{68 \text{ g}} = 80 \text{ g}$$

Example 2

The equation for the decomposition of calcium carbonate is $CaCO_3(s) \rightarrow CaO(s) + CO_2(g)$. How many moles of calcium carbonate are required for the production of 3.5 g of calcium oxide?

$$(C = 12, O = 16, Ca = 40)$$

Solution

We will convert 3.5 g of CaO to moles and then multiply the result by the mole ratio of reactants to products.

$$n = \frac{m}{M}$$

$$m = 3.5 \text{ g}$$

$$M = (40 + 16) \text{ g mol}^{-1} = 56 \text{ g mol}^{-1}$$

$$m = ?$$

Substituting, we have

$$n = \frac{3.5 \text{ g}}{56 \text{ g}} \times 1 \text{ mol} = 0.0625 \text{ mol}$$

$$\begin{array}{lccc} & CaCO_3(s) & \rightarrow \quad CaO(s) & + \quad CO_2(g) \\ \text{Moles} & 1 & 1 & 1 \\ \text{Mole ratio} & 1 & : \quad 1 & : \quad 1 \end{array}$$

Since the mole ratio of the reactant to the product is 1:1, it follows that the required number of moles of $CaCO_3$ is also 0.0625 mol.
Another way of setting up the calculation is as follows.
Let the required amount of $CaCO_3$ be n. According to the equation, 1 mol of $CaCO_3$ produces 1 mol of CaO, therefore, we can write:

$$\frac{1 \text{ mol of } CaCO_3}{1 \text{ mol of } CaO} = \frac{n}{0.0625 \text{ mol of } CaO}$$

$$\therefore n = 0.0625 \text{ mol}$$

Alternatively, we could multiply the given mass by the mass ratio of reactants to products and then convert the result to moles. We leave this method for you to verify as a practice exercise.

Chemical Equations and Stoichiometry

Example 3

Consider the following reaction:

$$4FeS(s) + 7O_2(g) \rightarrow 2Fe_2O_3(s) + 4SO_2(g)$$

Determine:

(a) the amount of oxygen, in moles, required to burn 1.5 mol of iron(II) sulfide completely;
(b) the mass of iron(II) sulfide required to react completely with 2.70 mol of oxygen.

$$(O = 16, S = 32, Fe = 56)$$

Solution

(a) We only need to multiply the mole ratio of O_2 to FeS by the given amount of FeS.

	$4FeS(s)$	+	$7O_2(g)$	$\rightarrow$	$2Fe_2O_3(s)$	+	$4SO_2(g)$
Moles	4		7		2		4
Mole ratio	4	:	7	:	2	:	4

So, the mole ration of O_2 to FeS = $\frac{7}{4}$.

Since the amount of FeS supplied is 1.5 mol, it follows that the amount, n, of O_2 required is:

$$n = \frac{7}{4} \times 1.5 \text{ mol} = 2.6 \text{ mol}$$

Alternatively, we can set up the calculation as follows.
   Let n be the required amount of O_2. From the equation, 4 mol of FeS requires 7 mol of O_2 for complete reaction; hence,

$$\frac{4 \text{ mol of FeS}}{7 \text{ mol of } O_2} = \frac{1.5 \text{ mol of FeS}}{n}$$

So

$$n \times 4 \text{ mol} = 1.5 \text{ mol} \times 7 \text{ mol}$$

$$\therefore n = \frac{1.5 \text{ mol} \times 7 \text{ mol}}{4 \text{ mol}} = 2.6 \text{ mol}$$

(b) If the mass of FeS is required, we must multiply the given mass of O_2 by the mass ratio of FeS to O_2. The very first step is to convert 2.70 mol of O_2 to mass, as follows.

$$n = \frac{m}{M}$$

$$\therefore n = m \times M$$

$$n = 2.70 \text{ mol}$$

$$M = (2 \times 16) \text{ g mol}^{-1} = 32 \text{ g mol}^{-1}$$

$$m = ?$$

Substituting, we have

$$m = 2.70 \text{ mol} \times \frac{32 \text{ g}}{1 \text{ mol}} = 86.4 \text{ g}$$

	4FeS(s)	+	$7\text{O}_2\text{(g)}$	$\rightarrow$	$2\text{Fe}_2\text{O}_3\text{(s)}$	+	$4\text{SO}_2\text{(g)}$
Moles	4		7		2		4
Mass	(4×88) g		(7×32) g		(2×160) g		(4×64) g
Mass ratio	352	:	224	:	320	:	256

So, the mass ratio of FeS to $O_2 = \dfrac{352}{224}$.

It follows that the mass, m, of FeS required to react completely with 86.4 g of O_2 is:

$$m = \frac{352}{224} \times 86.4 \text{ g} = 136 \text{ g}$$

We can also set up the calculation as follows.
Let m be the required mass of FeS. According to the equation, 352 g of FeS requires 224 g of O_2 for complete reaction. Therefore,

$$\frac{352 \text{ g of FeS}}{224 \text{ g of O}_2} = \frac{m}{86.4 \text{ g of O}_2}$$

So

$$m \times 224 \text{ g} = 352 \text{ g} \times 86.4 \text{ g}$$

$$\therefore m = \frac{352 \text{ g} \times 86.4 \text{ g}}{224 \text{ g}} = 136 \text{ g}$$

Example 4

Methane burns in air according to the following equation.

$$\text{CH}_4\text{(g)} + 2\text{O}_2\text{(g)} \rightarrow \text{CO}_2\text{(g)} + 2\text{H}_2\text{O(l)}$$

What volume of carbon dioxide would be produced from the complete burning of 1.5 kg of methane at STP?

$$\left(H = 1,\ C = 12,\ O = 16;\ V_m = 22.4 \text{ dm}^3 \text{ mol}^{-1}\right)$$

Chemical Equations and Stoichiometry

Solution

You will recall that the number of moles of a gas at STP is given by

$$n = \frac{V}{22.4 \text{ dm}^3 \text{ mol}^{-1}}$$

$$\therefore V = n \times 22.4 \text{ dm}^3 \text{ mol}^{-1}$$

The amount of CO_2 must, of course, be determined using stoichiometry. The first step is to convert the given amount of CH_4 to moles.

$$n = \frac{m}{M}$$

$$m = 1.5 \text{ kg} = 1500 \text{ g}$$

$$M = (12 + \{1 \times 4\}) \text{ g mol}^{-1} = 16 \text{ g mol}^{-1}$$

$$n = ?$$

We now substitute to obtain

$$n = \frac{1500 \text{ g}}{16 \text{ g}} \times 1 \text{ mol} = 93.75 \text{ mol}$$

$$\begin{array}{lcccc}
 & CH_4(g) + & 2O_2(g) \rightarrow & CO_2(g) + & 2H_2O(l) \\
\text{Mole} & 1 & 2 & 1 & 2 \\
\text{Mole ratio} & 1 & : 2 & : 1 & : 2
\end{array}$$

So, Mole ratio of CO_2 to CH_4 = 1:1.
∴ Number of moles of CO_2 produced at STP = 1 × 93.75 mol = 93.75 mol
As usual, we can set up the calculation as follows.
Let the amount of CO_2 produced at STP be n. According to the equation, 1 mol of CH_4 will produce 1 mol of CO_2. Therefore,

$$\frac{1 \text{ mol of } CH_4}{1 \text{ mol of } CO_2} = \frac{93.75 \text{ mol of } CH_4}{n}$$

So

$$n = 93.75 \text{ mol}$$

The final step is to calculate the volume of 93.7 mol of CO_2 at STP.

$$V = 93.75 \text{ mol} \times \frac{22.4 \text{ dm}^3}{1 \text{ mol}} = 2100 \text{ dm}^3$$

Example 5

Determine the mass of sodium sulfate that would be produced when 50.0 cm³ of a 0.10 mol dm⁻³ solution of sulfuric acid is mixed with excess sodium hydroxide solution. The equation for the reaction is: $H_2SO_4(aq) + 2NaOH(aq) \rightarrow Na_2SO_4(aq) + 2H_2O(l)$.

$$(H = 1, O = 16, Na = 23, S = 32)$$

Solution

The first step is to determine the number of moles of sulfuric acid present in the solution. You will recall that

$$n = C \times V$$

$$C = 0.10 \text{ mol dm}^{-3}$$

$$V = 50.0 \text{ cm}^3 = 0.0500 \text{ dm}^3$$

$$n = ?$$

Substituting, we have

$$n = \frac{0.10 \text{ mol}}{1 \text{ dm}^3} \times 0.0500 \text{ dm}^3 = 0.005 \text{ mol}$$

The next step is to convert 0.005 mol of the acid to mass.

$$n = \frac{m}{M}$$

$$\therefore m = n \times M$$

$$M = (\{1 \times 2\} + 32 + \{16 \times 4\}) \text{ g mol}^{-1} = 98 \text{ g mol}^{-1}$$

$$n = 0.005 \text{ mol}$$

$$m = ?$$

Substituting, we have

$$m = 0.005 \text{ mol} \times \frac{98 \text{ g}}{1 \text{ mol}} = 0.49 \text{ g}$$

Finally, we have to multiply this mass of H_2SO_4 by the mass ratio of Na_2SO_4 to H_2SO_4.

	H_2SO_4 (aq)	+	2NaOH(aq)	→	Na_2SO_4 (aq)	+	$2H_2O$ (l)
Mole	1		2		1		2
Mass	(1×98) g		(2×40) g		(1×142) g		(2×18) g
Mass ratio	98	:	80	:	142	:	36

So, the mole ratio of Na_2SO_4 to $H_2SO_4 = \frac{142}{98}$.

Since the mass of H_2SO_4 supplied is 0.49 g, it follows that the mass, m, of Na_2SO_4 produced is:

$$m = \frac{142}{98} \times 0.49 \text{ g} = 0.71 \text{ g}$$

Alternatively, we can set up the calculation as follows.
   Let the mass of Na_2SO_4 produced from 0.49 g of H_2SO_4 be m. Since the equation shows that 98 g of H_2SO_4 will produce 142 g of Na_2SO_4, it follows that

Chemical Equations and Stoichiometry

$$\frac{98 \text{ g of } H_2SO_4}{142 \text{ g of } Na_2SO_4} = \frac{0.49 \text{ g of } H_2SO_4}{m}$$

So

$$m \times 98\,g = 0.49\,g \times 142\,g$$

$$\therefore m = \frac{0.49\,g \times 142\,g}{98} = 0.71\,g$$

Example 6

Determine the volume of a 0.25 mol dm^{-3} hydrochloric acid solution required for the complete reaction with the calcium oxide produced from the full decomposition of 4.5 g of calcium carbonate. The balanced equations of the reactions are as follows:

$$CaCO_3(s) \rightarrow CaO(s) + CO_2(g)$$

$$CaO(s) + 2HCl(aq) \rightarrow CaCl_2(aq) + H_2O(l)$$

$$(C = 12, O = 16, Ca = 40)$$

Solution

We will begin by calculating the volume of HCl(aq).

$$n = C \times V$$

$$\therefore V = \frac{n}{C}$$

The first step is to calculate the amount of CaO that would be produced from the decomposition of 4.5 g of CaCO$_3$. To do this, we first convert the given mass of CaCO$_3$ to moles.

$$n = \frac{m}{M}$$

$$m = 4.5\,g$$

$$M = \left(40 + 12 + \{3 \times 16\}\right) \text{ g mol}^{-1} = 100 \text{ g mol}^{-1}$$

$$n = ?$$

Now, we substitute the values to obtain

$$n = \frac{4.5\,g}{100\,g} \times 1 \text{ mol} = 0.045 \text{ mol}$$

$$CaCO_3(s) \rightarrow CaO(s) + CO_2(g)$$

Mole	1	1	1
Mole ratio	1 :	1 :	1

So, the mole ratio of CaO to CaCO$_3$ = 1:1.

Since the number of moles of $CaCO_3$ supplied is 0.045 mol, it follows that the number of moles, n, of CaO produced is 0.045 mol.

Alternatively, let the amount of CaO produced from 0.045 mol of $CaCO_3$ be n. Since, according to the equation, 1 mol of $CaCO_3$ produces 1 mol of CaO, it follows that

$$\frac{1 \text{ mol of } CaCO_3}{1 \text{ mol of } CaO} = \frac{0.045 \text{ mol of } CaCO_3}{n}$$

So

$$n = 0.045 \text{ mol}$$

The next step is to determine the number of moles of HCl(aq) required for the complete reaction of 0.045 mol of CaO.

$$CaO(s) \;+\; 2HCl(aq) \;\rightarrow\; CaCl_2(aq) \;+\; H_2O(l)$$

Moles	1	2	1	1
Mole ratio	1 :	2 :	1 :	1

So, the mole ratio of HCl to CaO = 2:1.

Since the amount of CaO supplied is 0.045 mol, it follows that the required amount, n, of HCl(aq) is:

$$n = 2 \times 0.045 \text{ mol} = 0.090 \text{ mol}$$

Alternatively, let the amount of HCl(aq) required for the complete reaction of 0.045 mol of CaO be n. Since, according to the equation, 1 mol of CaO requires 2 mol of HCl(aq), it follows that

$$\frac{2 \text{ mol of } HCl(aq)}{1 \text{ mol of } CaO} = \frac{n}{0.045 \text{ mol of } CaO}$$

So

$$n = 2 \times 0.045 \text{ mol} = 0.090 \text{ mol}$$

The last step is to calculate the volume of the acid solution, which contains 0.090 mol of the acid.

$$V = \frac{n}{C}$$

$$n = 0.090 \text{ mol}$$

$$C = 0.25 \text{ mol dm}^{-3}$$

$$V = ?$$

Finally, we now substitute to obtain

$$V = \frac{0.09 \text{ mol}}{0.25 \text{ mol}} \times 1 \text{ dm}^3 = 0.36 \text{ dm}^3$$

Chemical Equations and Stoichiometry

PRACTICE PROBLEMS

1. Determine the number of moles of chlorine required for the complete reaction of 5.5 mol of iron. The equation for the reaction is: $2Fe(s) + 3Cl_2(g) \rightarrow 2FeCl_3(s)$.
2. Determine the number of moles of hydrogen required for the production of 4.0 mol of ammonia gas. The equation of the reaction is: $N_2(g) + 3H_2(g) \rightleftharpoons 2NH_3(g)$.
3. Consider the following reaction:

$$AlCl_3(s) + 3NaOH(aq) \rightarrow Al(OH)_3(s) + 3NaCl(aq).$$

   Determine the mass of aluminium chloride required for the complete reaction of 1.25 mol of sodium hydroxide.

$$(Al = 27, \; Cl = 35.5)$$

4. 2.5 mol of potassium permanganate was mixed with excess hydrochloric acid solution. Determine the mass of potassium chloride produced. The equation of the reaction is:

$$2KMnO_4(aq) + 16HCl(aq) \rightarrow 2KCl(aq) + 2MnCl_2(aq) + 8H_2O(l) + 5Cl_2(g)$$

$$(Cl = 35.5, \; K = 39)$$

5. 4.5 g of magnesium is mixed with an excess solution of hydrochloric acid. Calculate the mass of magnesium chloride produced. The equation of the reaction is: $Mg(s) + 2HCl(aq) \rightarrow MgCl_2(aq) + H_2(g)$.

$$(Mg = 24, \; Cl = 35.5)$$

6. Sodium peroxide reacts with water at room temperature as follows:

$$2Na_2O_2(aq) + 2H_2O(l) \rightarrow 4NaOH(aq) + O_2(g)$$

   Determine the volume of oxygen that would be produced when 5.8 g of sodium peroxide reacts completely with water at STP.

$$(O = 16, \; Na = 23; \; V_m = 22.4 \; dm^3 mol^{-1})$$

7. On vigorous heating, zinc burns in oxygen to form zinc oxide—a white solid—as shown by the reaction:

$$2Zn(s) + O_2(g) \rightarrow 2ZnO(s)$$

   What mass of zinc oxide would be produced when 50.0 cm³ of oxygen completely reacts with zinc at RTP?

$$(O = 16, \; Zn = 65; \; V_m = 24 \; dm^3 \; mol^{-1})$$

8. One of the chemical properties of ammonia is that it reacts with acids to form salts. An example is given below.

$$2NH_3(g) + H_2SO_4(aq) \rightarrow (NH_4)_2 SO_4(aq)$$

Calculate the mass of ammonium sulfate expected from the complete reaction of 50.0 cm³ of a 0.15 mol dm⁻³ sulfuric acid solution with ammonia gas.

$$(H = 1, \ N = 14, \ O = 16, \ S = 32)$$

9. Chlorine gas dissolves in water to form hypochlorous and hydrochloric acids as follows:

$$H_2O(l) + Cl_2(g) \rightarrow HClO(aq) + HCl(aq)$$

On exposure to sunlight, the hypochlorous acid decomposes as follows:

$$HClO(aq) \rightarrow 2HCl(aq) + O_2(g)$$

Calculate the maximum volume of oxygen that would be produced from the decomposition of the hypochlorous acid produced when 250 cm³ of chlorine completely reacts with water at STP.

$$(V_m = 22.4 \ dm^3 \ mol^{-1})$$

10. 1.5 mol of hydrogen reacts completely with nitrogen to form ammonia gas. What mass of ammonium chloride would be produced if the ammonia produced reacts completely with hydrochloric acid? The equation of the last reaction is: $NH_3(g) + HCl(aq) \rightarrow NH_4Cl(aq)$.

$$(H = 1, \ N = 14, \ Cl = 35.5)$$

5.2.1 Limiting Reagent

When two or more reactants are mixed, the amounts of products formed depend on the amount of one reactant unless the reactants are supplied in stoichiometric proportions—i.e., they are mixed exactly in the right ratio as specified by the equation of reaction.

The *limiting reagent* is the reactant that determines the maximum amount of products that can be formed in a reaction. Typically, it is the reactant that produces the least amount of any product. A reaction stops once all the limiting reagent has been consumed.

Example 1

Burning phosphorus reduces nitrogen monoxide to nitrogen according to the following equation:

$$P_4(s) + 10NO(g) \rightarrow 2P_2O_5(s) + 5N_2(g)$$

Determine:

(a) the limiting reagent when 3.5 mol of phosphorus is mixed with 5.0 mol of nitrogen monoxide;
(b) the amount of nitrogen produced in the above reaction.

Solution

(a) As stated earlier, the limiting reagent is the one that leads to the production of the least amount of a particular product. Thus, we need to calculate the amount of any product, such as N_2, that would be produced by each of the two reactants.

Chemical Equations and Stoichiometry

$$P_4(s) + 10NO(g) \rightarrow 2P_2O_5(s) + 5N_2(g)$$

Moles	1	10	2	5
Mole ratio	1 :	10 :	2 :	5

The amount, n, of N_2 produced from 3.5 mol of phosphorus is calculated as follows. The mole ratio of N_2 to $P_4 = 5:1$.

$$\therefore n = 5 \times 3.5 \text{ mol} = 17.5 \text{ mol}$$

The amount, n, of N_2 produced from 5.0 mol of nitrogen(II) oxide is calculated as follows.

$$\text{Mole ratio of } N_2 \text{ to } NO = \frac{5}{10} = \frac{1}{2}$$

$$\therefore n = \frac{1}{2} \times 5.0 \text{ mol} = 2.5 \text{ mol}$$

The reactants, P and NO, lead to the production of 17.5 mol and 2.5 mol of N_2, respectively. Since the smaller amount was obtained from NO, it is the limiting reagent.

(b) The maximum amount of a product that can be produced in a chemical reaction is the amount produced by the limiting reagent. Consequently, the amount of nitrogen produced is 2.5 mol.

Example 2

Metals that are higher up the Activity Series displace those lower down the series from solutions of their salts. An example is given by the equation:

$$Cu(s) + 2AgNO_3(aq) \rightarrow Cu(NO_3)_2(aq) + Ag(s)$$

Suppose we mix 75.0 g of copper with 88 g of silver nitrate. Determine:

(a) the limiting reagent;
(b) the mass of any reactant left.

$$(H = 1, N = 14, O = 16, Cu = 63.5, Ag = 108)$$

Solution

(a) We will determine the limiting reagent by calculating the mass of silver that would be formed from each of the reactants.

	Cu(s)	+ 2AgNO$_3$(aq)	$\rightarrow$ Cu(NO$_3$)$_2$(aq)	+ Ag(s)
Moles	1	2	1	1
Mass	(1×63.5) g	(2×170) g	(1×187.5) g	(1×108) g
Mass ratio	63.5 :	340 :	187.5 :	108

The mass, m, of silver that would be produced from 75.0 g of copper is calculated as follows.

$$\text{Mass ratio of Ag to Cu} = \frac{108}{63.5}$$

$$\therefore m = \frac{108}{63.5} \times 75.0 \text{ g} = 128 \text{ g}$$

The mass, m, of silver that would be produced from 88 g of $AgNO_3$ is calculated as follows.

$$\text{Mass ratio of Ag to AgNO}_3 = \frac{108}{340}$$

$$\therefore m = \frac{108}{340} \times 88 \text{ g} = 28 \text{ g}$$

The reactants, Cu and $AgNO_3$, lead to the production of 128 g and 28 g of Ag, respectively. Since the smaller amount was obtained from $AgNO_3$, it is the limiting reagent.

(b) The reaction stops once all the given mass of the limiting reactant, $AgNO_3$, has been used up. The mass, m, of Cu that would react with $AgNO_3$ is obtained as follows.

$$\text{Mass ratio of Cu to AgNO}_3 = \frac{63.5}{340}$$

Since the mass of $AgNO_3$ supplied (used up) is 88 g, it follows that

$$m = \frac{63.5}{340} \times 88 \text{ g} = 16.435 \text{ g}$$

Finally, the mass, m, of excess Cu—i.e., the mass of Cu left after the reaction—is obtained by deduction:

$$m = (75.0 - 16.435) \text{ g} = 58.6 \text{ g}$$

PRACTICE PROBLEMS

1. Nitrogen dioxide oxidises red-hot metals and is itself reduced to nitrogen in the process. An example is the reaction below:

$$4Cu(s) + 2NO_2(g) \rightarrow 4CuO(s) + N_2(g)$$

   Determine the limiting reagent when 10 mol of copper is mixed with 15 mol of nitrogen dioxide.

2. Calculate the mass of ammonia that would be produced when 5.0 mol of nitrogen is mixed with 4.5 mol of hydrogen. The equation of reaction is: $N_2(g) + 3H_2(g) \rightleftharpoons 2NH_3(g)$.

$$(H = 1, \ N = 14)$$

3. Calcium carbide is produced industrially by heating calcium oxide with coke (carbon) as shown below:

$$CaO(s) + 3C(s) \rightarrow CaC_2(s) + CO(g)$$

   Calculate the mass of any excess reactant when 45 g of calcium oxide is heated with 35 g of coke.

$$(C = 12, \ O = 16, \ Ca = 40)$$

Chemical Equations and Stoichiometry

4. Insoluble carbonates are precipitated by adding sodium carbonate solution to a solution of an appropriate metallic salt. An example is given by the equation below:

$$CuSO_4(aq) + Na_2CO_3(aq) \rightarrow CuCO_3(s) + Na_2SO_4(aq)$$

Calculate the mass of the salt formed after 150 g of copper(II) sulfate is mixed with 85 g of sodium carbonate.

$$(C = 12, \ O = 16, \ Na = 23, \ S = 32, \ Cu = 63.5)$$

5.2.2 Gas (Volume-Volume) Stoichiometry

According to Avogadro's law, which states that equal volumes of all gases at the same temperature and pressure contain the same number of molecules, the number of moles of a gas is proportional to its volume. Based on this principle, the number of moles (or stoichiometric coefficients) of gases in gas-phase reactions can be treated as volumes, provided that the reactions are under constant conditions of pressure and temperature. An example is given below.

$$2CO(g) \ + \ O_2(g) \ \rightarrow \ 2CO_2(g)$$

Moles	2	1	2
Volumes	2	1	2
Mole / volume ratio	2	: 1	: 2

In performing calculations in gas stoichiometry, it is usually assumed that temperature and pressure are constant if no mention is made of the conditions of the reaction.

Example 1

The combustion of ethene is represented by the following equation:

$$C_2H_4(aq) + 3O_2(g) \rightarrow 2CO_2(g) + 2H_2O(g).$$

What volume of oxygen is required to react with 50.0 cm³ of ethene?

Solution

The procedure is similar to other stoichiometric calculations.

$$C_2H_4(aq) \ + \ 3O_2(g) \ \rightarrow \ 2CO_2(g) \ + \ 2H_2O(g)$$

Mole	1	3	2	2
Mole / Volume ratio	1	: 3	: 2	: 2

So, the volume ratio of O_2 to C_2H_4 = 3:1.
 Since the volume of C_2H_4 supplied is 50.0 cm³, it follows that the volume, V, of O_2 required is:

$$V = 3 \times 50.0 \ cm^3 = 150 \ cm^3$$

Example 2

Propane burns in air according to the following equation:

$$C_3H_8(g) + 5O_2(g) \rightarrow 3CO_2(g) + 4H_2O(g)$$

Calculate the volume of carbon dioxide that would be produced from the complete burning of 25 cm³ of propane?

Solution

We will set up the calculation as follows.

$$C_3H_8(g) + 5O_2(g) \rightarrow 3CO_2(g) + 4H_2O(g)$$

	$C_3H_8(g)$	$5O_2(g)$	$3CO_2(g)$	$4H_2O(g)$
Moles	1	5	3	4
Mole / Volume ratio	1	: 5	: 3	: 4

So, the volume ratio of CO_2 to C_3H_8 = 3:1.
Since the volume of C_3H_8 supplied is 25 cm³, it follows that the volume, V, of CO_2 produced is:

$$V = 3 \times 25 \text{ cm}^3 = 75 \text{ cm}^3$$

Example 3

Butane burns in air according to the following equation:

$$2C_4H_{10}(g) + 13O_2(g) \rightarrow 8CO_2(g) + 10H_2O(g)$$

35 cm³ of propane is sparked with 150 cm³ of air. If all the gases are measured at STP and air contains 21% of oxygen, calculate:

(a) the total volume of all residual gases;
(b) the volume of residual gases after the mixture is passed through excess sodium hydroxide solution.

Solution

(a) Air is made up of 21% oxygen. Therefore, since the volume of air supplied is 150 cm³, it follows that the volume, V, of oxygen supplied is:

$$V = \frac{21}{100} \times 150 \text{ cm}^3 = 31.5 \text{ cm}^3$$

So, the volume of excess (unreacted) air is: (150 − 31.5) cm³ = 114.5 cm³
Since the reactants are not mixed in stoichiometric proportions, we need to determine the limiting reagent. We will determine the limiting reagent based on the volume of CO_2 produced.

$$2C_4H_{10}(g) + 13O_2(g) \rightarrow 8CO_2(g) + 10H_2O(g)$$

	$2C_4H_{10}(g)$	$13O_2(g)$	$8CO_2(g)$	$10H_2O(g)$
Mole	2	13	8	10
Mole / Volume ratio	2	: 13	: 8	: 10

Chemical Equations and Stoichiometry

The volume, V, of CO_2 that would be produced from 35 cm³ of C_4H_{10} is calculated as follows.

$$\text{Volume ratio of } CO_2 \text{ to } C_4H_{10} = \frac{8}{2} = 4:1$$

Since the available volume of C_4H_{10} is 35 cm³, it follows that

$$V = 4 \times 35 \text{ cm}^3 = 140 \text{ cm}^3$$

Similarly, the volume, V, of CO_2 that would be produced from 31.5 cm³ of O_2 is calculated as follows.

$$\text{Volume ratio of } CO_2 \text{ to } O_2 = \frac{8}{13}$$

Since the available volume of O_2 is 31.5 cm³, it follows that

$$V = \frac{8}{13} \times 31.5 \text{ cm}^3 = 19.4 \text{ cm}^3$$

∴ The limiting reagent is O_2. This implies that the volume of CO_2 produced is 19.4 cm³. The volume, V, of C_4H_{10} that would react with 19.4 cm³ of O_2 is calculated as follows.

$$\text{Volume ratio of } C_4H_{10} \text{ to } O_2 = \frac{2}{13}$$

$$\therefore V = \frac{2}{13} \times 31.5 \text{ cm}^3 = 4.85 \text{ cm}^3$$

Thus, the volume of excess (unreacted) C_4H_{10} is: $(35 - 4.85) \text{ cm}^3 = 30.15 \text{ cm}^3$
The next step is to determine the volume, V, of $H_2O(g)$ that would be produced from 31.5 cm³ of O_2.

$$\text{Volume ratio of } H_2O \text{ to } O_2 = \frac{10}{13}$$

$$\therefore V = \frac{10}{13} \times 31.5 \text{ cm}^3 = 24.23 \text{ cm}^3$$

The summary of the results is provided below.

	$2C_4H_{10}(g)$	+	$13O_2(g)$	→	$8CO_2(g)$	+	$10H_2O(g)$
Volume before reaction (cm³)	35		31.5		–		–
Volume after reaction (cm³)	30.15		–		19.4		24.23

The volume, V, of the residual gases is given by

V = volume of excess air + volume of excess C_4H_{10} + volume of CO_2 + volume of $H_2O(g)$
$= (114.5 + 30.15 + 19.4 + 24.23) \text{ cm}^3 = 188 \text{ cm}^3$

(b) Carbon dioxide will react completely with excess sodium hydroxide. Hence, the remaining volume, V, of residual gases is given by

V = volume of excess air + volume of excess C_4H_{10} + volume of $H_2O(g)$
$= (114.5 + 30.15 + 24.23) \text{ cm}^3 = 168.9 \text{ cm}^3$

PRACTICE PROBLEMS

1. Carbon monoxide is oxidised to carbon dioxide as follows:

$$2CO(g) + O_2(g) \rightarrow 2CO_2(g)$$

   What volume of oxygen is required to fully react with 50.00 cm³ of carbon monoxide?

2. Ammonia reacts with oxygen to produce nitrogen and water as follows:

$$4NH_3(g) + 3O_2(g) \rightarrow 2N_2(g) + 6H_2O(l)$$

   What volume of nitrogen would be produced when 150.00 cm³ of ammonia is sparked with 100.00 cm³ of oxygen, assuming the reaction takes place at constant temperature and pressure?

3. 80.00 cm³ of nitrogen monoxide is sparked with 150.00 cm³ of air. Determine the total volume of the resulting gas mixture if all the gases are measured at STP. Assume that air contains 21% of air. The equation of reaction is: $2NO(g) + O_2(g) \rightarrow 2NO_2(g)$.

4. 50.00 cm³ of ethyne is sparked with 200.00 cm³ of oxygen. Determine the total volume of theresulting mixture after it is passed through excess calcium hydroxide solution.

5.2.3 PERCENTAGE YIELD

In practice, the amounts of products obtained from chemical reactions are usually lower than those predicted by stoichiometric calculations. Some factors that could cause this discrepancy include the loss of reactants, side reactions (the formation of substances not shown by the reaction equation) and the presence of impurities. For instance, in the reaction: $CaCO_3(s) \rightarrow CaO(s) + CO_2(g)$, carbon monoxide may also be formed, lowering the amounts of the expected products.

The amount of a product obtained from stoichiometry calculations is called the *theoretical yield*, while the amount obtained in practice is called the *actual yield*. The percentage yield of a product is obtained by expressing its actual yield as a percentage of its theoretical yield. In other words,

$$\% \text{yield} = \frac{\text{Actual yield}}{\text{Theoretical yield}} \times 100\%$$

Example

Potassium chlorate dissociates into potassium chloride and oxygen as follows:

$$2KClO_3(s) \rightarrow 2KCl(s) + 3O_2(g)$$

Given that the actual volume of oxygen obtained from the decomposition of 1.5 g of potassium chlorate is 300.00 cm³ at RTP, determine the percentage yield of the reaction.

$$(O = 16, K = 39, Cl = 35.5; V_m = 24\,dm^3\,mol^{-1})$$

Solution

We have to apply the relation:

$$\% \text{yield} = \frac{\text{Actual yield}}{\text{Theoretical yield}} \times 100\%$$

Chemical Equations and Stoichiometry

The theoretical yield is obtained from stoichiometric calculations.

$$2KClO_3(s) \rightarrow 2KCl(s) + 3O_2(g)$$

Mole   2   2   3
Mole ratio   2   :   2   :   3

The number of moles of $KClO_3$ supplied is obtained as follows.

$$n = \frac{m}{M}$$

$$m = 1.5 \text{ g}$$

$$M = (39 + 35.5 + \{3 \times 16\}) \text{ g mol}^{-1} = 122.5 \text{ g mol}^{-1}$$

$$n = ?$$

Substituting, we have

$$n = \frac{1.5 \text{ g}}{122.5 \text{ g}} \times 1 \text{ mol} = 0.01224 \text{ mol}$$

The number of moles, n, of O_2 that produced from 0.01224 mol of $KClO_3$ is determined as follows.

$$\text{Mole ratio of } O_2 \text{ to } KClO_3 = \frac{3}{2}$$

$$\therefore n = \frac{3}{2} \times 0.01224 \text{ mol} = 0.01836 \text{ mol}$$

We must now convert 0.01836 mol to volume at RTP.

$$n = \frac{V}{24 \text{ dm}^3 \text{ mol}^{-1}}$$

$$\therefore V = n \times 24 \text{ dm}^3 \text{ mol}^{-1}$$

$$V = ?$$

Substituting, we have

$$V = 0.01836 \text{ mol} \times \frac{24 \text{ dm}^3}{1 \text{ mol}} = 0.44064 \text{ dm}^3$$

So

$$\text{Theoretical yield is} = 0.440064 \text{ dm}^3$$

$$\text{Actual yield} = 300.00 \text{ cm}^3 = 0.30 \text{ dm}^3$$

$$\% \text{ yield} = ?$$

Finally, we now substitute the values to obtain

$$\% \text{ yield} = \frac{0.30 \text{ dm}^3}{0.44064 \text{ dm}^3} \times 100\% = 68\%$$

PRACTICE PROBLEMS

1. 2.5 g of copper was obtained when 4.5 g copper(II) oxide was mixed with excess carbon monoxide. Determine the percentage yield of copper. The equation for the reaction is as follows:

$$CuO(s) + CO(g) \rightarrow Cu(s) + CO_2(g)$$

$$(O = 16, Cu = 63.5)$$

2. Insoluble bases are prepared by precipitation. An example is the reaction between copper(II) sulfate and potassium hydroxide to yield potassium sulfate and copper(II) hydroxide. Given that 0.25 g of copper(II) hydroxide was obtained when 50.0 cm³ of a 0.15 mol dm⁻³ solution of potassium hydroxide was mixed with 50.0 cm³ of a 0.20 mol dm³ solution of copper(II) sulfate, determine the percentage yield of the reaction. The equation for the reaction is as follows:

$$CuSO_4(aq) + 2KOH(aq) \rightarrow K_2SO_4(aq) + Cu(OH)_2(s).$$

$$(H = 1, O = 16, Cu = 63.5)$$

5.2.4 THE LAWS OF CHEMICAL COMBINATION

Chemical reactions are described by the following laws:

1. **The Law of Conservation of Mass**: this law states that matter is neither created nor destroyed during a chemical reaction but changes from one form to another. Consequently, mass is conserved in chemical reactions.
2. **The Law of Definite (Constant) Proportions**: this law states that all pure samples of a particular chemical compound contain the same elements combined in the same proportion by mass. For example, all pure samples of water will always contain 11.11% hydrogen and 88.89% oxygen, regardless of the source or method of preparation.
3. **The Law of Multiple Proportions**: this law states that if two elements, A and B, combine to form more than one chemical compound, the various masses of one element, A, which combine separately with a fixed mass of the other element, B, are in a simple multiple ratio. For instance, the different masses of copper that combine with a fixed mass of oxygen to form CuO and Cu_2O are in a simple multiple ratio of 2:1.

Example 1

The following data were obtained when two samples of copper(II) oxide, prepared using different methods, were analysed.

	A	B
Mass of CuO	10.0 g	7.4 g
Mass of Cu	8.0 g	5.7 g

Determine the mass percentage of copper in each sample and state which law of chemical combination your results confirm.

Chemical Equations and Stoichiometry

Solution

The mass percent of copper in each sample is determined from the relation:

$$\text{Mass percentage of Cu} = \frac{\text{Mass of Cu}}{\text{Mass of CuO}} \times 100\%$$

For sample A:

$$\text{Mass percentage of Cu} = \frac{8.0\,\text{g}}{10.0\,\text{g}} \times 100\% = 80\%$$

For sample B:

$$\text{Mass percentage of Cu} = \frac{5.9\,\text{g}}{7.4\,\text{g}} \times 100\% = 79.7\%$$

It can be observed from the results that the percentage of copper in the two samples of the oxide is approximately 80%, despite the different methods of preparation. This confirms the law of definite proportions.

Example 2

2.50 g of copper(I) chloride and 3.80 g of copper(II) chloride were reduced to constant masses of 2.20 g and 3.00 g, respectively. Determine:

(a) the mass of copper that would combine with 1.00 g of oxygen in each case;
(b) the law of chemical combination your results confirm.

Solution

(a) The mass of oxygen in each sample is obtained by deduction as shown below.

	Cu_2O	CuO
Mass of oxide	2.50 g	3.80 g
Mass of copper residue	2.20 g	3.00 g
Mass of oxygen	0.300 g	0.800 g

The mass, m, of Cu that would react with 1.00 g of O_2 in each case is calculated as follows.
For Cu_2O:

$$\text{Mass ratio of Cu to } O_2 = \frac{2.20}{0.300}$$

$$\text{Mass of } O_2 = 1.00\,\text{g}$$

$$\therefore m = \frac{2.20}{0.300} \times 1.00\,\text{g} = 7.33\,\text{g}$$

For CuO:

$$\text{Mass ratio of Cu to } O_2 = \frac{3.00}{0.800}$$

$$\therefore m = \frac{3.00}{0.800} \times 1.00\,g = 3.75\,g$$

(b) The results show that the masses of Cu that react separately with a fixed mass (1 g) of O_2 to form two different oxides of Cu are in the simple ratio of approximately 2:1. This confirms the law of multiple proportions.

PRACTICE PROBLEMS

1. The analysis of 1.51 g and 0.50 g pure samples of sodium chloride, prepared in different-ways, produced 0.91 g and 0.30 g of chlorine, respectively. Determine the mass percent of sodium in each sample and state which law your results verify.
2. The analysis of 1.50 g of iron(II) chloride and 0.85 g of iron(III) chloride produced 0.66 g and 0.29 g of iron, respectively. Determine the mass of iron that would combine with 5.00 g ofchlorine in each case and state which chemical law your results support.

FURTHER PRACTICE PROBLEMS

1. Balance the following chemical equations.
   (a) $Na(s) + O_2(g) \rightarrow Na_2O(s)$
   (b) $4NH_3(g) + O_2(g) \rightarrow H_2O(g) + N_2(g)$
   (c) $Al(NO_3)_3(s) \rightarrow Al_2O_3(s) + O_2(g) + 12NO_2(g)$
   (d) $Al(OH)_3(s) + H_2SO_4(aq) \rightarrow Al_2(SO_4)_3(aq) + H_2O(l)$
   (e) $H_2S(g) + O_2(g) \rightarrow 2H_2O(l) + SO_2(g)$

2. Chlorine displaces bromine gas from sodium bromide solution as follows:

$$Cl_2(g) + 2NaBr(aq) \rightarrow 2NaCl(aq) + Br_2(l)$$

   What mass of bromine would be produced from the complete reaction of 0.50 mol of sodium bromide?

$$(Na = 23,\ Br = 79.9)$$

3. Consider the following reaction:

$$2Pb(NO_3)_2(s) \rightarrow 2PbO(s) + 4NO_2(g) + O_2(g)$$

   Calculate the mass of lead(II) nitrate required to produce 0.25 mol of lead(II) oxide.

$$(N = 14,\ O = 16,\ Pb = 207)$$

4. What volume of hydrogen gas would be produced when excess magnesium reacts with 25.0 cm³ of a 0.15 mol dm³ solution of hydrochloric acid at STP? The equation for the reaction is as follows:

$$Mg(s) + 2HCl(aq) \rightarrow MgCl_2(aq) + H_2(g)$$

$$(V_m = 22.4\,dm^3\,mol^{-1})$$

Chemical Equations and Stoichiometry

5. Sodium nitrate dissociates as follows:

$$2NaNO_3(s) \rightarrow 2NaNO_2(s) + O_2(g)$$

Calculate the volume of oxygen that would be produced from the complete decomposition of 8.5 g of sodium nitrate.

$$(N = 14,\ O = 16,\ Na = 23;\ V_m = 22.4\,dm^3\,mol^{-1})$$

6. 5.0 g of magnesium were mixed with 150 cm³ of carbon dioxide at STP. The equation of reaction is as follows:

$$2Mg(s) + CO_2(g) \rightarrow 2MgO(s) + C(s)$$

Determine the mass of magnesium oxide produced.

$$(O = 16,\ Mg = 24;\ V_m = 22.4\,dm^3\,mol^{-1})$$

7. 150.0 cm³ of a 0.15 mol dm⁻³ solution of sodium hydroxide was mixed with 50.0 cm³ of a 0.10 mol dm⁻³ solution of sulfuric acid. Calculate the mass of sodium nitrate produced.

$$(N = 14,\ = O = 16,\ Na = 23)$$

8. The reduction of iron(III) oxide by hydrogen is given by the following equation:

$$Fe_2O_3(s) + 3H_2(g) \rightarrow 2Fe(s) + 3H_2O(g)$$

Determine the limiting reagent when 5.0 g of iron(III) oxide is mixed with 1.0 g of hydrogen. Hence, determine the mass FOR excess reactant:

$$(H = 1,\ O = 16,\ Fe = 56)$$

9. Determine the mass of excess reactant when 0.5 g of lead(II) sulfide is mixed with 25.00 cm³ of oxygen at STP. The equation FOR the reaction is as follows:

$$2PbS(s) + 3O_2(g) \rightarrow 2PbO(s) + 2SO_2(g)$$

$$(S = 32,\ Pb = 207;\ V_m = 22.4\,dm^3\,mol^{-1})$$

10. 10.0 cm³ of sulfur dioxide was sparked with 25.00 cm³ of oxygen to produce sulfur trioxide. Calculate the volume of the resulting mixture if all gases were measured at STP.

11. 50.00 cm³ of hydrogen was sparked with 150.00 cm³ of air at 1 atm and 150°C. Calculate the volume of the resulting mixture at the same temperature and pressure. Assume that air contains 21% of oxygen.

12. 0.50 g of magnesium hydroxide was produced when 25.0 cm³ of a 0.25 mol dm⁻³ sodium hydroxide solution reacted completely with magnesium sulfate. Determine the percentage yield of the base. The equation for the reaction is as follows:

$$2NaOH(aq) + MgSO_4(aq) \rightarrow Mg(OH)_2(s) + Na_2SO_4(aq)$$

$$(H = 1,\ O = 16,\ Mg = 24)$$

13. 30.0 cm³ of carbon dioxide was produced when 25.00 cm³ of a 0.10 mol dm⁻³ solution of sodium carbonate was mixed with 30.00 cm³ of a 0.15 mol dm⁻³ solution of nitric acid at STP. Determine the percentage yield.

$$\left(V_m = 22.4 \, dm^3 \, mol^{-1}\right)$$

14. The analysis of two different samples of zinc oxide produced the following results.

	A	B
Mass of the sample	1.5 g	2.0 g
Mass of the zinc residue	1.2 g	1.6 g

Show that the results support the law of definite proportions.

15. The analysis of 0.50 g of tin(II) oxide and 0.85 g of tin(IV) oxide produced 0.44 g and 0.67 g of tin, respectively. Determine the mass of tin that would react with 1.0 g of oxygen in each case, and state the law of chemical combination your results support.

6 Redox Reactions

6.1 INTRODUCTION

Oxidation-reduction (redox) reactions are among the most important types of reactions encountered in chemistry. Some examples of redox reactions include burning, rusting and the displacement of metals from salts. As we shall see shortly, redox reactions are now defined in terms of electron transfer. However, the concept of oxidation and reduction dates back to long before the discovery of the electron. The older definitions of oxidation and reduction are based on oxygen transfer, hydrogen transfer and the transfer of electropositive or electronegative elements.

As you will observe from the equations given below, oxidation and reduction cannot occur independently of each other. In fact, oxidation and reduction are *opposite* yet *complementary processes*. In other words, the oxidation of a substance is always accompanied by the reduction of some other substance and vice versa.

6.1.1 OXIDATION AND REDUCTION IN TERMS OF OXYGEN TRANSFER

The concept of oxidation and reduction was originally based on oxygen transfer. According to this definition, oxidation is the *addition of oxygen* to a substance, while reduction is the *removal of oxygen* from a substance.

For example,

$$2ZnO(s) + C(s) \rightarrow 2Zn(s) + CO_2(g)$$

(Reduction: ZnO → Zn; Oxidation: C → CO$_2$)

In the above reaction, carbon has been oxidised to carbon monoxide, while zinc oxide has been reduced to zinc. Zinc oxide is called the *oxidising agent*, while carbon is the *reducing agent*.

6.1.2 OXIDATION AND REDUCTION IN TERMS OF HYDROGEN TRANSFER

Oxidation is the *removal of hydrogen* from a substance, while reduction is the *addition of hydrogen* to a substance. In other words, the definitions of oxidation and reduction in terms of hydrogen are the opposite of those based on oxygen transfer.

Let us consider the reaction below.

$$H_2S(g) + Cl_2(g) \rightarrow 2HCl(g) + S(g)$$

(Reduction: Cl$_2$ → HCl; Oxidation: H$_2$S → S)

In this reaction, hydrogen sulfide has been oxidised to sulfur, while chlorine has been reduced to hydrogen chloride. Hydrogen sulfide is the *reducing agent*, while chlorine is the *oxidising agent*.

6.1.3 Oxidation and Reduction in Terms of Electropositive or Electronegative Elements

In terms of transfer of electronegative elements, oxidation is defined as the *addition of electronegative elements* to a substance, while reduction is the *removal of electronegative elements* from a substance.

For example, consider the reaction below.

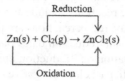

In this reaction, zinc has been oxidised to zinc chloride, while chlorine has been reduced to zinc chloride. Zinc is the *reducing agent*, while chlorine is the *oxidising agent*.

In terms of transfer of electropositive elements, oxidation is the *removal of electropositive elements* from a substance, while reduction is the *addition of electropositive elements* to a substance.

For example,

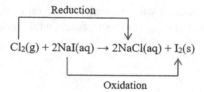

In the above reaction, chlorine has been reduced to sodium chloride, while sodium iodide has been oxidised to iodine. Chlorine is the *oxidising agent*, while sodium iodide is the *reducing agent*.

6.2 MODERN DEFINITIONS OF OXIDATION AND REDUCTION

In the 20th century, chemists discovered that *electron transfer* is common to all redox reactions. In fact, all the old definitions of oxidation and reduction discussed above involve *electron transfer*. Thus, oxidation and reduction are now defined in terms of electron transfer. Oxidation is defined as *a process involving electron loss*, while reduction is *a process involving electron gain*. An oxidising agent is *an electron acceptor*, while a reducing agent is *an electron donor*.

For example, in the redox reaction: $Zn(s) + Cl_2(g) \rightarrow ZnCl_2(s)$, Zn has been oxidised to Zn^{2+} (zinc ion) by losing two electrons, while each of the two atoms in Cl_2 has been reduced to Cl^- (chloride ion) by gaining an electron as shown below.

$$Zn - 2e^- \rightarrow Zn^{2+} \left(\text{oxidation}\right)$$

$$Cl_2 + 2e^- \rightarrow 2Cl^- \left(\text{reduction}\right)$$

There are many helpful mnemonics for remembering the definitions of oxidation and reduction. A common one is '**OIL RIG**': **O**xidation **I**s **L**oss of electron, **R**eduction **I**s **G**ain of electrons (Figure 6.1).

Redox Reactions

	Oxidation	Reduction
Oxygen	Gain	Loss
Hydrogen	Loss	Gain
Electronegative element	Gain	Loss
Electropositive element	Loss	Gain
Electron	Loss	Gain

FIGURE 6.1 Definitions of oxidation and reduction.

6.2.1 Oxidation Number

The oxidation number of an element represents *the apparent electrical charge an atom has in a compound or ion, determined by a set of arbitrary rules*. The rules for assigning oxidation numbers to elements are as follows.

1. The oxidation number of all elements in their uncombined (free) state is zero. For example, in Cl_2, Na and S_2, the oxidation number is 0.
2. The oxidation number of all Group 1 elements (e.g., Na, K, Li and H) is +1, except in metal hydrides (e.g., NaH), where hydrogen has an oxidation number of −1.
3. The oxidation number of all Group 2 elements (e.g., Mg, Ca, Ba and Be) is +2.
4. Oxygen usually has an oxidation number of −2, except in peroxides (e.g., Na_2O_2 and H_2O_2), where it is −1
5. The oxidation number of halogens (e.g., F, Br, Cl and I) is usually −1, except when bonded to oxygen or a more electronegative element. Fluorine, being the most electronegative element, always has an oxidation number of −1.
6. The oxidation number of simple ions corresponds to their charge. For example, the oxidation numbers of Fe^{2+}, Al^{3+}, Na^+ and Cl^- are +2, +3, +1 and −1, respectively.
7. The sum of oxidation numbers in a polyatomic ion equals its overall charge. For example, in OH^-, the oxidation numbers of oxygen and hydrogen must sum to −1.
8. The sum of oxidation numbers in a compound is always zero. For example, in H_2O, the oxidation numbers of oxygen and hydrogen must sum to 0.

It is important to emphasise that, unlike onic charge, where the '+' or '−' sign follows a number, an oxidation number must be preceded by the '+' or '−' sign follows the number (e.g., Al^{3+}), oxidation numbers must be written with the sign first (e.g., +3 for Al^{3+}).

We will now illustrate these rules with the following examples.

Example 1

Determine the oxidation number of sulfur in H_2SO_4.

Solution

Let the oxidation number of sulfur be x.

$$H = +1$$
$$O = -2$$
$$H_2SO_4 = 0$$
$$S = x$$

So

$$(1 \times 2) + x + (-2 \times 4) = 0$$
$$2 + x - 8 = 0$$
$$\therefore x = 8 - 2 = +6$$

Example 2

Determine the oxidation number of manganese in $KMnO_4$.

Solution

Let the oxidation number of manganese be x.

$$K = +1$$
$$O = -2$$
$$KMnO_4 = 0$$
$$Mn = x$$

So

$$1 + x + (-2 \times 4) = 0$$
$$1 + x - 8 = 0$$
$$\therefore x = 8 - 1 = +7$$

Example 3

Determine the oxidation number of chromium in $Cr_2O_7^{2-}$.

Solution

Let the oxidation number of chromium be x.

$$O = -2$$

Redox Reactions

$$Cr_2O_7^{2-} = -2$$
$$Cr = x$$

So
$$(x \times 2) + (-2 \times 7) = -2$$
$$2x - 14 = -2$$
$$2x = -2 + 14 = +12$$
$$\therefore x = \frac{+12}{2} = +6$$

Example 4

Determine the oxidation number of platinum in [Pt(NH$_3$)$_2$Cl$_4$].

Solution

Let the oxidation number of platinum be x.

$$Cl = -1$$
$$\left[Pt(NH_3)_2 Cl_4\right] = 0$$
$$Pt = x$$

So
$$x + (0 \times 2) + (-1 \times 4) = 0$$
$$x - 4 = 0$$
$$\therefore x = +4$$

PRACTICE PROBLEMS

1. Determine the oxidation number of copper in each of the following substances.
    (a) Cu^+
    (b) Cu^{2+}
2. Determine the oxidation number of nitrogen in each of the following substances.
    (a) HNO_3
    (b) HNO_2
    (c) NH_4Cl
    (d) N_2O_4
    (e) N_2O
    (f) NO_2
    (g) NO_3^-
3. Determine the oxidation number of the elements in bold.
    (a) [**Fe**(OH$_2$)$_6$]$^{3+}$
    (b) H$_3$**P**O$_4$

(c) $KClO_3$
(d) HCO_3^-
(e) Na_2SiO_3

6.2.1.1 Use of Oxidation Number in Defining and Identifying Redox Reactions

When a substance loses electrons, its *oxidation number increases*. Conversely, when a substance gains electrons, its *oxidation number decreases*. Thus, oxidation and reduction can also be defined in terms of oxidation numbers. Oxidation is *a process involving an increase in oxidation number*, while reduction is *a process involving a decrease in oxidation number*.

Consequently, a redox reaction is any reaction in which there is *a change in oxidation number*. An oxidising agent is a substance *whose oxidation number reduces*, meaning it gains elctrons. A reducing agent is a substance *whose oxidation number increases*, meaning it loses electrons.

This is illustrated by the following equations.

$$Mg(s) - 2e^- \rightarrow Mg^{2+}$$
$$(0) \qquad\qquad (+2) \text{ (oxidation)}$$

$$1/2 O_2 + 2e^- \rightarrow O^{2-}$$
$$(0) \qquad\qquad (-2) \text{ (reduction)}$$

Note: the oxidation number of each substance is written below it in parentheses.

As we shall see shortly, oxidation numbers play a key role in assigning systematic names to certain inorganic compounds.

Examples

Identify the redox reactions in the following list.

1. $NaOH(aq) + HCl(aq) \rightarrow NaCl(aq) + H_2O(l)$
2. $CaCO_3(s) \rightarrow CaO + CO_2(g)$
3. $2FeCl_2(aq) + Cl_2(g) \rightarrow 2FeCl_3(aq)$
4. $2KClO_3(s) \rightarrow 2KCl(aq) + 3O_2(g)$
5. $C(s) + ZnO(s) \rightarrow CO_2(g) + 2Zn(s)$

Solution

We need to determine whether any of the elements in each reaction undergoes a change in oxidation number from reactants to products. The oxidation number of each element is specified below it in parentheses.

1. $NaOH(aq) \quad + \quad HCl(aq) \quad \rightarrow \quad NaCl(aq) \quad + \quad H_2O(l)$
   $(+1)(-2)(+1) \qquad (+1)(-1) \qquad\qquad (+1)(-1) \qquad\quad (+1)(-2)$

We observe that there is no change in oxidation number; this is an acid-base reaction, not a redox reaction.

2. $CaCO_3(s) \quad \rightarrow \quad CaO \quad + \quad CO_2(g)$
   $(+2)(+4)(-2) \qquad\quad (+2)(-2) \qquad (+4)(-2)$

There is no change in oxidation number; this is a thermal decomposition reaction, not a redox reaction.

3. $2FeCl_2(aq) \quad + \quad Cl_2(g) \quad \rightarrow \quad 2FeCl_3(aq)$
   $(+2)(-1) \qquad\qquad (0) \qquad\qquad (+3)(-1)$

Redox Reactions

Iron's oxidation number increases from +2 to +3 (oxidation), while chlorine's oxidation number reduces from 0 to −1. This is a redox reaction.

4. $2KClO_3(s) \quad + \quad 2KCl(aq) \quad \rightarrow \quad 3O_2(g)$
$(+1)(+5)(-2) \quad\quad (+1)(-1) \quad\quad\quad (0)$

Chlorine's oxidation number reduces from +5 to −1 (reduction), and oxygen's oxidation number increases from −2 to 0 (oxidation). This is a redox reaction.

5. $C(s) \quad + \quad ZnO(s) \quad \rightarrow \quad CO_2(g) \quad + \quad 2Zn(s)$
$(0) \quad\quad\quad (+2)(-2) \quad\quad (+4)(-2) \quad\quad (0)$

Carbon's oxidation number increases 0 to +4 (oxidation), while zinc's oxidation number reduces from +2 to 0 (reduction). This is a redox reaction.

PRACTICE PROBLEMS

Determine whether the following reactions are redox reactions.
1. $2HNO_3(aq) + K_2CO_3(aq) \rightarrow 2KNO_3(aq) + H_2O(l) + CO_2(g)$
2. $CH_4(g) + 2O_2(g) \rightarrow CO_2(g) + 2H_2O(l)$
3. $NH_4Cl(s) \rightarrow NH_3(g) + HCl(g)$
4. $H_2S(g) + I_2(g) \rightarrow 2HI(g) + S(s)$
5. $2NH_3(g) + H_2SO_4(aq) \rightarrow (NH_4)_2SO_4(aq)$

6.3 NOMENCLATURE OF POLYATOMIC IONS AND INORGANIC COMPOUNDS

Nomenclature refers to a system of naming. In chemistry, the body responsible for establishing the conventions or rules for naming compounds is the **I**nternational **U**nion of **P**ure and **A**pplied **C**hemists (IUPAC). The system of naming compounds based on the set of rules stipulated by the IUPAC is called *IUPAC Nomenclature*. In this naming system, a unique name is assigned to each substance in such a way that its composition or chemical formula can be easily deduced from the name. This type of name is often called a *systematic name*.

Nevertheless, the traditional or common names of many compounds—established long before the existence of the IUPAC—are still widely used alongside their IUPAC names. This persistence is mainly due to historical reasons and intent to facilitate learning.

6.3.1 BINARY IONIC COMPOUNDS

Binary ionic compounds consist of only two elements. Examples include NaCl, NaI, CuO and KI. The first step in naming an ionic compoundis to state the name the metallic element (which is usually the first element in the formula), as it appears in the Periodic Table. This is followed by its oxidation number, written in Roman numerals within parentheses, except when the element has only one possible oxidation state. Finally, the non-metallic element is named by modifying its name to end in '–ide.' Some examples are given in Table 6.1.

6.3.2 BINARY COVALENT (MOLECULAR) COMPOUNDS

Binary molecular compounds, also known as binary covalent compounds, are composed mainly of two non-metals. Examples include NH_3, H_2O, CO_2, CO, HCl, and N_2O.

There are two versions of IUPAC nomenclature for binary molecular compounds. For example, the gas CO_2 can be called either carbon dioxide or carbon(IV) oxide, depending on regional conventions or the examination board.

In the first version, the first element is named as it appears in the Periodic Table, followed by the name of the second element, modified to end in '–ide.' Additionally, a Greek prefix is used to

TABLE 6.1
IUPAC Names of Binary Ionic Compounds

Compound	IUPAC Name
NaCl	Sodium chloride
CaO	Calcium oxide
NaH	Sodium hydride
CaC_2	Calcium carbide
Cu_2O	Copper(I) oxide
CuO	Copper(II) oxide
$CoCl_2$	Cobalt(II) chloride
$FeCl_2$	Iron(II) chloride
$FeCl_3$	Iron(III) chloride
KI	Potassium iodide

TABLE 6.2
IUPAC Names of Binary Molecular Compounds

Compound	Version I	Version II
CO	Carbon monoxide	Carbon(II) oxide
CO_2	Carbon dioxide	Carbon(IV) oxide
SO_3	Sulfur trioxide	Sulfur(VI) oxide
N_2O	Dinitrogen monoxide	Nitrogen(I) oxide
NO_2	Nitrogen dioxide	Nitrogen(IV) oxide
P_4O_6	Tetraphosphorus hexaoxide	Phosphorous(III) oxide
P_4O_{10}	Tetraphosphorus decaoxide	Phosphorous(V) oxide
SF_4	Sulfur tetrafluoride	Sulfur(IV) fluoride
$AlCl_3$	Aluminium trichloride	Aluminium chloride

indicate the number of atoms of each element in the chemical formula: *mono-* (1), *di-* (2), *tri-* (3), *tetra-* (4), *penta-* (5), *hexa-* (6), *hepta-* (7), *octa-* (8), *nona-* (9), *deca-*(10), etc. However, the prefix *mono-* is never used for the first element. This is the official IUPAC naming convention for binary molecular compounds and is by far the most widely used system. It is the naming convention adopted in this book.

In the second system, binary molecular compounds are named similarly to binary ionic compounds. Some examples of compound names based on both conventions are provided in Table 6.2.

It is important to note that NH_3 (ammonia) and H_2O (water) are exceptions to both naming conventions. Their traditional names are retained in the IUPAC system of nomenclature.

6.3.3 OXYANIONS

Oxyanions are a group of polyatomic ions containing oxygen, such as SO_4^{2-}, CO_3^{2-}, PO_4^{3-}, CO_3^{-}, BrO_3^{-}, IO_3^{-} and NO_3^{-}. The general formula of oxyanions is $A_xO_y^{z-}$, where A is a non-metal called the central atom.

There are two versions of the IUPAC nomenclature of oxyanions in use. For example, SO_4^{2-} may be called either the sulfate ion or the tetraoxosulfate(VI) ion, depending on regional conventions or the examination board.

TABLE 6.3
IUPAC Names of Polyatomic Ions

Oxyanion	Version I	Version II
CO_3^{2-}	Carbonate	Carbonate(IV)
HCO_3^-	Hydrogencarbonate	Hydrogencarbonate(IV)
SO_4^{2-}	Sulfate	Tetraoxosulfate(VI)
SO_3^{2-}	Sulfite	Trioxosulfate(IV)
$S_2O_3^{2-}$	Thiosulfate	Trioxosulfate(II)
$S_2O_2^{2-}$	Thiosulfite	Dioxosulfate(I)
HSO_4^-	Hydrogensulfate	Hydrogentetraoxosulfate(VI)
ClO^-	Hypochlorite	Oxochlorate(I)
ClO_2^-	Chlorite	Dioxochlorate(III)
ClO_3^-	Chlorate	Trioxochlorate(V)
ClO_4^-	Perchlorate	Tetraoxochlorate(VII)
PO_4^{3-}	Phosphate	Tetraoxophosphate(V)
NO_3^-	Nitrate	Trioxonitrate(V)
NO_2^-	Nitrite	Dioxonitrate(III)
$H_2PO_4^-$	Dihydrogenphosphate	Hydrogentetraoxophosphate(V)
BrO_3^-	Bromate	Trioxobromate(V)

6.3.3.1 Version I

The most widely used and official IUPAC naming convention for oxyanions involves indicating the relative number of oxygen atoms in an ion by modifying the name of the central atom to end in *–ate* or *–ite*. This is the method adopted in this book.

For oxyanions where the central atom is not a halogen, the name of the central atom is modified to end in *–ate* if the ion contains the maximum number of oxygen atoms. If an ion contains contains one fewer oxygen atom than the *–ate* ion, its name is modified to end in *–ite*.

When the central atom is a halogen, the ion can have a maximum of four oxygen atoms and a minimum of one. This means halogens can form four different oxyanions: AO_4^-, AO_3^-, AO_2^- and AO^-, where A represents the halogen. An oxyanion of the form AO_4^- is named by adding the prefix *per-* to the halogen's name, which is then modified to end in *–ate*. An oxyanion of the form AO_3^- is named by simply modifying the halogen's name to end in *–ate*. An oxyanion of the form AO_2^- is named by modifying the halogen's name to end in *–ite*. The oxyanion with the fewest oxygen atom (AO^-) is named by adding the prefix *hypo-* to the halogen's name, which is then modified to end in *–ite*.

Additionally, the name of an oxyanion must be modified appropriately when it reacts with another element to form a new ion. The prefix *thio-* is added if one of the oxygen atoms in an oxyanion has been replaced with sulfur atom. When an oxyanion bonds with a hydrogen atom, its name must be prefixed with hydrogen. If more than one hydrogen atom is present, a Greek prefix is used to indicate the number of hydrogen atoms.

Examples of oxyanions and other polyatomic ions illustrating these rules are given in Table 6.3.

6.3.3.2 Version II

An alternative version of the IUPAC naming convention for oxyanions, used in a handful of educational systems, specifies the oxidation number of the central atom (in Roman numerals, in parentheses), and uses Greek prefixes to indicate the number of oxygen atoms in the ion. This method often produces systematic names that are somewhat complex and less commonly used.

TABLE 6.4
IUPAC Names of Compounds of Polyatomic Ions

Compound	Version I	Version II
$AgNO_3$	Silver nitrate	Silver trioxonitrate(V)
$Al_2(SO_4)_3$	Aluminium sulfate	Aluminium tetraoxosulfate(VI)
$(NH_4)_2CO_3$	Ammonium carbonate	Ammonium trioxocarbonate(IV)
$Ca(OH)_2$	Calcium hydroxide	Calcium hydroxide
$FeSO_4$	Iron(II) sulfate	Iron(II) tetraoxosulfate(VI)
$Fe_2(SO_4)_3$	Iron(III) sulfate	Iron(III) tetraoxosulfate(VI)
$NaNO_3$	Sodium nitrate	Sodium trioxonitrate(V)
$KMnO_4$	Potassium permanganate	Potassium heptaoxomanganate(VII)
$CuSO_4$	Copper(II) sulfate	Copper(II) tetraoxosulfate(VI)

In this version, oxygen is named first by modifying its name to *oxo*. If more than one oxygen atom is present, a Greek prefix is added before *oxo* to indicate the number of oxygen atoms. The central atom is then named with its name modified to end in *–ate*. Finally, the oxidation number of the central atom is written in Roman numerals, in parentheses.

When hydrogen combines with an oxyanion to form a new ion, the word *hydrogen* is prefixed to the original name of the ion.

Examples of oxyanions and other polyatomic ions illustrating these rules are provided in Table 6.3. In both naming systems, the inclusion of the word *ion* in the name of an oxyanion is optional, and most texts omit it.

It is important to note that these naming rules do not apply to some commonly encountered polyatomic ions such as OH^-, NH_4^+ and CN^-, which are known by their common names: hydroxide, ammonium and cyanide. It is recommended that students memorise these names.

6.3.4 COMPOUNDS OF POLYATOMIC IONS

The compounds containing polyatomic ions are named similarly to binary ionic compounds, with the polyatomic ion treated as a single unit. It is important to note that these compounds can be named using two different nomenclature systems, depending on the chosen convention for the polyatomic ion. Examples are given in Table 6.4.

6.3.5 OXOACIDS

Oxoacids are compounds containing oxyanions and hydrogen as the only electropositive element. Some examples include H_2SO_4, HNO_3, $HClO$, H_3PO_4 and H_2CO_3. The names of oxoacids are based on the names of their oxyanions, and there are two possible naming conventions, depending on the system adopted for the oxyanion.

The most widely used naming system—and the official IUPAC nomenclature—is that based on Version I of the naming convention for oxyanions, as outlined above. In this version, an acid containing an *-ate* oxyanion is named by modifying the name of the ion to end in *-ic*, followed by the word *acid*. An acid containing an *-ite* oxyanion is named by modifying the name of the ion to end in *-ous*, followed by the word *acid*.

In the second version, the word *acid* is simply added to the name of an oxyanion based on Version II of the oxyanion nomenclature.

These rules are illustrated in Table 6.5.

Redox Reactions

TABLE 6.5
IUPAC Names of Oxoacids

Oxoacid	Version I	Version II
HNO_3	Nitric acid	Trioxonitrate(V) acid
HNO_2	Nitrous acid	Dioxonitrate(III) acid
H_2SO_4	Sulfuric acid	Tetraoxosulfate(VI) acid
H_2CO_3	Carbonic acid	Trioxosulfate(IV) acid
$HClO$	Hypochlorous acid	Trioxocarbonate(IV) acid
H_3PO_4	Phosphoric acid	Oxochlorate(I) acid
$HBrO_3$	Bromic acid	Tetraoxophosphate(V) acid

TABLE 6.6
IUPAC Names of Oxoacids

Hydrohalic Acid	IUPAC Name
HCl(aq)	Hydrochloric acid
HF(aq)	Hydrofluoric acid
HBr(aq)	Hydrobromic acid
H_2S(aq)	Hydrosulfuric acid
HI(aq)	Hydroiodic acid

6.3.6 Aqueous Solutions of Hydrogen Halides

Hydrogen halides are compounds formed between hydrogen and the halogens. Examples include HCl, HBr, HF and HI. These compounds form acids—known as hydrohalic acids—when dissolved in water. A hydrohalic acid is named by adding the prefix *hydro-* to the name of the halogen, modifying the ending to *-ic*, and then adding the word *acid*. Examples are given in Table 6.6.

6.3.7 Hydrated Salts

Hydrated salts are salts that contain one or more molecules of water of crystallisation. Examples include $CuSO_4.5H_2O$, $Na_2CO_3.10H_2O$, $Na_2CO_3.H_2O$ and $FeSO_4.7H_2O$. A hydrated salt is named by first naming the anhydrous salt (the salt without water of crystallisation) as usual, followed by the word *hydrate*, which is prefixed by the Greek numeral indicating the number of water molecules. Examples are given in Table 6.7.

6.3.8 Double Salts (Alums)

Double salts, commonly called alums, are hydrated salts containing three different ions: two cations and an anion. Examples include $KCr(SO_4)_2.12H_2O$, $(NH_4)_2Fe(SO_4)_2.6H_2O$ and $KAl(SO_4)_2.12H_2O$.

To name a double salt, the less electropositive cation is named first, followed by the second cation. If a cation has more than one possible oxidation state, its oxidation number should be specified using Roman numerals in parentheses. Finally, the anion is named, followed by the number of molecules of water, as done for other hydrated salts. Examples are given in Table 6.8.

TABLE 6.7
IUPAC Names of Hydrated Salts

Hydrated Salt	Version I	Version II
$CuSO_4 \cdot H_2O$	Copper(II) sulfate monohydrate	Copper(II) tetraoxosulfate(VI) monohydrate
$CuSO_4 \cdot 5H_2O$	Copper(II) sulfate pentahydrate	Copper(II) tetraoxosulfate(VI) pentahydrate
$Na_2CO_3 \cdot 10H_2O$	Sodium carbonatedecahydrate	Sodium trioxocarbonate(IV) decahydrate
$FeSO_4 \cdot 7H_2O$	Iron(II) sulfate heptahydrate	Iron(II) tetraoxosulfate(VI) heptahydrate
$CoCl_2 \cdot 6H_2O$	Cobalt(II) chloride hexahydrate	Cobalt(II) chloride hexahydrate

TABLE 6.8
IUPAC Names of Double Salts

Double Salt	Version I	Version II
$(NH_4)_2Fe(SO_4)_2 \cdot 6H_2O$	Ammonium iron(II) sulfate hexahydrate	Ammonium iron(II) tetraoxosulfate(VI) hexahydrate
$KAl(SO_4)_2 \cdot 12H_2O$	Aluminium potassium sulfate dodecahydrate	Aluminium potassium tetraoxosulfate(VI) dodecahydrate
$KCr(SO_4)_2 \cdot 12H_2O$	Chromium(III) potassium sulfate dodecahydrate	Chromium(III) potassium tetraoxosulfate(VI) dodecahydrate

PRACTICE PROBLEMS

State the IUPAC names of the following substances.
1. (a) PCl_5
   (b) SiF_4
   (c) CS_2
   (d) P_4O_6
   (e) Al_2O_3
2. (a) H_3PO_3
   (b) K_2CrO_4
   (c) $CoCl_2$
   (d) $Ca(H_2PO_4)_2$
   (e) N_2O_5
3. (a) IO_2^-
   (b) SiO_3^{2-}
   (c) $Zn(NO_3)_2 \cdot 6H_2O$
   (d) HIO_2
   (e) SnO_2

6.4 BALANCING REDOX EQUATIONS

A simple way of balancing redox equations is by the method of half-equations, where separate equations are written for oxidation and reduction. This method follows these steps:

1. Deduce the oxidising and reducing agents along with their respective products.
2. Write the half-equations for oxidation and reduction, then balance them as follows:

Redox Reactions

- Use appropriate numerical coefficients to balance the atoms and charges.
- Balance oxygen atoms using H_2O molecules.
- Balance hydrogen atoms using H^+ in acidic media and OH^- in basic media.
- Ensure that the number of electrons lost by the reducing agent equals the number of electrons gained by the oxidising agent.

3. Combine the two half-equations, cancel out the electrons, and write the overall balanced equation.

This procedure is illustrated with the following examples.

Example 1

Break the equation $Zn(s) + Cu^+(aq) \rightarrow Zn^+(aq) + Cu(s)$ into two half-equations—one for reduction and one for oxidation.

Solution

From the changes in oxidation number, we can see that the reducing agent (the substance that is oxidised) is $Zn(s)$, which is oxidised to Zn^{2+}. Thus, the oxidation half-equation is:

$$Zn(s) \rightarrow Zn^{2+}(aq)$$

The Zn atoms on both sides are already balanced. However, the charges must be balanced, as the total charge on the LHS is 0, while the RHS has a charge of +2. To balance the charges, we add two electrons ($2e^-$) to the RHS:

$$Zn(s) \rightarrow Zn^{2+}(aq) + 2e^-$$

Similarly, the oxidising agent (the substance that is reduced) is Cu^{2+}, which is reduced to Cu. Thus, the reduction half-equation is:

$$Cu^{2+}(aq) \rightarrow Cu(s)$$

The Cu atoms on both sides are already balanced. To balance the charges, we add the two electrons transferred from the reducing agent to the oxidising agent:

$$Cu^{2+}(aq) + 2e^- \rightarrow Cu(s)$$

Example 2

Write the balanced ionic equation for the redox reaction between potassium iodide, KI, solution and iron(III) sulfate, $Fe_2(SO_4)_3$, solution.

Solution

The reducing agent is I^-, which is oxidised to I_2. Thus, the oxidation half-equation is written as:

$$I^-(aq) \rightarrow I_2(aq)$$

Balancing the atoms and charges, we add two electrons ($2e^-$) to the RHS:

$$2I^-(aq) \rightarrow I_2(aq) + 2e^-$$

The oxidising agent is Fe^{3+}, which is reduced to Fe^{2+}. Thus, the reduction half-equation is written as:

$$Fe^{3+}(aq) \rightarrow Fe^{2+}(aq)$$

To balance the charges, we add one electron (e⁻) to the LHS:

$$Fe^{3+}(aq) + e^- \rightarrow Fe^{2+}(aq)$$

The number of electrons transferred from the reducing agent must equal the number of electrons gained by the oxidising agent. In other words, the number of electrons in both half-equations must be the same. To achieve this, we multiply the reduction half-equation by 2:

$$2Fe^{3+}(aq) + 2e^- \rightarrow 2Fe^{2+}(aq)$$

Now, we combine the two half-equations:

$$2I^-(aq) + 2Fe^{3+}(aq) + 2e^- \rightarrow I_2(aq) + Fe^{2+}(aq) + 2e^-$$

Finally, we cancel out the electrons on both sides to obtain the overall balanced equation:

$$2I^-(aq) + 2Fe^{3+}(aq) \rightarrow I_2(aq) + Fe^{2+}(aq)$$

Example 3

Write the ionic equation for the redox reaction between potassium permanganate and iron(II) sulfate in an acidic medium.

Solution

The reducing agent is Fe^{2+}, which is oxidised to Fe^{3+}. Thus, the oxidation half-equation is:

$$Fe^{2+}(aq) \rightarrow Fe^{3+}(aq) + e^-$$

The oxidising agent is MnO_4^-, which is reduced to Mn^{2+}. Thus, the reduction half-equation is:

$$MnO_4^-(aq) \rightarrow Mn^{2+}(aq)$$

Next, we balance the oxygen atoms by adding 4 H_2O molecules to the RHS:

$$MnO_4^-(aq) \rightarrow Mn^{2+}(aq) + 4H_2O(l)$$

Since the reaction occurs in an acidic medium, we balance the newly introduced hydrogen atoms by adding 8 H⁺ to the LHS:

$$MnO_4^-(aq) + 8H^+(aq) \rightarrow Mn^{2+}(aq) + 4H_2O(l)$$

Now, we balance the charges. The total charge on the LHS is +7 (−1 from MnO_4^- and +8 from H⁺), while the total charge on the RHS is +2 from Mn^{2+}. To balance this, we add five electrons (5e⁻) to the LHS:

$$MnO_4^-(aq) + 8H^+(aq) + 5e^- \rightarrow Mn^{2+}(aq) + 4H_2O(l)$$

Now, we ensure that the number of electrons transferred by the reducing agent equals the number gained by the oxidising agent. To do this, we multiply the oxidation half-equation by 5:

$$5Fe^{2+}(aq) \rightarrow 5Fe^{3+}(aq) + 5e^-$$

Now, we combine both half-equations to obtain

$$MnO_4^-(aq) + 8H^+(aq) + 5Fe^{2+}(aq) + 5e- \rightarrow Mn^{2+}(aq) + 4H_2O(l) + 5Fe^{3+}(aq) + 5e^-$$

Finally, we cancel out the electrons on both sides to obtain the balanced redox equation:

$$MnO_4^-(aq) + 8H^+(aq) + 5Fe^{2+}(aq) \rightarrow Mn^{2+}(aq) + 4H_2O(l) + 5Fe^{3+}(aq)$$

PRACTICE PROBLEMS

1. Break the following redox reactions into two half-equations:
   (a) $Cl_2(g) + 2I^-(aq) \rightarrow 2Cl^-(aq) + I_2(aq)$
   (b) $Cu(s) + 2Ag^+(aq) \rightarrow Cu^{2+}(aq) + 2Ag(s)$
   (c) $Fe(s) + 2Ag^+(aq) \rightarrow Fe^{2+}(aq) + 2Ag(s)$

2. Balance the redox reaction between potassium dichromate and iron(II) sulfate in an acidic medium.

FURTHER PRACTICE PROBLEMS

1. Determine the oxidation numbers of the atoms in bold.
   (a) $H_2\mathbf{Si}O_3$
   (b) $Na_2\mathbf{S}_2O_3$
   (c) $H_2\mathbf{P}O_4^-$
   (d) $Al_2\mathbf{Si}_2O_7$
   (e) $K_2[\mathbf{Ni}(CN)_4]$

2. Determine which of the following equations are redox reactions.
   (a) $N_2(g) + 3H_2(g) \rightarrow 2NH_3(g)$
   (b) $PCl_5(s) \rightleftharpoons PCl_3(l) + Cl_2(g)$
   (c) $Cu(s) + N_2O(g) \rightarrow N_2(g) + CuO(s)$
   (d) $Fe_2O_3(s) + 3H_2(g) \rightarrow 2Fe(s) + 3H_2O(g)$
   (e) $2KMnO_4(aq) + 16HCl(aq) \rightarrow 2KCl(aq) + 2MnCl_2(aq) + 8H_2O(l) + 5Cl_2(g)$

3. State the IUPAC names of the following:
   (a) H_2SiO_3
   (b) $Na_2S_2O_3$
   (c) SnO
   (d) Hg_2SO_4
   (e) HIO_3
   (f) P_4O_{10}
   (g) $Cr_2O_7^{2-}$
   (h) $Cu(NO_3)_2.3H_2O$
   (i) $KMgCl_3.6H_2O$
   (j) BrO_2^-

4. Split the following redox equations into half-equations.
    (a) $2Fe(s) + 3Cl_2(g) \rightarrow 2Fe^{3+}(aq) + 6Cl^-(aq)$
    (b) $I_2(aq) + 2S_2O_3^{2-}(aq) \rightarrow 2I^-(aq) + S_4O_6^{2-}(aq)$
    (c) $2MnO_4^-(aq) + 16H^+(aq) + 5C_2O_4^{2-}(aq) \rightarrow 2Mn^{2+}(aq) + 8H_2O(l) + 10CO_2(g)$
    (d) $8I^-(aq) + MnO_4^{2-}(aq) + 16H^+(aq) \rightarrow 4I_2(g) + 2Mn^{2+}(aq) + 8H_2O(l)$
    (e) $5SO_2(g) + 2H_2O(l) + 2MnO_4^-(aq) \rightarrow 2Mn^{2+}(aq) + 4H^+(aq) + 5SO_4^{2-}(aq)$

5. Write a balanced ionic equation for the following reactions.
    (a) The redox reaction between chromium(III) hydroxide and potassium hypochlorite in a basic medium.
    (b) The redox reaction between potassium permanganate and hydrogen peroxide in an acidic medium.

7 Atomic Structure

7.1 ORBITALS AND ELECTRONIC CONFIGURATION

Electrons are found around the nucleus of an atom in regions called *shells* or *energy levels*. The energy of each shell is quantised—i.e., restricted to a particular value. A shell with the lowest possible energy in an atom is called the *ground state*. When an atom gains sufficient energy, an electron moves to a higher energy level in a process called *excitation*, and the electron is then said to be in an *excited state*.

In Chapter 1, we took a cursory look at electronic configuration—or the arrangement of electrons around the nucleus of an atom—using the s, p, d, f notation. Here we will examine the details of how this works.

The state of electrons in an atom is defined by four quantum numbers: the principal quantum number n, the subsidiary (azimuthal or angular momentum) quantum number l, the magnetic quantum number m_l and the spin quantum number m_s.

7.1.1 THE PRINCIPAL QUANTUM NUMBER

The principal quantum number n describes the shell or energy level of an atom. It has integral (whole-numbers) values ranging from 1 to infinity. The ground-state electron of an atom has $n = 1$, the next higher energy level has $n = 2$, and so on. Thus, all electrons in the same shell have the same n value. As stated in Chapter 1, the maximum possible number of electrons in a shell is given by the formula $2n^2$.

7.1.2 THE SUBSIDIARY (AZIMUTHAL OR ANGULAR MOMENTUM) QUANTUM NUMBER

The subsidiary quantum number l defines *energy sublevels* within a shell. It has integral values ranging from 0 to $n - 1$, where n is the principal quantum number. The number of energy sublevels in a shell corresponds to the number of possible l values.

For instance, if $n = 4$, then l can take values = 0, 1, 2 and 3, meaning there are four sublevels in that shell. These sublevels are designated by the letters s, p, d and f, respectively. All electrons within the same energy sublevel have the same value of l.

The energy of an orbital increases with n for values below 4 in the order 1s < 2s < 2p < 3s < 3p < 3d. However, for values of n from 4 and above, exceptions occur, following this order: $(n + 1)$s < nd and $(n + 1)$p < $(n + 2)$s < nf.

For example, a 3d orbital has a higher energy than a 4s orbital, meaning 4s < 3d. Similarly, a 4f orbital has a higher energy than a 6s or 5p orbital, meaning 5p < 6s < 4f. Thus, orbitals are arranged in increasing energy order as follows: 1s < 2s < 2p < 3s < 3p < 4s < 3d < 4p < 5s < 4d < 5p < 6s < 4f < 5d < 6p < 7s < 5f < 6d < 7p. A simple way to rememer this order is to follow the arrows in the energy-level diagram shown below.

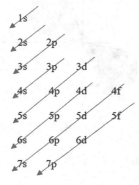

7.1.3 THE MAGNETIC QUANTUM NUMBER

The magnetic quantum number m_l determines the number of orbitals within a given sublevel, which represent the most likely locations of an electron at any given time. It has integral values ranging from $-l$ to $+l$, where l is the subsidiary quantum number. Each value of m_l corresponds to a specific orbital within the subshell. For instance, in the p subshell ($l = 1$), the possible values are $m_l = -1, 0$ and $+1$, corresponding to p orbitals: p_x, p_y and p_z. Similarly, the d subshell ($l = 2$) has five possible values of m_l: $-2, -1, 0, +1$ and $+2$, corresponding to five orbitals: d_{xy}, d_{yz}, d_{xz}, $d_{x^2-y^2}$ and d_{z^2}.

In electronic configurations, each orbital can hold a maximum of two electrons, as dictated by the Pauli exclusion principle (see Subsection 7.1.5.2). Thus, the maximum number of electrons in the s, p, d and f subshells is 2, 6, 10 and 14 electrons, respectively.

The shapes of s, p and d orbitals are illustrated in Figures 7.1–7.3. The s orbital is *spherically symmetrical*, with its size increasing with the principal quantum number, n. The p_x, p_y and p_z orbitals have a *dumb bell shape*, each symmetrical about the x-, y- and z-axes, respectively. The d_{xy}, d_{yz}, d_{xz} and $d_{x^2-y^2}$ orbitals have four lobes arranged in a plane between the axes, while the d_{z^2} orbital consists of two lobes with a doughnut-shaped region between them. The shapes of f orbitals are more complex and will not be discussed here.

7.1.4 THE SPIN QUANTUM NUMBER$_s$

Electrons possess an intrinsic quantum property called spin, represented by the spin quantum number, m_s. This property gives rise to a magnetic field that interacts with other magnetic fields.

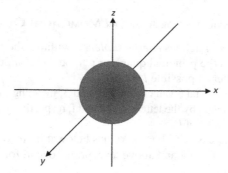

FIGURE 7.1 The shape of the *s* orbital.

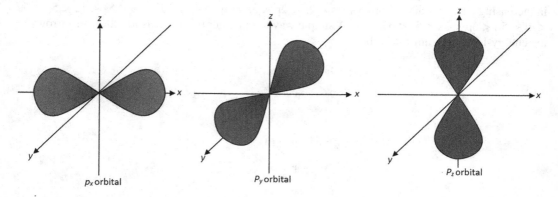

FIGURE 7.2 The shapes of the p orbitals.

Atomic Structure

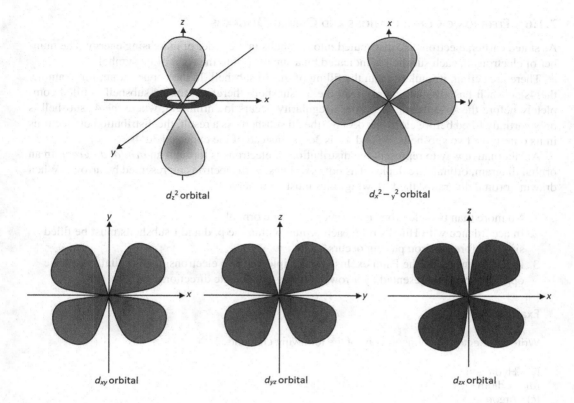

FIGURE 7.3 The shapes of the *d* orbitals.

When two electrons occupy the same orbital, they must have opposite spins, generating magnetic fields that are equal in magnitude but opposite in direction. As a result, the orbital becomes magnetically neutral.

The opposite spins of the electrons are indicated by the spin quantum number, m_s, which has values of $-\frac{1}{2}$ and $+\frac{1}{2}$.

7.1.5 Principles for Filling Orbitals with Electrons

The following principles govern how electrons are distributed among orbitals when writing electronic configurations. Thus, lower-energy orbitals fill before higher-energy ones.

7.1.5.1 Aufbau or Building Up Principle

The Aufbau (from the German word for *building up*) principle states that electrons occupy orbitals in order of increasing energy.

7.1.5.2 Pauli Exclusion Principle

The Paul exclusion principle states that no two electrons in an atom can have the same set of values for all four quantum numbers. This means that even electrons in the same orbital, which share the same values of n, l and m_l, must have opposite spins, distinguishing them by their m_s values.

7.1.5.3 Hund's Rule

Hund's rule states that the most stable arrangement of electrons within a subshell is the one that results in the highest possible number of unpaired electrons with parallel spins. This means that electrons occupy empty orbitals singly before pairing occurs. As we will see shortly, this rule is crucial for correctly drawing orbital diagrams.

7.1.6 Electronic Configurations and Orbital Diagrams

As stated earlier, electrons are distributed into subshells in the order of increasing energy. The number of electrons in each subshell is indicated by a superscript to the right of its symbol.

There are certain irregularities in the filling of the 3d subshell. In the copper atom, for example, the 4s subshell has a higher energy than the 3d subshell; therefore, the 3d subshell is filled completely before the 4s subshell. A similar irregularity occurs in chromium, where the 4s subshell is only partially filled before electrons occupy the 3d subshell. As a result, the distribution of electrons in its outermost two subshells, 3d and 4s, is $3d^54s^1$ instead of the expected $3d^44s^2$.

An alternative way to represent the distribution of electrons is through an *orbital diagram*. In an orbital diagram, orbitals are depicted as boxes or lines, with electrons represented by arrows. When drawing orbital diagrams, the following rules must be observed:

1. No more than two electrons may occupy a single orbital.
2. In accordance with Hund's rule, each orbital within the p, d and f subshells must be filled singly before electron pairing occurs.
3. In accordance with the Pauli exclusion principle, the two electrons in an orbital must have opposite spins, represented by arrows pointing in opposite directions

Example 1

Write the electronic configurations of the following elements.

(a) Hydrogen
(b) Sodium
(c) Argon
(d) Calcium
(e) Scandium
(f) Chromium
(g) Copper
(h) Chloride ion
(i) Sodium ion

Solutions

As stated above, the s, p, d and f can accommodate a maximum of 2, 6, 10 and 14 electrons, respectively. Using this, along with the principles discussed earlier, the electronic configurations of the elements are written as follows.

(a) H: 1 electron	$1s^1$
(b) Na: 11 electrons	$1s^22s^22p^63s^1 = [Ne]3s^1$
(c) Ar: 18 electrons	$1s^22s^22p^63s^23p^6$
(d) Ca: 20 electrons	$1s^22s^22p^63s^23p^64s^2 = [Ar]4s^2$
(e) Sc: 21 electrons	$1s^22s^22p^63s^23p^64s^23d^1 = [Ar]4s^23d^1$
(f) Cr: 24 electrons	$1s^22s^22p^63s^23p^64s^13d^5 =[Ar]4s^13d^5$
(g) Cu: 29 electrons	$1s^22s^22p^63s^23p^63d^{10}4s^1 = [Ar]3d^{10}4s^1$
(h) Cl⁻: 18 electrons	$1s^22s^22p^63s^23p^6$
(i) Na⁺: 10 electrons	$1s^22s^22p^6$

Note: The shorthand notation for writing electronic configurations expresses the electronic configuration of an element in terms of the symbol of the nearest preceding noble gas, enclosed in square brackets. For example, the electronic configuration of sodium can be written as $[Ne]3s^1$.

Atomic Structure

Example 2: Draw the Orbital Diagrams of the Following Elements

(a) Hydrogen
(b) Helium
(c) Carbon
(d) Nitrogen
(e) Chlorine

Solution

Using the rules stated above, the corresponding orbital diagrams are drawn as follows.

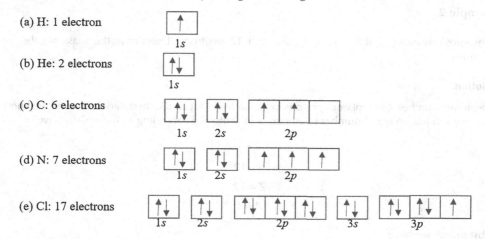

PRACTICE PROBLEMS

Write the electronic configurations and orbital diagrams of the following elements.
1. Oxygen
2. Potassium

7.2 MASS NUMBER

As mentioned in Chapter 1, the atom consists of a dense nucleus where most of its mass is concentrated. The nucleus contains protons and neutrons, and the number of protons in an atom of an element is called its *atomic number*, Z. The *mass number*, A, is the total number of protons and neutrons in an atom. In other words,

$$A = n + Z$$

where n is the number of neutrons in the atom.

An atom of an element is typically represented by writing its atomic and mass numbers as subscript and superscript, respectively, to the left of its symbol, X, i.e., $^{A}_{Z}X$. For example, an oxygen atom with the mass number of 16, O-16, is written as $^{16}_{8}O$.

Example 1

Sodium has an atomic number of 11. Determine the mass number of a sodium atom that contains 12 neutrons.

Solution

We know that the mass number, A, is given by

$$A = n + Z$$

$$n = 12$$
$$Z = 11$$
$$A = ?$$

Substituting, we have

$$A = 12 + 11 = 23$$

Example 2

A magnesium atom contains 12 electrons and 12 neutrons. Determine the mass number of the atom.

Solution

The atomic number, or number of protons in the atom, is not explicitly stated. However, we know that an atom has an equal number of electrons as protons. Proceeding as before, we have

$$A = n + Z$$
$$n = 12$$
$$Z = 12$$
$$A = ?$$

Substituting, we have

$$A = 12 + 12 = 24$$

Example 3

An atom of silicon has a mass number of 28. Determine the number of neutrons in the atom, given that the atomic number of silicon is 14.

Solution

As usual, the mass number is given by

$$A = n + Z$$
$$\therefore n = A - Z$$
$$A = 28$$
$$Z = 14$$
$$n = ?$$

Substituting these values, we have

$$n = 28 - 14 = 14$$

PRACTICE PROBLEMS

1. An atom of an element contains 8 protons and 8 neutrons. What is the mass number of the atom?
2. Determine the mass number of an atom containing 10 electrons and 12 neutrons.

Atomic Structure

3. Determine the number of neutrons in the atom $^{238}_{92}U$.
4. An atom of an element contains 144 neutrons. How many electrons does the atom contain if its mass number is 234?

7.3 CALCULATING RELATIVE ATOMIC MASSES

In Chapter 1, we stated that the relative atomic masses of most elements are reported as the weighted average of their isotopic masses. We will illustrate how this works with the following examples.

Example 1

Chlorine has two isotopes with masses 35 and 37, with relative abundances of 75% and 25%, respectively. Determine the relative atomic mass of the element.

Solution

The relative atomic mass, A_r, of the element is obtained from the isotopic masses and their relative abundances as follows.

$$Cl\text{-}35 = 75\%$$
$$Cl\text{-}37 = 25\%$$
$$\therefore A_r = \left(\frac{75}{100} \times 35\right) + \left(\frac{25}{100} \times 37\right) = 35.5$$

Example 2

A mass spectrometer shows that magnesium has three isotopes of masses 24, 25 and 26, with the relative abundances of 79%, 10% and 11%, respectively. Determine the relative atomic mass of magnesium.

Solution

As usual, the relative atomic mass of the element is calculated from its isotopic masses, and their relative abundances, as shown below.

$$Mg\text{-}24 = 79\%$$
$$Mg\text{-}25 = 10\%$$
$$Mg\text{-}26 = 11\%$$
$$\therefore A_r = \left(\frac{79}{100} \times 24\right) + \left(\frac{10}{100} \times 25\right) + \left(\frac{11}{100} \times 26\right) = 24$$

Example 3

Copper, with a relative atomic mass of 63.54, has two isotopes with masses 63 and 65. Determine the relative abundance of the isotope with mass 65.

Solution

Let the relative abundance of Cu-65 be x. Since there are only two isotopes, then the relative abundance of Cu-63 is $(100\% - x)$.

$$Cu\text{-}65 = x$$
$$Cl\text{-}63 = (100\% - x)$$

$$\therefore 63.54 = \left(\frac{x}{100} \times 65\right) + \left(\frac{100-x}{100} \times 63\right)$$

So

$$63.54 = \frac{6300 + 2x}{100}$$

Cross-multiplying, we have

$$6354 = 6300 + 2x$$

$$2x = 54$$

$$\therefore x = \frac{54}{2} = 27\%$$

PRACTICE PROBLEMS

1. Determine the relative atomic masses of the elements with the following isotopes.
   (a) ^{79}Br, 50.69%; ^{81}Br, 49.31%
   (b) ^{6}Li, 7.59%; ^{7}Li, 92.41%
   (c) ^{28}Si, 92.23%; ^{29}Si, 4.68%; ^{30}Si, 3.09%

2. Silver has two isotopes with masses 107 and 109. Determine the relative abundance of the isotope with mass 107, given that the relative atomic mass of silver is 107.96.

FURTHER PRACTICE PROBLEMS

1. An atom has 1 neutron. Determine its mass number if it has an atomic number of 1.

2. Calculate the mass number of an atom if it contains four neutrons and three electrons.

3. Determine the number of neutrons in the following atoms.
   (a) $^{28}_{14}$Si
   (b) $^{31}_{15}$P
   (c) $^{14}_{7}$N
   (d) $^{1}_{1}$H
   (e) $^{52}_{24}$Cr

4. Write the electronic configurations and orbital diagrams of the following substances.
   (a) Aluminium
   (b) Sodium ion

5. Write the electronic configuration of the following atoms.
   (a) Titanium
   (b) Chloride ion

6. Determine the relative atomic masses of the elements whose isotopes are given below.
   (a) ^{12}C, 98.93%; ^{13}C, 1.07%
   (b) ^{121}Sb, 57.21%; ^{123}Sb, 42.79%
   (c) ^{39}K, 93.26%; ^{40}K, 0.012%; ^{41}K, 6.73%
   (d) ^{16}O, 99.76%; ^{17}O, 0.040%; ^{18}O, 0.20%
   (e) ^{84}Sr, 0.56%; ^{86}Sr, 9.86%; ^{87}Sr, 7%; ^{88}Sr, 82.58%

7. Boron has an atomic mass of 10.81. The element has two isotopes of masses 10 and 11. Determine the relative abundance of B-11.

8 Gas Laws

8.1 MOLE FRACTION

It is a common practice to express the composition of gaseous mixtures in terms of the *mole fractions* of their components. The mole fraction, χ, of a gas in a mixture is the ratio of the number of moles of that gas to the total number of moles of all gases in the mixture:

$$\chi_i = \frac{n_i}{n_t}$$

where

χ_i = mole fraction of a gas i
n_i = number of moles of gas i
n_t = total number of moles of gases in the mixture

Since mole fraction is a ratio, it has no unit. Additionally, the sum of the mole fractions of all components in a gaseous mixture should always equal 1 when calculated correctly.

Example 1

A gaseous mixture contains 1.50 g of argon, 2.50 g of helium and 5.00 g of neon. Determine the mole fraction of each gas.

$$(He = 4,\ Ne = 20,\ Ar = 40)$$

Solution

The first step is to calculate the number of moles of each component.

$$n = \frac{m}{M}$$

For argon:

$$m = 1.50\ g$$

$$M = 40\ g\ mol^{-1}$$

$$n = ?$$

Substituting, we have

$$n = \frac{1.5\ g}{40\ g} \times 1\ mol = 0.0375\ mol$$

For helium:

$$m = 2.50\ g$$

$$M = 4 \text{ g mol}^{-1}$$
$$n = ?$$

Substituting, we have

$$n = \frac{2.5 \text{ g}}{4 \text{ g}} \times 1 \text{ mol} = 0.625 \text{ mol}$$

For neon:

$$m = 5.00 \text{ g}$$
$$M = 20 \text{ g mol}^{-1}$$
$$n = ?$$

Substituting, we have

$$n = \frac{5.00 \text{ g}}{20 \text{ g}} \times 1 \text{ mol} = 0.25 \text{ mol}$$

Now, we calculate the mole fraction of each component as follows.

$$\chi_i = \frac{n_i}{n_t}$$

$$n_t = (0.0375 + 0.625 + 0.25) = 0.9125 \text{ mol}$$

$$\chi_{Ar} = ?$$
$$\chi_{He} = ?$$
$$\chi_{Ne} = ?$$

Substituting, we have

$$\chi_{Ar} = \frac{0.0375 \text{ mol}}{0.9125 \text{ mol}} = 0.0411$$

$$\chi_{He} = \frac{0.625 \text{ mol}}{0.9125 \text{ mol}} = 0.685$$

$$\chi_{Ne} = \frac{0.25 \text{ mol}}{0.9125 \text{ mol}} = 0.274$$

The mole fractions sum up to 1: $0.0411 + 0.685 + 0.274 = 1.00$
 Note that in this type of calculation, the last mole fraction can also be obtained by subtracting the sum of the other mole fractions from 1. However, slight differences may arise due to rounding In this specific case,

$$\chi_{Ne} = 1.00 - (0.0411 + 0.685) = 0.274$$

Gas Laws

Example 2

The mole fraction of argon in a gaseous mixture is 0.20. Calculate the number of moles of argon in the mixture if the total number of moles of gases in the mixture is 4.51 mol.

Solution

Using the formula for mole fraction:

$$\chi_{Ar} = \frac{n_{Ar}}{n_t}$$

$$\therefore n_{Ar} = \chi_{Ar} \times n_t$$

$$\chi_{Ar} = 0.20$$

$$\chi_t = 4.51 \text{ mol}$$

$$\chi_{Ar} = ?$$

We will now substitute to obtain

$$n_{Ar} = 0.20 \times 4.51 \text{ mol} = 0.90 \text{ mol}$$

As stated previously, the last mole fraction can be determined by deduction. We leave it to you to verify this as practice.

PRACTICE PROBLEMS

1. A gas mixture contains 2.00 g of methane, CH_4, 10.00 g of benzene, C_6H_6, 3.05 g of ethane, C_2H_6, and 10.00 g of helium. Determine the mole fraction of each gas.

$$\left(H = 1, \ He = 4, \ C = 12 \right)$$

2. The mole fraction of helium in a gas mixture is 0.110. Determine the mass of helium in the mixture if the total number of moles of gases in the mixture is 5.50 mol.

$$\left(He = 4 \right)$$

8.2 DALTON'S LAW OF PARTIAL PRESSURES

Dalton's law of partial pressures states that the total pressure exerted by a mixture of gases equals the sum of the partial pressures of the individual gases that make up the mixture.

The pressure of each gas is called its partial pressure. Suppose we have a mixture of n gases, then, according to Dalton's law of partial pressures, the total pressure, P_t, of the mixture is given by.

$$P_t = P_1 + P_2 + P_3 + \ldots + P_n$$

The partial pressure, P_i, of a gas in a gaseous mixture is expressed in terms of its mole fraction as follows:

$$P_i = \chi_i P_t = \frac{n_i}{n_t} P_t$$

The actual pressure exerted by a wet gas, such as that collected over water, is the difference between the measured pressure of the gas and the saturated vapour pressure of water at the prevailing temperature.

$$P_g = P - P_{s.v.}$$

where
P_g = actual pressure of the gas
P = measured pressure of the gas
$P_{s.v.}$ = saturated vapour pressure of water

Example 1

A cylinder contains neon, argon and helium. Calculate the total pressure exerted by the mixture if the partial pressures of the gases are 250 mmHg, 300 mmHg and 400 mmHg, respectively.

Solution

We have to apply Dalton's law as follows.

$$P_t = P_{Ne} + P_{Ar} + P_{He}$$

$$P_{Ne} = 250 \text{ mmHg}$$

$$P_{Ar} = 300 \text{ mmHg}$$

$$P_{He} = 400 \text{ mmHg}$$

$$P_t = ?$$

Substituting we have

$$P_t = (250 + 300 + 400) \text{ mmHg} = 950 \text{ mmHg}$$

Example 2

A mixture containing nitrogen, carbon dioxide and oxygen is stored over water and has a pressure of 855 mmHg at 10°C. If the saturated vapour pressure of water at 10°C is 9.2 mmHg, determine the pressure exerted by nitrogen if the partial pressures of carbon dioxide and oxygen are 320 mmHg and 250 mmHg, respectively.

Solution

According to Dalton's law,

$$P_t = P_{Ne} + P_{CO_2} + P_{O_2} + P_{s.v.}$$

$$P_t = 855 \text{ mmHg}$$

$$P_2 = 320 \text{ mmHg}$$

$$P_3 = 250 \text{ mmHg}$$

$$P_{s.v.} = 9.2 \text{ mmHg}$$

$$P_1 = ?$$

Gas Laws

Substituting, we have

$$855 \text{ mmHg} = P_{N_2} + 320 \text{ mmHg} + 250 \text{ mmHg} + 9.2 \text{ mmHg}$$

$$\therefore P_{N_2} = (855 - 579.2) \text{ mmHg} = 275.8 \text{ mmHg}$$

Example 3

A mixture of 0.20 mol of neon, 0.25 mol of helium and 0.35 mol of nitrogen exerts a total pressure of 2.5 atm. Determine the partial pressure of helium in the mixture.

Solution

We have to calculate the partial pressure of nitrogen from its mole fraction as follows.

$$P_{N_2} = \chi_{N_2} = P_t = \frac{n_{N_2}}{n_t} P_t$$

$$n_{N_2} = 0.25 \text{ mol}$$

$$n_t = (0.20 + 0.25 + 0.35) \text{ mol} = 0.80 \text{ mol}$$

$$P_t = 2.5 \text{ atm}$$

$$P_{N_2} = ?$$

Substituting we have

$$P_{N_2} = \frac{0.25 \text{ mol}}{0.80 \text{ mol}} \times 2.5 \text{ atm} = 0.78 \text{ atm}$$

PRACTICE PROBLEMS

1. A gaseous mixture contains four gases whose pressures are 450 mmHg, 900 mmHg, 600 mmHg and 300 mmHg. Calculate the total pressure exerted by the mixture.
2. A mixture of three gases A, B and C exerts a pressure of 850 mmHg. Determine the pressure of C if the pressures of A and B are 250 mmHg and 350 mmHg, respectively.
3. Determine the partial pressure of argon in a mixture containing 0.15 mol of nitrogen, 0.10 mol of argon, 0.25 mol of helium and 0.20 mol of neon, given that the total pressure exerted by the mixture is 1.5 atm.

8.3 BOYLE'S LAW

Boyle's law states that the pressure of a fixed mass of a gas is inversely proportional to its volume at constant temperature. Mathematically, Boyle's law is expressed as:

$$P \alpha \frac{1}{V}$$

So

$$P = \frac{k}{V}$$

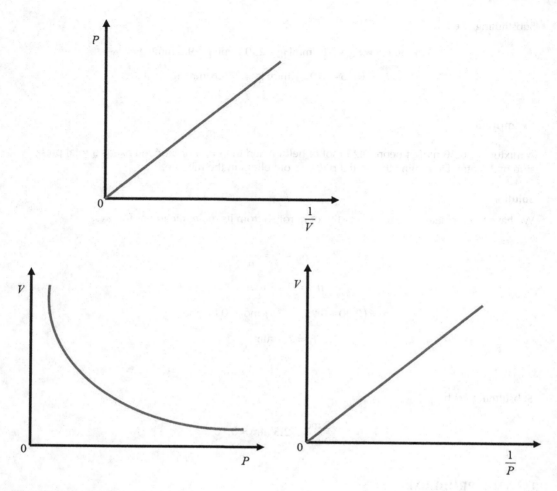

FIGURE 8.1 Graphs of Boyle's Law.

where k is a constant.

$$\therefore PV = k \text{ or } P_1V_1 = P_2V_2$$

The graphical representations of Boyle's law are shown in Figure 8.1.

Example 1

A gas occupies a volume of 250 cm³ at a pressure of 760 mmHg. At what pressure would the volume reduce to 220 cm³ if the temperature remains constant?

Solution

Applying Boyle's law, we have

$$P_1V_1 = P_2V_2$$

$$\therefore P_2 = \frac{P_1V_1}{V_2}$$

Gas Laws

$$P_1 = 760 \text{ mmHg}$$
$$V_1 = 250 \text{ cm}^3$$
$$V_2 = 220 \text{ cm}^3$$
$$P_2 = ?$$

Substituting we have

$$P_2 = \frac{760 \text{ mmHg} \times 250 \text{ cm}^3}{220 \text{ cm}^3} = 863 \text{ mmHg}$$

Example 2

A gas occupies a volume of 380 cm³ at 760 mmHg. Calculate the volume of the gas at 850 mmHg if the temperature remains constant.

Solution

We apply the relation:

$$P_1 V_1 = P_2 V_2$$

$$\therefore V_2 = \frac{P_1 V_1}{P_2}$$

$$P_1 = 760 \text{ mmHg}$$
$$P_2 = 850 \text{ mmHg}$$
$$V_1 = 380 \text{ cm}^3$$
$$P_2 = ?$$

We now substitute to obtain

$$V_2 = \frac{760 \text{ mmHg} \times 380 \text{ cm}^3}{850 \text{ mmHg}} = 350 \text{ cm}^3$$

PRACTICE PROBLEMS

1. A gas has a volume of 150 cm³ at 1 atm. Calculate the volume of the gas if the pressure is increased to 1.7 at constant temperature.
2. 2.5 dm³ of a gas was found to exert a pressure of 760 mmHg. Calculate the pressure of the gas if its volume is reduced by 500 cm³ at constant temperature.

8.4 CHARLES' LAW

Charles' law states that the volume of a fixed mass of a gas is directly proportional to its absolute temperature (measured in kelvin) at constant pressure.

Mathematically, Charles' law is expressed as:

$$V \propto T$$

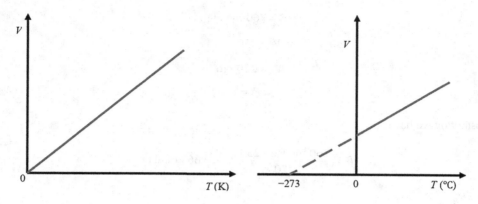

FIGURE 8.2 Graphs of Charles' Law.

So

$$V = kT$$

where k is a constant.

$$\therefore \frac{V}{T} = k \quad \text{or} \quad \frac{V_1}{T_1} = \frac{V_2}{T_2}$$

The graphical representations of Charles' law are shown in Figure 8.2.

Example 1

A gas occupies a volume of 55 cm³ at 25°C and 800 mmHg. Determine the volume of the gas if the temperature is increased to 30°C at constant pressure.

Solution

We apply Charles's law as follows.

$$\frac{V_1}{T_1} = \frac{V_2}{T_2}$$

$$\therefore V_2 = \frac{V_1 T_2}{T_1}$$

$$V_1 = 55 \text{ cm}^3$$

$$T_1 = 25°C = (25 + 273) \text{ K} = 298 \text{ K}$$

$$T_2 = 33°C = (33 + 273) \text{ K} = 306 \text{ K}$$

$$V_2 = ?$$

Substituting, we have

$$V_2 = \frac{55 \text{ cm}^3 \times 306 \text{K}}{298 \text{K}} = 56 \text{ cm}^3$$

Gas Laws

Example 2

The volume of a gas is 550.0 cm³ at STP. At what temperature would the volume of the gas increase to 570.0 cm³ at constant pressure.

Solution

We have to apply the relation:

$$\frac{V_1}{T_1} = \frac{V_2}{T_2}$$

So

$$V_2 T_1 = V_1 T_2$$

$$\therefore T_2 = \frac{V_2 T_1}{V_1}$$

$$V_1 = 550.0 \text{ cm}^3$$

$$T_1 = \text{standard temperature} = 298 \text{ K}$$

$$V_2 = 570.0 \text{ cm}^3$$

$$T_2 = ?$$

We now substitute to obtain

$$T_2 = \frac{570.0 \text{ cm}^3 \times 298 \text{ K}}{550.0 \text{ cm}^3} = 309 \text{ K}$$

PRACTICE PROBLEMS

1. A sample of a gas occupies a volume of 1.8 dm³ at 55°C. What volume would it occupy at 27°C at constant pressure?
2. The volume of a gas is 150 cm³ at 100°C. Calculate the temperature at which the volume would rise to 350 cm³ at the same pressure.

8.5 THE GENERAL (COMBINED) GAS LAW

The combination of Boyle's and Charles' laws gives the general or combined gas law, expressed as follows:

$$\frac{PV}{T} = k \quad \text{or} \quad \frac{P_1 V_1}{T_1} = \frac{P_2 V_2}{T_2}$$

Example 1

A gas occupies a volume of 250 cm³ at 32°C and 550 mmHg. Calculate its volume at STP.

Solution

We apply the general gas equation:

$$\frac{P_1 V_1}{T_1} = \frac{P_2 V_2}{T_2}$$

$$\therefore V_2 = \frac{P_1 V_1 T_2}{P_2 T_1}.$$

$P_1 = 550 \text{ mmHg}$

$V_1 = 250 \text{ cm}^3$

$T_1 = 32°C = (32+273) \text{ K} = 305 \text{ K}$

$P_2 = \text{Standard pressure} = 760 \text{ mmHg}$

$T_2 = \text{Standard temperature} = 273 \text{ K}$

$V_2 = ?$

Substituting, we have

$$V_2 = \frac{550 \text{ mmHg} \times 250 \text{ cm}^3 \times 273 \text{ K}}{760 \text{ mmHg} \times 305 \text{ K}} = 160 \text{ cm}^3$$

Example 2

The volume of a gas is 5.0 dm³ at 25°C and 840 mmHg. Determine the temperature at which the volume of the gas would rise to 5.5 dm³ at 740 mmHg.

Solution

Using the general gas law, we have

$$\frac{P_1 V_1}{T_1} = \frac{P_2 V_2}{T_2}$$

$$\therefore T_2 = \frac{P_2 V_2 T_1}{P_1 V_1}$$

$P_1 = 840 \text{ mmHg}$

$V_1 = 5.0 \text{ dm}^3$

$T_1 = 25°C = (25+273) \text{ K} = 298 \text{ K}$

$P_2 = 740 \text{ mmHg}$

$V_2 = 5.5 \text{ dm}^3$

$T_2 = ?$

We now substitute the values, to obtain

$$T_2 = \frac{740 \text{ mmHg} \times 5.5 \text{ dm}^3 \times 298 \text{ K}}{840 \text{ mmHg} \times 5.0 \text{ dm}^3} = 290 \text{ K}$$

Gas Laws

Example 3

A certain mass of a gas occupies a volume of 350 cm³ at −10°C. Determine the initial pressure of the gas if its volume rises to 420 cm³ at STP.

Solution

Using the general gas equation, we have

$$\frac{P_1 V_1}{T_1} = \frac{P_2 V_2}{T_2}$$

$$\therefore P_1 = \frac{P_2 V_2 T_1}{V_1 T_2}$$

$$V_1 = 350 \text{ cm}^3$$

$$T_1 = -10°C = (-10 + 273) \text{ K} = 263 \text{ K}$$

$$P_2 = 760 \text{ mmHg}$$

$$V_2 = 420 \text{ cm}^3$$

$$T_2 = 273 \text{ K}$$

$$P_1 = ?$$

We now substitute to obtain

$$P_1 = \frac{760 \text{ mmHg} \times 420 \text{ cm}^3 \times 263 \text{ K}}{350 \text{ cm}^3 \times 273 \text{ K}} = 880 \text{ mmHg}$$

PRACTICE PROBLEMS

1. Calculate the volume of a gas at 42°C and 1 atm if it occupies a volume of 450 cm³ at 30°C and 1.2 atm.
2. A gas occupies a volume of 2.0 dm³ at STP. Calculate the pressure at which its volume would reduce to 750 cm³ at −5°C.
3. Calculate the temperature at which the volume of a gas at 650 mmHg and 50°C would reduce to two-thirds its initial value when the pressure increases to 750 mmHg.

8.6 THE IDEAL GAS EQUATION (LAW)

The ideal gas equation (or law) is obtained by combining Avogadro's law with the general gas law. The ideal gas law for n moles of a gas is expressed as:

$$PV = nRT$$

where R is the universal gas constant.

The value of R depends on the unit of pressure, as shown in Table 8.1.

TABLE 8.1
Values of R at Different Temperatures

Unit of Pressure	Value of R
Atm	0.08206 dm³ atm mol⁻¹ K⁻¹
kPa	8.314 m³ kPa mol⁻¹ K⁻¹ (8.314 J mol⁻¹ K⁻¹)
Torr	62.4 dm³ Torr mol⁻¹ K⁻¹

The ideal gas equation is an example of an *equation of state*—a mathematical relation that connects the four properties of a gas: volume, pressure, temperature and number of moles.

It is important to note that, when using the ideal gas laws for calculations, volume and temperature must always be expressed in their SI units: dm³ for volume and K for temperature.

Example 1

Calculate the amount, in moles, of an ideal gas if it occupies a volume of 550 cm³ at 5.5 kPa and 25°C.

$$(R = 8.314 \text{ J mol}^{-1}\text{K}^{-1})$$

Solution

Using the ideal gas law, we have

$$PV = nRT$$

$$\therefore n = \frac{PV}{RT}$$

$$P = 5.5 \text{ kPa}$$

$$V = 550 \text{ cm}^3 = 0.55 \text{ dm}^3$$

$$T = 25°C = (25 + 273) \text{ K} = 298 \text{ K}$$

$$R = 8.314 \text{ J mol}^{-1} \text{ K}^{-1} = 8.314 \text{ dm}^3 \text{ kPa mol}^{-1} \text{ K}^{-1}$$

$$n = ?$$

Substituting, we have

$$n = \frac{5.5 \text{ kPa} \times 0.55 \text{ dm}^3 \times 1 \text{ K} \times 1 \text{ mol}}{8.314 \text{ dm}^3 \text{ kPa} \times 298 \text{ K}} = 0.0012 \text{ mol}$$

Example 2

Calculate the volume occupied by 2.5 mol of an ideal gas at 1.5 atm and 27°C.

$$(R = 0.08206 \text{ dm}^3 \text{ atm mol}^{-1}\text{K}^{-1})$$

Solution

Applying the ideal gas law, we have

$$PV = nRT$$

$$\therefore V = \frac{nRT}{P}$$

$$P = 1.5 \text{ atm}$$

$$T = 27°C = (27 + 273) \text{ K} = 300 \text{ K}$$

$$n = 2.5 \text{ mol}$$

$$R = 0.08206 \text{ dm}^3 \text{ atm mol}^{-1} \text{ K}^{-1}$$

$$V = ?$$

We now substitute to obtain

$$V = \frac{1.5 \text{ mol} \times 0.08206 \text{ dm}^3 \text{ atm} \times 300 \text{ K}}{1.5 \text{ atm} \times 1 \text{ mol} \times 1 \text{ K}} = 25 \text{ dm}^3$$

Example 3

3.5 mol of an ideal gas occupies a volume of 10.2 dm³ at 2.5 atm. Calculate the temperature of the gas.

$$\left(R = 0.08206 \text{ dm}^3 \text{ atm mol}^{-1} \text{ K}^{-1}\right)$$

Solution

As usual, when three state properties are given, we apply the ideal gas law:

$$PV = nRT$$

$$\therefore T = \frac{PV}{nR}$$

$$P = 2.5 \text{ atm}$$

$$V = 10.2 \text{ dm}^3$$

$$n = 3.5 \text{ mol}$$

$$R = 0.08206 \text{ dm}^3 \text{ atm mol}^{-1} \text{ K}^{-1}$$

$$T = ?$$

We now substitute to obtain

$$T = \frac{2.5 \text{ atm} \times 10.2 \text{ dm}^3 \times 1 \text{ mol} \times 1 \text{ K}}{3.5 \text{ mol} \times 0.08206 \text{ dm}^3 \text{ atm}} = 88 \text{ K}$$

Example 4

At what pressure would 5.5 g of oxygen occupy a volume of 2500 cm³ at 256°C, assuming the gas behaves like an ideal gas?

$$(O = 16;\ R = 8.314\ J\ mol^{-1}\ K^{-1})$$

Solution

Using the ideal gas law, we have

$$PV = nRT$$

$$\therefore P = \frac{nRT}{V}$$

Since the number of moles is given by

$$n = \frac{m}{M}$$

Substituting this into the equation, we have

$$P = \frac{\frac{m}{M}RT}{V}$$

$$T = 256°C = (256 + 273)\ K = 532\ K$$

$$V = 2500\ cm^3 = 2.5\ dm^3$$

$$R = 8.314\ dm^3\ kPa\ mol^{-1}\ K^{-1}$$

$$m = 5.5\ g$$

$$M = (16 \times 2)\ g\ mol^{-1} = 32\ g\ mol^{-1}$$

$$P = ?$$

We now substitute to obtain

$$P = \frac{\frac{5.5\ g}{32\ g} \times 8.314\ dm^3\ kPa \times 300\ K \times 1\ mol}{2.5\ dm^3 \times 1\ mol \times 1\ K} = 170\ kPa$$

Example 5

1.02 g of an ideal gas occupies a volume of 12.5 dm³ at 25°C and 4.58 kPa. Determine the molar mass of the gas.

$$(R = 8.314\ J\ mol^{-1}\ K^{-1})$$

Solution

We apply the ideal gas law:

$$PV = nRT$$

Gas Laws

Since the number of moles is given by

$$n = \frac{m}{M},$$

then

$$PV = \frac{m}{M}RT$$

$$\therefore M = \frac{mRT}{PV}$$

$P = 4.58$ kPa

$V = 12.5$ dm^3

$T = 25°C = (25+273)$ K $= 298$ K

$m = 1.02$ g

$R = 8.314$ dm^3 kPa mol^{-1} K^{-1}

$M = ?$

Substituting, we have

$$M = \frac{1.02 \text{ g} \times 8.314 \text{ dm}^3 \text{ kPa} \times 298 \text{ K}}{4.58 \text{ kPa} \times 12.5 \text{ dm}^3 \times 1 \text{ mol} \times 1 \text{ K}} = 44 \text{ g mol}^{-1}$$

Example 6

An ideal gas occupies a volume of 2.5 dm^3 at $-45°C$ and 2.1 atm. Calculate the number of moles of the gas if the volume is reduced to 1.5 dm^3 at $-10°C$.

$$\left(R = 0.08206 \text{ dm}^3 \text{ atm mol}^{-1} \text{ K}^{-1}\right)$$

Solution

The number of moles of the gas must be calculated under the final conditions of volume, pressure and temperature. The final pressure is obtained using the combined gas law as follows.

$$\frac{P_1V_1}{T_1} = \frac{P_2V_2}{T_2}$$

$$\therefore P_2 = \frac{P_1V_1T_2}{V_2T_1}$$

$P_1 = 2.1$ atm

$V_1 = 2.5$ dm^3

$T_1 = -45°C = (-45+273)$ K $= 228$ K

$V_2 = 1.5$ dm^3

$T_2 = -10°C = (-10+273)$ K $= 263$ K

$P_2 = ?$

Substituting, we have

$$P_2 = \frac{2.1 \text{ atm} \times 2.5 \text{ dm}^3 \times 263 \text{ K}}{1.5 \text{ dm}^3 \times 228 \text{ K}} = 4.037 \text{ atm}$$

Now, we calculate the number of moles of the gas using the ideal gas law:

$$PV = nRT$$

$$\therefore n = \frac{PV}{RT}$$

$$P = 4.037 \text{ atm}$$

$$V = 1.5 \text{ dm}^3$$

$$T = 263 \text{ K}$$

$$R = 0.08206 \text{ dm}^3 \text{ atm mol}^{-1} \text{ K}^{-1}$$

$$n = ?$$

Substituting, we have

$$n = \frac{4.037 \text{ atm} \times 1.5 \text{ dm}^3 \times 1 \text{ mol} \times 1 \text{ K}}{0.08206 \text{ dm}^3 \text{ atm} \times 263 \text{ K}} = 0.28 \text{ mol}$$

PRACTICE PROBLEMS

1. Calculate the volume of 0.55 mol of an ideal gas at 218 kPa and 100°C.
$$\left(R = 8.314 \text{ J mol}^{-1} \text{ K}^{-1}\right)$$

2. 1.5 mol of an ideal gas occupies a volume of 5.5 dm³ at 2.98 atm. Determine the temperature of the gas.
$$\left(R = 0.08206 \text{ dm}^3 \text{ atm mol}^{-1} \text{ K}^{-1}\right)$$

3. Calculate the volume of 1.5 g of hydrogen chloride at 33°C and 155 kPa assuming the gas behaves like an ideal gas.
$$\left(H = 1, \; Cl = 35.5; \; R = 8.314 \text{ J mol}^{-1} \text{ K}^{-1}\right)$$

4. 5.5 dm³ of a gas weighs 0.253 g at 2.7 kPa and −15°C. Determine the molar mass of the gas.
$$\left(R = 8.314 \text{ J mol}^{-1} \text{ K}^{-1}\right)$$

8.7 VAPOUR DENSITY AND RELATIVE MOLECULAR MASS

Vapour density, V.D., of a gas or vapour is the number of times a given volume of the gas or vapour is heavier than the same volume of hydrogen under identical conditions of temperature and pressure.

Mathematically, vapour density is expressed as:

$$\text{V.D.} = \frac{\text{Mass of } x \text{ volume of a gas/vapour}}{\text{Mass of } x \text{ volume of hydrogen}} = \frac{\text{Mass of } n \text{ molecules of a gas/vapour}}{\text{Mass of } n \text{ molecules of hydrogen}}$$

For 1 molecule of a gas or vapour we have

$$\text{V.D.} = \frac{\text{Mass of 1 molecule of gas/vapour}}{\text{Mass of 1 molecule of hydrogen}}$$

Gas Laws

Since hydrogen is a diatomic gas, we modify the expression as follows:

$$\text{V.D.} = \frac{\text{Mass of 1 molecule of a gas / vapour}}{2 \times \text{mass of 1 atom of hydrogen}}$$

On the hydrogen scale,

$$M_r = \frac{\text{Mass of 1 molecule of a gas / vapour}}{\text{Mass of 1 atom of hydrogen}}$$

$$\therefore \text{V.D.} = \frac{M_r}{2} \text{ or } M_r = 2 \times \text{V.D.}$$

Example 1

The vapour density of a gas is 22. Determine its relative molecular mass.

Solution

We must apply the relationship between vapour density and relative molecular mass:

$$M_r = 2 \times \text{V.D.}$$
$$\text{V.D.} = 22$$
$$M_r = ?$$

Substituting, we have

$$M_r = 2 \times 22 = 44$$

Example 2

Determine the vapour density of hydrogen chloride, HCl.

$$(H = 1,\ Cl = 35.5)$$

Solution

We apply the relation:

$$\text{V.D.} = \frac{M_r}{2}$$
$$M_r = 1 + 35.5 = 36.5$$
$$\text{V.D.} = ?$$

Substituting, we have

$$\text{V.D.} = \frac{36.5}{2} = 18.3$$

PRACTICE PROBLEMS

1. Determine the relative molecular mass of a gas whose vapour density is 15.
2. Calculate the molar mass of a gas whose vapour density is 40.

3. Determine the vapour densities of the following gases.
   (a) CH_4
   (b) SO_2

$$(H = 1, \ C = 12, \ O = 16, \ S = 32)$$

8.8 GRAHAM'S LAW OF DIFFUSION OR EFFUSION

Graham's law of diffusion or effusion states that the rate of diffusion or effusion of a gas is inversely proportional to the square root of its density, provided that temperature and pressure remain constant.

This means that the lower the density or molar mass of a gas, the faster it will diffuse or effuse. Mathematically, Graham's law is expressed as:

$$R \alpha \frac{1}{\sqrt{d}}$$

where

R = rate of diffusion
d = gas density

Since the density of a gas is proportional to its molar mass, we can also write the proportionality as follows:

$$R \alpha \frac{1}{\sqrt{M_r}}$$

For two gases (1 and 2) diffusing under the same conditions, the relative rates of diffusion or effusion are given as:

$$\frac{R_1}{R_2} = \sqrt{\frac{d_2}{d_1}} = \sqrt{\frac{M}{M}}$$

Since the rate of diffusion is inversely proportional to the time taken, we can also express the relationship as:

$$\frac{R_1}{R_2} = \frac{t_2}{t_1} = \sqrt{\frac{d_2}{d_1}} = \sqrt{\frac{M_2}{M_1}}$$

where t_1 and t_2 represent the time required for equal volume of the two gases to diffuse or effuse under identical conditions.

The volumetric rate of diffusion of a gas is given by

$$R = \frac{V}{t}$$

where

V = volume of gas
t = the time required for a given volume of gas to flow through a specified area

Gas Laws

Example 1

25 cm³ of a gas diffuses through a porous plug in 1.5 min. Determine the rate of diffusion of the gas in cm³ s⁻¹.

Solution

We apply the relation:

$$R = \frac{V}{t}$$

$$V = 25 \text{ cm}^3$$

$$t = 1.5 \text{ min} = (1.5 \times 60) \text{ s} = 90 \text{ s}$$

$$R = ?$$

Substituting, we have

$$R = \frac{25 \text{ cm}^3}{90 \text{ s}} = 0.28 \text{ cm}^3 \text{ s}^{-1}$$

Example 2

The rate of diffusion of a gas is 1.5 cm³ s⁻¹. How long would it take 50.5 cm³ of the gas to diffuse under the same conditions?

Solution

The time is calculated from the rate of diffusion using the relation:

$$R = \frac{V}{t}$$

$$\therefore t = \frac{V}{R}$$

$$V = 50.5 \text{ cm}^3$$

$$R = 1.5 \text{ cm}^3 \text{ s}^{-1}$$

$$t = ?$$

Substituting, we have

$$t = \frac{50.5 \text{ cm}^3 \times 1 \text{s}}{1.5 \text{ cm}^3} = 34 \text{ s}$$

Example 3

A gas diffuses 0.213 times as fast as hydrogen. Calculate the molar mass of the gas.

Solution

$$(H = 1)$$

We apply Graham's law:

$$\frac{R_{H_2}}{R_2} = \sqrt{\frac{M_2}{M_{H_2}}}$$

$$R_{H_2} = 1$$
$$R_2 = 0.213$$
$$M_{H_2} = (2 \times 1) \text{ g mol}^{-1} = 2 \text{ g mol}^{-1}$$
$$M_2 = ?$$

Substituting, we have

$$\frac{1}{0.213} = \sqrt{\frac{M_2}{2 \text{ g mol}^{-1}}}$$

To clear the square root from the equation, we square both sides:

$$\frac{1}{0.045369} = \frac{M_2}{2 \text{ g mol}^{-1}}$$

Cross-multiplying, we have

$$0.045369 M_2 = 2 \text{ g mol}^{-1}$$

$$M_2 = \frac{2 \text{ g mol}^{-1}}{0.045369} = 44 \text{ g mol}^{-1}$$

$$\therefore M = 44 \text{ g mol}^{-1}$$

Example 4

A container contains methane, CH_4, and krypton. How many times as fast as krypton would the methane diffuse?

$$(H = 1, \; C = 12, \; Kr = 84)$$

Solution

Applying Graham's law, we proceed as follows.

$$\frac{R_{CH_4}}{R_{Kr}} = \sqrt{\frac{M_{Kr}}{M_{CH_4}}}$$

$$\therefore M_{Kr} = 84 \text{ g mol}^{-1}$$

$$M_{CH_4} = (12 + \{4 \times 1\}) \text{ g mol}^{-1} = 16 \text{ g mol}^{-1}$$

$$\frac{R_{CH_4}}{R_{Kr}} = ?$$

Substituting, we have

$$\frac{R_{CH_4}}{R_{Kr}} = \sqrt{\frac{84 \text{ g mol}^{-1}}{16 \text{ g mol}^{-1}}} = 2.29$$

Gas Laws

Example 5

250 cm³ of nitrogen diffuses through an orifice in 35 s, while it takes 37.4 s for the same volume of an unknown gas to diffuse through the orifice under the same conditions. Determine the molar mass of the unknown gas.

$$(N = 14)$$

Solution

Applying Graham's law, we have

$$\frac{t_{N_2}}{t_2} = \sqrt{\frac{M_{N_2}}{M_2}}$$

$$t_{N_2} = 35 \text{ s}$$

$$t_2 = 37.4 \text{ s}$$

$$M_{N_2} = (14 \times 2) \text{ g mol}^{-1} = 28 \text{ g mol}^{-1}$$

$$M_2 = ?$$

Substituting, we have

$$\frac{35 \text{ s}}{37.5 \text{ s}} = \sqrt{\frac{28 \text{ g mol}^{-1}}{M_2}}$$

Squaring both sides, we obtain

$$\frac{1225}{1398.76} = \frac{28 \text{ g mol}^{-1}}{M_2}$$

Cross-multiplying, we have

$$M_2 \times 1225 = 39165.28 \text{ g mol}^{-1}$$

$$\therefore M_2 = \frac{39165.28 \text{ g mol}^{-1}}{1225} = 32 \text{ g mol}^{-1}$$

Example 6

350 cm³ of nitrogen diffuses through an orifice in 15 s. How long would it take 550 cm³ of ammonia to diffuse under the same conditions?

$$(H = 1, N = 14)$$

Solution

We apply Graham's law:

$$\frac{t_{N_2}}{t_{NH_3}} = \sqrt{\frac{M_{N_2}}{M_{NH_3}}}$$

The time, t_{N_2}, taken for 550 cm³ of nitrogen (equal volume as ammonia) to diffuse under the same conditions is calculated as follows.

$$\frac{350 \text{ cm}^3}{15 \text{ s}} = \frac{550 \text{ cm}^3}{t_{N_2}}$$

So

$$350 \text{ cm}^3 \times t_{N_2} = 550 \text{ cm}^3 \times 15 \text{ s}$$

$$t_{N_2} = \frac{550 \text{ cm}^3 \times 15 \text{ s}}{350 \text{ cm}^3} = 23.5714 \text{ s}$$

$$M_{N_2} = (14 \times 2) \text{ g mol}^{-1} = 28 \text{ g mol}^{-1}$$

$$M_{NH_3} = (14 + \{1 \times 3\}) \text{ g mol}^{-1} = 17 \text{ g mol}^{-1}$$

$$t_{NH_3} = ?$$

We now substitute to obtain

$$\frac{23.57 \text{ s}}{t_{NH_3}} = \sqrt{\frac{28 \text{ g mol}^{-1}}{17 \text{ g mol}^{-1}}}$$

Squaring both sides, we have

$$\frac{555.5449 \text{ s}}{t^2_{NH_3}} = \frac{28}{17}$$

So

$$t^2_{NH_3} \times 28 = 9444.2633$$

$$\therefore t_{NH_3} = \sqrt{\frac{9444.2633 \text{ s}}{28}} = 18 \text{ s}$$

Alternatively, we can begin by calculating the rate of diffusion of nitrogen.

$$R_{N_2} = \frac{V_{N_2}}{t_{N_2}}$$

$$V_{N_2} = 350 \text{ cm}^3$$

$$t_{N_2} = 15 \text{ s}$$

$$R_{N_2} = ?$$

Substituting, we have

$$R_{N_2} = \frac{350 \text{ cm}^3}{15 \text{ s}} = 23.33 \text{ cm}^3 \text{ s}^{-1}$$

Gas Laws

The next step is to determine the rate of diffusion of ammonia using Graham's law.

$$\frac{R_{N_2}}{R_{NH_3}} = \sqrt{\frac{M_{NH_3}}{M_{N_2}}}$$

$$M_{N_2} = 28 \text{ g mol}^{-1}$$

$$M_{NH_3} = 17 \text{ g mol}^{-1}$$

$$R_{NH_3} = ?$$

Substituting, we have

$$\frac{23.33 \text{ cm}^3 \text{ s}^{-1}}{R_{NH_3}} = \sqrt{\frac{17 \text{ g mol}^{-1}}{28 \text{ g mol}^{-1}}}$$

Squaring both sides, we have

$$\frac{544.2889 \text{ cm}^6 \text{ s}^{-2}}{R^2_{NH_3}} = \frac{17}{28}$$

So

$$R^2_{NH_3} \times 17 = 15240.0892 \text{ cm}^6 \text{ s}^{-2}$$

$$\therefore R_{NH_3} = \sqrt{\frac{15240.0892 \text{ cm}^6 \text{ s}^{-2}}{17}} = 29.941 \text{ cm}^3 \text{ s}^{-1}$$

Finally, we can determine the time required for 550 cm³ of ammonia to diffuse as follows.

$$R_{NH_3} = \frac{V_{NH_3}}{t_{NH_3}}$$

$$\therefore t_{NH_3} = \frac{V_{NH_3}}{t_{NH_3}}$$

$$V_{NH_3} = 550 \text{ cm}^3$$

$$t_{NH_3} = ?$$

Substituting the values, we obtain

$$R_{NH_3} = \frac{550 \text{ cm}^3}{29.941 \text{ cm}^3} \times 1 \text{ s} = 18 \text{ s}$$

Example 7

250 cm³ of chlorine diffuses through an orifice in 25 s. What volume of nitrogen would diffuse through the same orifice in 10.5 s under similar conditions?

$$(N = 14, \text{ Cl} = 35.5)$$

Solution

We will begin by calculating the rate of diffusion of chlorine using the relation:

$$R_{Cl_2} = \frac{V_{Cl_2}}{t_{Cl_2}}$$

$$V_{Cl_2} = 250 \text{ cm}^3$$

$$t_{Cl_2} = 25 \text{ s}$$

$$R_{Cl_2} = ?$$

Substituting, we obtain

$$R_{Cl_2} = \frac{250 \text{ cm}^3}{25 \text{ s}} = 10 \text{ cm}^3 \text{ s}^{-1}$$

The next step is to calculate the rate of diffusion of nitrogen by applying Graham's law.

$$\frac{R_{Cl_2}}{R_{N_2}} = \sqrt{\frac{M_{N_2}}{M_{Cl_2}}}$$

$$M_{Cl_2} = (35.5 \times 2) \text{ g mol}^{-1} = 71 \text{ g mol}^{-1}$$

$$M_{N_2} = (14 \times 2) \text{ g mol}^{-1} = 28 \text{ g mol}^{-1}$$

$$R_{N_2} = ?$$

Substituting the values, we obtain

$$\frac{10 \text{ cm}^3 \text{ s}^{-1}}{R_{N_2}} = \sqrt{\frac{28 \text{ g mol}^{-1}}{71 \text{ g mol}^{-1}}}$$

Squaring both sides, we obtain

$$\frac{100 \text{ cm}^6 \text{ s}^{-2}}{R^2_{N_2}} = \frac{28}{71}$$

So

$$R^2_{N_2} \times 28 = 7100 \text{ cm}^6 \text{ s}^{-2}$$

$$\therefore R_{N_2} = \sqrt{\frac{7100 \text{ cm}^6 \text{ s}^{-2}}{28}} = 15.92 \text{ cm}^3 \text{ s}^{-1}$$

The next step is to determine the time required for 250 cm³ of nitrogen to diffuse under the same conditions as chlorine.

$$R_{N_2} = \frac{V_{N_2}}{t_{N_2}}$$

$$\therefore t_{N_2} = \frac{V_{N_2}}{R_{N_2}}$$

$$V_{N_2} = 250 \text{ cm}^3$$
$$t_{N_2} = ?$$

Substituting, we have

$$t_{N_2} = \frac{250 \text{ cm}^3}{15.924 \text{ cm}^3} \times 1\text{ s} = 15.7 \text{ s}$$

Finally, we can calculate the volume, V_{N_2}, of nitrogen that would diffuse in 10.5 s under the same conditions as chlorine.

$$\frac{250 \text{ cm}^3}{15.7 \text{ s}} = \frac{V_{N_2}}{10.5 \text{ s}}$$

So

$$250 \text{ cm}^3 \times 10.5\text{ s} = 15.7\text{ s} \times V_{N_2}$$

$$\therefore V_{N_2} = \frac{250 \text{ cm}^3 \times 10.5 \text{ s}}{15.7 \text{ s}} = 170 \text{ cm}^3$$

PRACTICE PROBLEMS

1. 50.5 cm³ of a gas diffuses in 15 s at a particular temperature and pressure. Determine the rate of diffusion of the gas under the given conditions.
2. 250 cm³ of hydrogen diffuses through a porous pot in 15 s, while it takes 60 s for the same volume of an unknown gas to diffuse through the same pot under identical conditions. Determine the molar mass of the unknown gas.
$$(H = 1)$$
3. 150 cm³ of chlorine diffuses through an orifice in 65 s. Determine the rate of diffusion of the same volume of oxygen under similar conditions.
$$(O = 16, Cl = 35.5)$$
4. It takes 13 s for 50.5 cm³ of nitrogen to diffuse through a porous pot. How long would it take for the same volume of carbon dioxide to diffuse under similar conditions?
$$(C = 12, N = 14, O = 16)$$
5. 45 cm³ of a gas whose vapour density is 23 diffuses in 5.0 s. How long would it take the same volume of carbon monoxide to diffuse under similar conditions?
$$(C = 12, O = 16)$$

FURTHER PRACTICE PROBLEMS

1. A mixture of gases contains 2.40 g, 1.10 g, 1.51 g and 2.00 g of oxygen, carbon dioxide, neon and nitrogen, respectively. Determine the mole fraction of each gas.
$$(H = 1, C = 12, N = 14, O = 16, Ar = 40)$$
2. A mixture of three gases contains 10.50 g of oxygen, 6.25 g of neon and an unknown mass of helium. Determine the mass of helium in the mixture if the total number of moles of the gases is 2.50.
$$(H = 4, O = 16, Ne = 20)$$

3. A mixture of nitrogen, hydrogen and neon exerts a total pressure of 900 mmHg. Determine the partial pressure of hydrogen if neon and nitrogen both exert a total pressure of 500 mmHg

4. A gas was collected over water in the laboratory at 20°C. Calculate the pressure exerted by the dry gas.
$$\text{(Vapour pressure of water at 20°C} = 17\,\text{mmHg)}$$

5. Determine the volume of 350 cm³ of a gas when its pressure is halved at constant temperature.

6. A gas has a volume of 500 cm³ at STP. Calculate the volume of the gas when its pressure is increased to 850 mmHg at constant temperature.

7. A gas fills 250 cm³ of a container at 770 mmHg. By what value must the pressure be reduced to increase the volume of the gas to 320 cm³ at the same temperature?

8. Calculate the volume of a gas at 50°C and 550 mmHg if it occupies a volume of 55 cm³ at 27°C and 550 mmHg.

9. A gas has a volume of 105 cm³ at 25°C. Calculate the temperature at which the volume of the gas would rise to 170 cm³ at constant pressure.

10. A gas occupies a volume of 450 cm³ at 780 mmHg and 30°C. At what pressure would the volume of the gas reduce by 70 cm³ at 52°C?

11. A gas occupies a volume of 520 cm³ at 1.5 atm and 25°C. Determine the volume of the gas at 47°C and 1.8 atm.

12. The volume of a gas is 450 cm³ at 55°C and 780 mmHg. Determine the temperature at which its volume would reduce to two-fifths its initial value of 670 mmHg.

13. A 350 cm³ sample of oxygen has a mass of 2.5 g at 53°C. Determine the pressure of the gas if it behaves like an ideal gas.
$$\left(O = 32;\ R = 0.08206\,\text{dm}^3\,\text{atm}\,\text{mol}^{-1}\,\text{K}^{-1}\right)$$

14. An ideal gas occupies a volume of 510 cm³ at 33°C and 280 kPa. Determine the amount of gas, in moles.
$$\left(R = 8.314\,\text{J}\,\text{mol}^{-1}\,\text{K}^{-1}\right)$$

15. A 850 cm³ sample of sulfur trioxide is at −12°C and 16 kPa. Determine the mass of the gas if it behaves like an ideal gas.
$$\left(O = 16,\ S = 32;\ R = 8.314\,\text{J}\,\text{mol}^{-1}\,\text{K}^{-1}\right)$$

16. 1.5 dm³ of an ideal gas weighs 1.5 g at 1.11 atm. Determine the temperature of the gas if it has a vapour density of 22.
$$\left(R = 0.08206\,\text{dm}^3\,\text{atm}\,\text{mol}^{-1}\,\text{K}^{-1}\right)$$

17. 2.5 dm³ of hydrogen at 55°C has a mass of 0.50 g. Calculate the pressure of the gas if it behaves like an ideal gas.
$$\left(R = 8.314\,\text{J}\,\text{mol}^{-1}\,\text{K}^{-1}\right)$$

18. 382 cm³ of an ideal gas weighs 2.5 g at 25°C and 2.5 atm. Calculate the molar mass of the gas.
$$\left(R = 0.08206\,\text{dm}^3\,\text{atm}\,\text{mol}^{-1}\,\text{K}^{-1}\right)$$

Gas Laws

19. 13.8 dm³ of an ideal gas weighs 1.5 g at −25°C and 5.1 kPa. Calculate the vapour pressure of the gas.
$$\left(R = 8.314 \, \text{J mol}^{-1} \, \text{K}^{-1}\right)$$

20. The rate of diffusion of a gas is 0.55 cm³ s⁻¹. How long would it take 350 cm³ of the gas to diffuse under similar conditions?

21. 550 cm³ of oxygen diffuses through an orifice in 45 s, while it takes 33 s for the same volume of an unknown gas to diffuse under similar conditions. Calculate the vapour density of the gas.
$$(O = 16)$$

22. 150 cm³ of carbon dioxide diffuses through a porous pot in 62 s. Calculate the time required for 250 cm³ of carbon monoxide to diffuse under similar conditions.
$$(\text{Relative atomic masses}: C = 12, \, O = 16)$$

23. 850 cm³ of sulfur dioxide diffuses through a porous pot in 62 s. What volume of methane would diffuse through the same pot in 25 s under similar conditions?
$$(H = 1, \, C = 12, \, S = 32)$$

24. Determine the number of times the rate of diffusion of ethene is as fast as that that of sulfur dioxide under similar conditions?
$$(H = 1, \, C = 12, \, O = 16, \, S = 32)$$

23. A certain volume of hydrogen was found to diffuse 3.6 times as fast as an unknown gas. Calculate the vapour pressure of the unknown gas.
$$(H = 1)$$

9 Properties of Solutions

9.1 INTRODUCTION

A solution is a homogeneous mixture of two or more substances. It consists of two parts: a solvent and one or more solutes. A solvent is a substance that dissolves another substance, called a solute. For example, a sugar solution is made by dissolving sugar—a solute—in water—a solvent. A solvent or solute can be a solid, liquid or gas. A solution in which the solvent is water is called an *aqueous solution*. In this book, we deal exclusively with aqueous solutions.

Water is often called the *universal solvent* because it can dissolve many more substances than any other liquid.

9.2 CONCENTRATION

The strength of a solution is measured in terms of its *concentration*. A *weak or dilute solution* contains a small amount of solute relative to the volume of the solvent, while a *concentrated solution* contains a large amount of the solute relative to the volume of the solvent.

There are several ways to express concentration, the most common being molar concentration (molarity), mass concentration, molality (m), normality, mass percent, volume percent, parts per million (ppm), parts per billion (ppb) and mole fraction. Mole fraction was treated in Chapter 8. In addition to mole fraction, we shall consider molar concentration and mass concentration.

Molar concentration, C, is expressed as the number of moles of solute per cubic decimetre of a solution. Mathematically,

$$C = \frac{n}{V}$$

where

n = number of moles of the solute
V = volume of the solution (in dm³)

The unit of molar concentration is mol dm⁻³ (or mol/dm³), commonly abbreviated as M (pronounced *molar*).

Mass concentration, ρ, is expressed as the mass of solute per cubic decimetre of solution. Mathematically,

$$\rho = \frac{m}{V}$$

where

m = mass of the solute in grammes
V = volume of the solution (in dm³)

The unit of mass concentration is g dm⁻³ (grams per cubic decimetre).
The relationship between molar and mass concentrations is:

$$C = \frac{\rho}{M}$$

where M is the molar mass of the solute

Properties of Solutions

Example 1

A sugar solution was found to contain 0.15 mol of the solute in 250 cm³ of the solution. Determine the molar concentration of the solution.

Solution

To determine the molar concentration of the sugar solution, we use the relation:

$$C = \frac{n}{V}$$

$$n = 0.15 \text{ mol}$$

$$V = 250 \text{ cm}^3 = 0.25 \text{ dm}^3$$

$$C = ?$$

Substituting, we have

$$C = \frac{0.15 \text{ mol}}{0.25 \text{ dm}^3} = 0.60 \text{ mol dm}^{-3}$$

Alternatively, we can scale up the solution from 250 cm³ (0.25 dm³) to 1000 cm³ (1.0 dm³), keeping in mind that the molar concentration, C, is the number of moles of solute in 1000 cm³ of solution.

$$\frac{0.25 \text{ dm}^3}{0.15 \text{ mol}} = \frac{1.0 \text{ dm}^3}{C}$$

So

$$C \times 0.25 \text{ dm}^3 = 1.0 \text{ dm}^3 \times 0.15 \text{ mol}$$

$$\therefore C = \frac{1.0 \text{ dm}^3 \times 0.15 \text{ mol}}{0.25 \text{ dm}^3} = 0.60 \text{ mol}$$

Since 1.0 dm³ of solution contains 0.60 mol of the solute, it follows that the molar concentration of the solution is 0.60 mol dm⁻³.

Example 2

550 cm³ of a sodium chloride solution contains 1.5 g of the salt. Determine:

(a) the mass concentration of the solution;
(b) the molarity of the solution.

$$(Na = 23, Cl = 35.5)$$

Solution

(a) The mass concentration of a solution is given as:

$$\rho = \frac{m}{V}$$

$$m = 1.5 \text{ g}$$

$$V = 550 \text{ cm}^3 = 0.55 \text{ dm}^3$$
$$V = ?$$

Substituting, we have

$$\rho = \frac{1.5 \text{ g}}{0.55 \text{ dm}^3} = 2.7 \text{ g dm}^{-3}$$

As usual, we can also scale up the solution as follows.

$$\frac{1.5 \text{ g}}{0.55 \text{ dm}^3} = \frac{\rho}{1.0 \text{ dm}^3}$$

So

$$\rho \times 0.55 \text{ dm}^3 = 1.0 \text{ dm}^3 \times 1.5 \text{ g}$$

$$\therefore \rho = \frac{1.0 \text{ dm}^3 \times 1.5 \text{ g}}{0.55 \text{ dm}^3} = 2.7 \text{ g}$$

1.0 dm³ of the solution contains 2.7 g of the solute; hence, the mass concentration of the solution is 2.7 g dm⁻³.

(b) The molar concentration is calculated using the relation:

$$C = \frac{\rho}{M}$$

$$\rho = 2.7 \text{ g dm}^{-1}$$
$$M = (23 + 35.5) \text{ g mol}^{-1} = 58.5 \text{ g mol}^{-1}$$
$$C = ?$$

Substituting, we have

$$C = \frac{2.7 \text{ g} \times 1 \text{ mol}}{58.5 \text{ g} \times 1 \text{ dm}^3} = 0.046 \text{ mol dm}^{-3}$$

An alternative approach is to first convert the mass of NaCl to moles as follows.

$$n = \frac{m}{M}$$

$$m = 1.5 \text{ g}$$
$$M = 58.5 \text{ g mol}^{-1}$$
$$n = ?$$

$$n = \frac{1.5 \text{ g} \times 1 \text{ mol}}{58.5 \text{ g}} = 0.02564 \text{ mol}$$

Now, we determine the molar concentration of the solution.

$$C = \frac{n}{V}$$

Properties of Solutions

$$V = 550 \text{ cm}^3 = 0.55 \text{ dm}^3$$

$$C = ?$$

Substituting, we have

$$C = \frac{0.02564 \text{ mol}}{0.55 \text{ dm}^3} = 0.047 \text{ mol dm}^{-3}$$

Note that the slight difference is a result of rounding.
We can also obtain the molar concentration by scaling up the solution as follows.

$$\frac{0.02564 \text{ mol}}{0.55 \text{ dm}^3} = \frac{C}{1.0 \text{ dm}^3}$$

So

$$C \times 0.55 \text{ dm}^3 = 1.0 \text{ dm}^3 \times 0.02564$$

$$\therefore C = \frac{0.02564 \text{ mol} \times 1 \text{ dm}^3}{0.55 \text{ dm}^3} = 0.047 \text{ mol}$$

Since 1.0 dm³ of the solution contains 0.047 mol of the salt, the molar concentration of the salt is 0.047 mol dm⁻³.

Example 3

The concentration of a salt solution is 5.5 g dm³. What volume of the solution would contain 0.55 g of the salt?

Solution

We know that mass concentration is expressed as follows:

$$\rho = \frac{m}{V}$$

$$\therefore V = \frac{m}{\rho}$$

$$m = 0.55 \text{ g}$$

$$\rho = 5.5 \text{ g dm}^{-3}$$

$$V = ?$$

Substituting, we have

$$V = \frac{0.55 \text{ g} \times 1 \text{ dm}^3}{5.5 \text{ g}} = 0.10 \text{ dm}^3$$

We will also arrive at this value if we scale down the solution as follows.

$$\frac{5.5 \text{ g}}{1.0 \text{ dm}^3} = \frac{0.55 \text{ g}}{V}$$

So

$$1.0 \text{ dm}^3 \times 0.55 \text{ g} = V \times 5.5 \text{ g}$$

$$\therefore V = \frac{0.55 \text{ g} \times 1.0 \text{ dm}^3}{5.5 \text{ g}} = 0.10 \text{ dm}^3$$

Example 4

Determine the mass of sodium hydroxide in 25 cm³ of a 0.25 mol dm⁻³ solution.

$$(H = 1, Cl = 35.5)$$

Solution

The first step is to obtain the mass concentration of the base.

$$C = \frac{\rho}{M}$$

$$\therefore \rho = C \times M$$

$$C = 0.25 \text{ mol dm}^{-3}$$

$$M = (1 + 35.5) \text{ g mol}^{-1} = 36.5 \text{ g mol}^{-1}$$

$$\rho = ?$$

Substituting, we have

$$\rho = \frac{0.25 \text{ mol} \times 36.5 \text{ g}}{1 \text{ dm}^3 \times 1 \text{ mol}} = 9.125 \text{ g dm}^{-3}$$

We can now determine the mass of the salt as follows.

$$\rho = \frac{m}{V}$$

$$\therefore m = \rho \times V$$

$$V = 25 \text{ cm}^3 = 0.025 \text{ dm}^3$$

$$m = ?$$

Substituting, we have

$$m = \frac{9.125 \text{ g}}{1 \text{ dm}^3} \times 0.025 \text{ dm}^3 = 0.23 \text{ g}$$

We will obtain the same value if we scale down the mass of the solute in 1.0 dm³ of solution.

$$\frac{9.125 \text{ g}}{1.0 \text{ dm}^3} = \frac{m}{0.025 \text{ dm}^3}$$

Properties of Solutions

So

$$9.125 \text{ g} \times 0.025 \text{ dm}^3 = m \times 1.0 \text{ dm}^3$$

$$\therefore m = \frac{9.125 \text{ g} \times 0.025 \text{ dm}^3}{1.0 \text{ dm}^3} = 0.23 \text{ g}$$

An alternative way to solve this problem is to calculate the number of moles of the solute in 0.025 dm³ of solution and then convert the result to mass.

$$C = \frac{n}{V}$$

$$\therefore n = C \times V$$

$$C = 0.25 \text{ mol dm}^{-3}$$

$$V = 0.025 \text{ dm}^3$$

$$n = ?$$

Substituting, we have

$$n = \frac{0.25 \text{ mol}}{1 \text{ dm}^3} \times 0.025 \text{ dm}^3 = 0.00625 \text{ mol}$$

As always, rather than applying a formula, we can solve this problem by scaling down the solution.

$$\frac{0.25 \text{ mol}}{1.0 \text{ dm}^3} = \frac{n}{0.025 \text{ dm}^3}$$

So

$$0.25 \text{ mol} \times 0.025 \text{ dm}^3 = n \times 1.0 \text{ dm}^3$$

$$\therefore n = \frac{0.25 \text{ mol} \times 0.025 \text{ dm}^3}{1.0 \text{ dm}^3} = 0.00625 \text{ mol}$$

Finally, we can calculate the mass of the solute as follows.

$$n = \frac{m}{M}$$

$$\therefore m = n \times M$$

$$n = 0.00625 \text{ mol}$$

$$M = 36.5 \text{ g mol}^{-3}$$

$$m = ?$$

Substituting the values, we have

$$m = 0.00625 \text{ mol} \times \frac{36.5 \text{ g}}{1 \text{ mol}} = 0.23 \text{ g}$$

PRACTICE PROBLEMS

1. Determine the number of moles of sulfuric acid in 50.0 cm³ of a 25.5 g dm⁻³ solution.

$$(H = 1, \ O = 16, \ S = 32)$$

2. Determine the mass of sodium hydroxide in 25 cm³ of a 0.10 mol dm⁻³ solution.

$$(H = 1, \ O = 16, \ Na = 23)$$

3. A solution contains 12.5 g of glucose, $C_6H_{12}O_6$, in 150 cm³ of solution. Determine:
   (a) the mass concentration of the solution;
   (b) the molarity of the solution.

$$(H = 1, \ C = 12, \ O = 16)$$

4. The molar concentration of an acid solution containing 6.5 g of the acid in 450 cm³ of solution is 0.72 mol dm⁻³. Determine the molar mass of the acid.

9.3 DILUTION LAW

When more solvent is added to a solution, its volume increases while its concentration reduces. However, the number of moles of solute remains the same. In other words,

$$n_1 = n_2$$

where n_1 and n_2 are the number of moles of the solute before and after dilution, respectively.
Since $n = C \times V$, we can also express this relationship as:

$$C_1 V_1 = C_2 V_2$$

This equation is known as the *dilution law* or *dilution formula*. When using the dilution formula, ensure that the volume units are consistent.

Example 1

A 0.15 mol dm⁻³ sucrose solution was diluted to 150 cm³. Determine the initial volume of the solution if its new concentration is 0.10 mol dm⁻³.

Solution

We will apply the dilution formula: as follows.

$$C_1 V_1 = C_2 V_2$$

$$\therefore V_1 = \frac{C_2 V_2}{C_1}$$

$$C_1 = 0.15 \text{ mol dm}^{-3}$$

$$C_2 = 0.10 \text{ mol dm}^{-3}$$

Properties of Solutions

$$V_2 = 150 \text{ cm}^3$$
$$V_1 = ?$$

Substituting, we have

$$V_1 = \frac{0.10 \text{ mol dm}^{-3} \times 150 \text{ cm}^3}{0.15 \text{ mol dm}^{-3}} = 100 \text{ cm}^3$$

Example 2

What volume of distilled water must be added to 450 cm³ of a 0.25 mol dm⁻³ hydrochloric acid solution to obtain a 0.15 mol dm⁻³ solution?

Solution

The first step is to determine the final volume of the solution.

$$C_1 V_1 = C_2 V_2$$
$$\therefore V_2 = \frac{C_1 V_1}{C_2}$$

$$C_1 = 0.25 \text{ mol dm}^{-3}$$
$$V_1 = 450 \text{ cm}^3$$
$$C_2 = 0.15 \text{ mol dm}^{-3}$$
$$V_2 = ?$$

Substituting, we have

$$V_2 = \frac{0.25 \text{ mol dm}^{-3} \times 450 \text{ cm}^3}{0.15 \text{ mol dm}^{-3}} = 750 \text{ cm}^3$$

The required volume of water is the difference between the final and initial volumes of the solution.

$$\therefore V = (750 - 450) \text{ cm}^3 = 300 \text{ cm}^3$$

Example 3

10.5 dm³ of a 0.15 mol dm⁻³ solution was prepared by diluting 55 cm³ of a stock solution. Determine concentration of the stock solution.

Solution

We apply the dilution formula as follows.

$$C_1 V_1 = C_2 V_2$$
$$\therefore C_1 = \frac{C_2 V_2}{V_1}$$

$$C_2 = 0.15 \text{ mol dm}^{-3}$$
$$V_2 = 10.5 \text{ dm}^3$$
$$V_1 = 55 \text{ cm}^3 = 0.055 \text{ dm}^3$$
$$C_1 = ?$$

We now substitute to obtain

$$C_1 = \frac{0.15 \text{ mol dm}^{-3} \times 10.5 \text{ dm}^3}{0.055 \text{ dm}^3} = 29 \text{ mol dm}^{-3}$$

PRACTICE PROBLEMS

1. What volume of a 5.5 mol dm^{-3} nitric acid solution is required to prepare 250 cm^3 of a 0.10 mol dm^{-3} solution of the acid?
2. What volume of distilled water must be added to 45 cm^3 of a 0.20 mol dm^3 solution to make a 0.11 mol dm^{-3} solution?
3. 35 cm^3 of distilled water was added to 25 cm^3 of a 0.15 mol dm^{-3} potassium hydroxide solution. Calculate the new concentration of the solution.
4. The concentration of a 250 cm^3 solution was reduced from 0.15 mol dm^{-3} to 0.050 mol dm^{-3}. Determine the new volume of the solution.

9.4 PREPARATION OF STANDARD SOLUTIONS

A standard solution is a solution of known concentration. Standard solutions are prepared using a *volumetric flask*.

To prepare a standard solution of a solid solute, such as a base, an appropriate mass of the solid is weighed and dissolved in distilled water. The solution is then carefully transferred into a volumetric flask and diluted to the etched ring graduation mark of the flask.

For liquids, such as acids, the procedure is not as straightforward. The required volume of the *stock solution* required to prepare the desired solution must be first calculated using the density of the liquid and its composition in the stock solution. This volume is then carefully measured and transferred into a volumetric flask to obtain the required standard solution.

Example 1

10.0 g of sodium hydroxide was dissolved in distilled water and then diluted up to the 250 cm^3 graduation mark in a 250 cm^3 volumetric flask. Determine the mass concentration and molar concentration of the resulting solution.

$$(H = 1, O = 16, Na = 23)$$

Solution

This problem involves determining the concentration of a solution containing 10.0 g of solute in 250 cm^3 of solution. We proceed as follows.

$$\rho = \frac{m}{V}$$
$$m = 10 \text{ g}$$

Properties of Solutions

$$V = 250 \text{ cm}^3 = 0.25 \text{ dm}^3$$

$$\rho = ?$$

Substituting, we have

$$\rho = \frac{10.0 \text{ g}}{0.25 \text{ dm}^3} = 40 \text{ g dm}^{-3}$$

As usual, we can scale up the solution as follows.

$$\frac{10 \text{ g}}{0.25 \text{ dm}^3} = \frac{\rho}{1.0 \text{ dm}^3}$$

So

$$10 \text{ g} \times 1 \text{ dm}^3 = \rho \times 0.25 \text{ dm}^3$$

$$\therefore \rho = \frac{10 \text{ g} \times 1 \text{ dm}^3}{0.25 \text{ dm}^3} = 40 \text{ g}$$

Thus, the mass concentration of the solute is 40 g dm⁻³.
The molar concentration of the solution is obtained as follows.

$$C = \frac{\rho}{M}$$

$$\rho = 40 \text{ g dm}^{-3}$$

$$M = (23 + 16 + 1) \text{ g mol}^{-1} = 40 \text{ g mol}^{-1}$$

$$C = ?$$

Substituting, we have

$$C = \frac{40 \text{ g} \times 1 \text{ mol}}{40 \text{ g} \times 1 \text{ dm}^3} = 1 \text{ mol dm}^{-1}$$

Example 2

What mass of potassium hydroxide is required to prepare 25.5 dm³ of a 12.0 g dm³ solution of the base?

Solution

The required mass can be determined from volume and mass concentration as follows.

$$\rho = \frac{m}{V}$$

$$\therefore m = \rho \times V$$

$$\rho = 12.0 \text{ g dm}^{-3}$$

$$V = 25.5 \text{ dm}^3$$
$$m = ?$$

Substituting, we have

$$m = \frac{12.0 \text{ g}}{1 \text{ dm}^3} \times 25.5 \text{ dm}^3 = 306 \text{ g}$$

Alternatively, we can scale up the volume of the solution as follows.

$$\frac{12 \text{ g}}{1.0 \text{ dm}^3} = \frac{m}{25.5}$$

So

$$m \times 1.0 \text{ dm}^3 = 12.0 \text{ g} \times 25.5 \text{ dm}^3$$
$$\therefore m = 306 \text{ g}$$
$$m = \frac{12.0 \text{ g} \times 25.5 \text{ dm}^3}{1.0 \text{ dm}^3} = 306 \text{ g}$$

Example 3

Determine the volume of concentrated sulfuric acid required to prepare a 0.10 mol dm^{-3} solution of the acid, given that the density and percentage composition of the acid are 1.84 g dm^{-3} and 98%, respectively.

$$(H = 1, O = 16, S = 32)$$

Solution

The desired molar concentration of the solution, 0.10 mol dm^{-3}, means that 0.10 mol of sulfuric acid must be present in 1.0 dm^3 of the solution. We now convert this to amount in moles to mass as follows.

$$n = \frac{m}{M}$$
$$\therefore m = n \times M$$
$$n = 0.10 \text{ mol}$$
$$M = (\{1 \times 2\} + 32 + \{16 \times 4\}) \text{ g mol}^{-1} = 98 \text{ g mol}^{-1}$$
$$m = ?$$

Substituting, we have

$$m = 0.10 \text{ mol} \times \frac{98 \text{ g}}{1 \text{ mol}} = 9.8 \text{ g}$$

The next step is to obtain the volume of this mass of the pure acid from its density.

$$d = \frac{m}{V}$$

Properties of Solutions

$$\therefore V = \frac{m}{d}$$

$$d = 1.84 \text{ g cm}^{-3}$$

$$V = ?$$

Substituting, we have

$$V = \frac{9.8 \text{ g} \times 1 \text{ cm}^3}{1.84 \text{ g}} = 5.326 \text{ cm}^3$$

Alternatively, we can scale up the volume as follows.

$$\frac{1.0 \text{ cm}^3}{1.84 \text{ g}} = \frac{V}{9.8 \text{ g}}$$

So

$$1.0 \text{ cm}^3 \times 9.8 \text{ g} = V \times 1.84 \text{ g}$$

$$\therefore V = \frac{1.0 \text{ cm}^3 \times 9.8 \text{ g}}{1.84 \text{ g}} = 5.326 \text{ cm}^3$$

Finally, we need to determine the volume of the stock solution (the concentrated acid solution) that contains this volume of pure acid. Let V be the required volume of the pure acid. Since the percentage composition of 98% means that 1 cm³ of the concentrated acid contains 0.98 cm³ of pure acid, it then follows that

$$\frac{1.0 \text{ cm}^3}{0.98 \text{ cm}^3} = \frac{V}{5.326 \text{ cm}^3}$$

So

$$1.0 \text{ cm}^3 \times 5.326 \text{ cm}^3 = V \times 0.98 \text{ cm}^3$$

$$\therefore V = \frac{1.0 \text{ cm}^3 \times 5.326 \text{ cm}^3}{0.98 \text{ cm}^3} = 5.4 \text{ cm}^3$$

An alternative approach to solving this problem is to first determine the mass of 1.0 dm³ of pure acid, i.e., its mass concentration, using its density. We proceed as follows.

$$d = \frac{m}{V}$$

$$\therefore m = d \times V$$

$$d = 1.84 \text{ g cm}^{-3}$$

$$V = 1.0 \text{ dm}^3 = 1000 \text{ cm}^3$$

$$m = ?$$

Substituting, we have

$$m = \frac{1.84 \text{ g}}{1 \text{ cm}^3} \times 1000 \text{ cm}^3 = 1840 \text{ g}$$

Alternatively, we can scale up the solution as follows.

$$\frac{1.0 \text{ cm}^3}{1.84 \text{ g}} = \frac{1000 \text{ cm}^3}{m}$$

So

$$1.0 \text{ cm}^3 \times m = 1.84 \text{ g} \times 1000 \text{ cm}^3$$

$$\therefore m = \frac{1.84 \text{ g} \times 1000 \text{ cm}^3}{1.0 \text{ cm}^3} = 1840 \text{ g}$$

The composition of pure acid in the concentrated acid solution is 98%. Therefore, the mass of pure acid in 1.0 dm³ of the concentrated acid solution is:

$$m = 0.98 \times 1840 \text{ g} = 1803.2 \text{ g}$$

Thus,

$$\rho = 1803.2 \text{ g dm}^{-3}$$

The next step is to convert this mass concentration to molar concentration.

$$C = \frac{\rho}{M}$$

$$\rho = 1803.2 \text{ g dm}^{-3}$$

$$M = 98 \text{ g mol}^{-1}$$

$$C = ?$$

Substituting, we have

$$C = \frac{1803.2 \text{ g} \times 1 \text{ mol}}{98 \text{ g} \times 1 \text{ dm}^3} = 18.4 \text{ mol dm}^3$$

Finally, we calculate the required volume of the concentrated acid using the dilution formula as follows.

$$C_1 V_1 = C_2 V_2$$

$$\therefore V_1 = \frac{C_2 V_2}{C_1}$$

$$C_1 = 18.4 \text{ mol dm}^{-3}$$

$$C_2 = 0.10 \text{ mol dm}^{-3}$$

$$V_2 = 1.0 \text{ dm}^3$$

$$V_1 = ?$$

Substituting, we have

$$\therefore V_1 = \frac{0.10 \text{ mol dm}^{-3} \times 1.0 \text{ dm}^3}{18.4 \text{ mol dm}^{-3}} = 0.0054 \text{ dm}^3$$

Properties of Solutions

Example 4

A nitric acid solution was prepared by diluting 5.5 cm³ of a stock solution of the acid to 1.0 dm³. Given that the density of the acid and its composition in the stock solution are 1.52 g cm⁻³ and 68%, respectively, determine the concentration of the resulting acid solution.

$$(H = 1, N = 14, O = 16)$$

Solution

We begin by calculating the volume, V, of pure acid in 5.5 cm³ of the stock solution. Since the composition of pure acid in the stock solution is 68%, it follows that

$$V = 0.68 \times 5.5 \text{ cm}^3 = 3.74 \text{ cm}^3$$

Next, we calculate the mass of this volume of pure acid using its density as follows.

$$d = \frac{m}{V}$$

$$\therefore m = d \times V$$

$$d = 1.52 \text{ g cm}^{-3}$$

$$m = ?$$

Substituting the values, we obtain

$$m = \frac{1.52 \text{ g}}{1 \text{ cm}^3} \times 3.74 \text{ cm}^3 = 5.7 \text{ g}$$

Alternatively, we can simply scale up the volume of solution.

$$\frac{1.0 \text{ cm}^3}{1.52 \text{ g}} = \frac{3.74 \text{ cm}^3}{m}$$

So

$$m \times 1.0 \text{ cm}^3 = 3.74 \text{ cm}^3 \times 1.52 \text{ g}$$

$$\therefore m = \frac{3.74 \text{ cm}^3 \times 1.52 \text{ g}}{1.0 \text{ cm}^3} = 5.7 \text{ g}$$

Since 5.5 cm³ of the concentrated acid was diluted to 1.0 dm³ of solution, and the mass of pure acid in this volume is 5.7 g, it follows that the mass concentration of the resulting nitric acid solution is 5.7 g dm⁻³.

Next, we calculate the molar concentration of the solution as follows.

$$C = \frac{\rho}{M}$$

$$\rho = 5.7 \text{ g dm}^{-3}$$

$$M = (1 + 14 + \{16 \times 3\}) \text{ g mol}^{-1} = 63 \text{ g mol}^{-1}$$

$$C = ?$$

Substituting, we have

$$C = \frac{5.7 \text{ g} \times 1 \text{ mol}}{63 \text{ g} \times 1 \text{ dm}^3} = 0.090 \text{ mol dm}^{-3}$$

An alternative approach is to first calculate the mass of 1.0 dm³ of pure acid as follows.

$$d = \frac{m}{V}$$

$$\therefore m = d \times V$$

$$d = 1.52 \text{ g cm}^{-3}$$

$$V = 1.0 \text{ dm}^3 = 1000 \text{ cm}^3$$

$$m = ?$$

Substituting, we have

$$m = \frac{1.52 \text{ g}}{1 \text{ cm}^3} \times 1000 \text{ cm}^3 = 1520 \text{ g}$$

As usual, this mass can also be obtained by scaling up the solution as follows.

$$\frac{1.0 \text{ cm}^3}{1.52 \text{ g}} = \frac{1000 \text{ cm}^3}{m}$$

So

$$1.0 \text{ cm}^3 \times m = 1000 \text{ cm}^3 \times 1.52 \text{ g}$$

$$\therefore m = \frac{1000 \text{ cm}^3 \times 1.52 \text{ g}}{1 \text{ cm}^3} = 1520 \text{ g}$$

Since the composition of pure acid in the concentrated acid solution is 68%, the mass, m, of pure acid contained in 1520 g of the concentrated acid solution is:

$$m = 0.68 \times 1520 \text{ g} = 1033.6 \text{ g}$$

This is the mass of pure acid in 1 dm³ of the concentrated acid solution. Therefore, the mass concentration of the concentrated acid solution is 1033.6 g dm⁻³. Next, the molar concentration is obtained as follows.

$$C = \frac{\rho}{M}$$

$$M = 63 \text{ g mol}^{-3}$$

$$C = ?$$

Substituting, we have

$$C = \frac{1033.6 \text{ g} \times 1 \text{ mol}}{63 \text{ g} \times 1 \text{ dm}^3} = 16.41 \text{ mol dm}^{-3}$$

Finally, we can determine the concentration of the dilute acid solution as follows.

$$C_1 V_1 = C_2 V_2$$

Properties of Solutions

$$\therefore C_2 = \frac{C_1 V_1}{V_2}$$

$$C_1 = 16.41 \text{ mol dm}^{-3}$$

$$V_1 = 5.5 \text{ cm}^3 = 0.0055 \text{ dm}^3$$

$$V_2 = 1.0 \text{ dm}^3$$

$$C_2 = ?$$

We now substitute to obtain

$$C_2 = \frac{16.41 \text{ mol dm}^{-3} \times 0.0055 \text{ dm}^3}{1.0 \text{ dm}^3} = 0.090 \text{ mol dm}^{-3}$$

PRACTICE PROBLEMS

1. A teacher intends to prepare 10.0 dm³ of a 0.250 mol dm⁻³ sodium carbonate solution. What mass of the compound does he need to prepare the solution?

$$(C = 12, \ O = 16, \ Na = 23)$$

2. The percentage composition of hydrochloric acid in a stock solution is 40%. Calculate the volume of the stock solution required to prepare a 0.15 mol dm⁻³ solution of the acid, given that its density is 1.198 g cm⁻³.

$$(H = 1, \ Cl = 35.5)$$

3. Calculate the molarity of an ammonia solution prepared by diluting 5.0 cm³ of a stock solution of the base to 1 dm³. The density of the ammonia solution is 0.89 g cm⁻³, and its composition in the stock solution is 80%.

$$(H = 1, \ N = 14)$$

9.5 SOLUBILITY

The solubility, S, of a substance is the maximum amount of that substance that will dissolve in a specified quantity of the solvent at a particular temperature. A solution that contains the maximum amount of a solute it can dissolve at a particular temperature is called a *saturated solution*.

Solubility is mostly expressed in terms of the mass of solute required to saturate 100 g of solvent. Another common way of expressing solubility is in the mass or number of moles of solute required to saturate 1.0 dm³ of solvent. The measurement of solubility in terms of the number of moles of solutes in 1.0 dm³ of a saturated solution is called *molar solubility*. Similarly, solubility can also be expressed in terms of the mass of solute in 1.0 dm³ of a saturated solution, which is referred to as mass solubility.

Solubility varies with *temperature*, the *nature of the solute and solvent, pressure* and *pH*. Generally, the solubility of solids increases with an increase in temperature, while the solubility of gases decreases with an increase in temperature.

Polar substances, such as ionic compounds, are soluble in polar solvents such as water, trichloromethane and ethanol. Non-polar substances, such as covalent compounds, are soluble in non-polar solvents such as benzene, cyclohexane, tetrachloromethane and toluene.

The solubility of ionic compounds containing basic ions increases with a decrease in pH, while pH has no effect on the solubility of other substances.

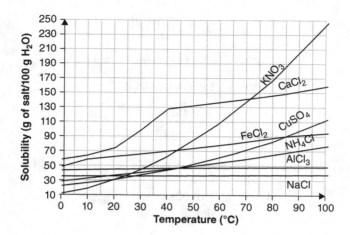

FIGURE 9.1 Solubility curves of different salts.

According to *Henry's Law*, the solubility of gases is directly proportional to pressure. In contrast, pressure has no effect on the solubility of solids.

When the solubility of a substance is plotted against temperature, the resulting graph is called a *solubility curve*. A solubility curve allows the solubility of a substance at a specific temperature to be easily determined. The solubility curves of some salts are shown in Figure 9.1. From the diagram, the solubility of calcium chloride is approximately 130 g/100 g of H_2O at 40°C.

Example 1

250 cm³ of a saturated solution of calcium hydroxide, $Ca(OH)_2$, contains 0.80 g of the base at 25°C. At this temperature, determine the solubility of the base in:
(a) g/100 g H_2O;
(b) g dm⁻³;
(c) mol dm⁻³.

Solution

(a) To obtain the solubility, S, of the base at a specified temperature, we need to scale down the volume of the saturated solution to 100 cm³. It is also important to know that, since the density of water is 1 g cm⁻³, the mass and volume of water are equivalent.

$$\frac{250 \text{ g}}{100 \text{ g}} = \frac{0.80 \text{ g}}{S}$$

So

$$250 \text{ g} \times S = 100 \text{ g} \times 0.80 \text{ g}$$

$$\therefore S = \frac{100 \text{ g} \times 0.80 \text{ g}}{250 \text{ g}} = 0.32 \text{ g}/100 \text{ g } H_2O$$

(b) Here, we have to scale up the solution to 1.0 dm³ as follows.

$$\frac{0.25 \text{ dm}^3}{0.80 \text{ g}} = \frac{1.0 \text{ dm}^3}{S}$$

Properties of Solutions

So

$$0.25 \text{ dm}^3 \times S = 1.0 \text{ dm}^3 \text{ g} \times 0.80 \text{ g}$$

$$\therefore S = \frac{1.0 \text{ dm}^3 \times 0.80 \text{ g}}{0.25 \text{ dm}^3} = 3.2 \text{ g}$$

Since 3.2 g of the solute is present in 1.0 dm³ of the saturated solution, the solubility is:

$$S = 3.2 \text{ g dm}^{-3}$$

Since the solubility of a substance in g dm^{-3} represents the mass concentration of its saturated solution, we can also apply the formula for mass concentration as follows.

$$S = \rho = \frac{m}{V}$$

$$m = 0.80 \text{ g}$$

$$V = 250 \text{ cm}^3 = 0.25 \text{ dm}^3$$

$$S = ?$$

Substituting, we have

$$S = \frac{0.80 \text{ g}}{0.25 \text{ dm}^3} = 3.2 \text{ g dm}^3$$

(c) The molar solubility of a substance is the molar concentration of its saturated solution.

$$S = C = \frac{\rho}{M}$$

$$\rho = 3.2 \text{ g dm}^{-3}$$

$$M = \left(40 + \{16 \times 2\} + \{1 \times 2\}\right) \text{ g mol}^{-1} = 74 \text{ g mol}^{-1}$$

$$S = ?$$

Substituting, we have

$$S = \frac{3.2 \text{ g} \times 1 \text{ mol}}{74 \text{ g} \times 1 \text{ dm}^3} = 0.043 \text{ mol dm}^{-3}$$

Alternatively, we can begin by calculating the number of moles of the solute in the saturated solution.

$$n = \frac{m}{M}$$

$$m = 0.80 \text{ g}$$

$$M = 74 \text{ g mol}^{-1}$$

$$n = ?$$

Substituting, we have

$$n = \frac{0.80 \text{ g} \times 1 \text{ mol}}{74 \text{ g}} = 0.0108 \text{ mol}$$

The solubility in mol dm^{-3} is obtained by scaling up the solution as follows.

$$\frac{0.25 \text{ dm}^3}{0.0108 \text{ mol}} = \frac{1.0 \text{ dm}^3}{n}$$

So

$$0.25 \text{ dm}^3 \times S = 1.0 \text{ dm}^3 \times 0.0108 \text{ mol}$$

$$\therefore S = \frac{1.0 \text{ dm}^3 \times 0.0108 \text{ mol}}{0.25 \text{ dm}^3} = 0.043 \text{ mol}$$

Since 1.0 dm^3 of the saturated solution contains 0.043 mol of the solute, the solubility is:

$$S = 0.043 \text{ mol dm}^{-3}$$

Alternatively, since molar solubility is the molar concentration of a saturated solution, we can also apply the formula as follows.

$$S = C = \frac{n}{V}$$

$$n = 0.0108 \text{ mol}$$

$$V = 0.25 \text{ dm}^3$$

$$n = ?$$

Substituting, we have

$$S = \frac{0.0108 \text{ mol}}{0.25 \text{ dm}^3} = 0.043 \text{ mol dm}^{-3}$$

Example 2

The solubility of sodium nitrate, NaNO$_3$, is 148 g/100 g H$_2$O at 80°C. Determine the mass of the salt required to saturate 65.0 g of water at 80°C.

Solution

The required mass, m, of NaNO$_3$, can be determined by scaling down the volume of solution as follows.

$$\frac{100 \text{ g}}{148 \text{ g}} = \frac{65.0 \text{ g}}{m}$$

So

$$100 \text{ g} \times m = 65 \text{ g} \times 148 \text{ g}$$

$$\therefore m = \frac{65.0 \text{ g} \times 148 \text{ g}}{100 \text{ g}} = 96.2 \text{ g}$$

Example 3

The solubility of potassium chlorate, KClO$_3$, is 0.011 mol dm^{-3} at a certain temperature. Determine the mass of the salt that would dissolve in 50.5 cm^3 of water at the same temperature.

$$(O = 16, Cl = 35.5, K = 39)$$

Properties of Solutions

Solution

We will begin by calculating the number of moles, n, of the solute that will dissolve in 50.5 cm³ of water.

$$\frac{1.0 \text{ dm}^3}{0.011 \text{ mol}} = \frac{0.0505 \text{ dm}^3}{n}$$

So

$$1.0 \text{ dm}^3 \times n = 0.0505 \text{ dm}^3 \times 0.011 \text{ mol}$$

$$\therefore n = \frac{0.0505 \text{ dm}^3 \times 0.011 \text{ mol}}{1.0 \text{ dm}^3} = 0.0005555 \text{ mol}$$

We can also use the formula for molar concentration as follows.

$$S = \frac{n}{V}$$

$$\therefore n = S \times V$$

$$S = 0.011 \text{ mol dm}^{-3}$$

$$V = 50.5 \text{ cm}^3 = 0.0505 \text{ dm}^3$$

$$n = ?$$

Substituting, we have

$$n = \frac{0.011 \text{ mol}}{1 \text{ dm}^3} \times 0.0505 \text{ dm}^3 = 0.0005555 \text{ mol}$$

The final step is to convert 0.0005555 mol of $KClO_3$ to mass.

$$n = \frac{m}{M}$$

$$\therefore m = n \times M$$

$$M = (39 + 35.5 + \{16 \times 3\}) \text{ g mol}^{-1} = 122.5 \text{ g mol}^{-1}$$

$$m = ?$$

Substituting, we have

$$m = 0.0005555 \text{ mol} \times \frac{122.5 \text{ g}}{1 \text{ mol}} = 0.068 \text{ g}$$

Alternatively, the required mass can be obtained from the mass concentration of the saturated solution.

$$S = C = \frac{\rho}{M}$$

$$\therefore \rho = C \times M$$

$$C = 0.011 \text{ mol}$$

$$M = 122.5 \text{ g mol}^{-1}$$

$$\rho = ?$$

Substituting, we have

$$\rho = \frac{0.011 \text{ mol}}{1 \text{ dm}^3} \times \frac{122.5 \text{ g}}{1 \text{ mol}} = 1.3475 \text{ g dm}^{-1}$$

The final step is to determine the mass of the solute in 0.0505 dm³ of the saturated solution.

$$\frac{1.0 \text{ dm}^3}{1.3475 \text{ g}} = \frac{0.0505 \text{ dm}^3}{m}$$

So

$$1.0 \text{ dm}^3 \times m = 0.0505 \text{ dm}^3 \times 1.3475 \text{ g}$$

$$\therefore m = \frac{0.0505 \text{ dm}^3 \times 1.3475 \text{ g}}{1.0 \text{ dm}^3} = 0.068 \text{ g}$$

The mass can also be determined from the formula for mass concentration as follows.

$$\rho = \frac{m}{V}$$

$$\therefore m = \rho \times V$$

$$\rho = 1.3475 \text{ g}$$

$$V = 0.0505 \text{ dm}^3$$

$$m = ?$$

Substituting, we have

$$m = \frac{1.3475 \text{ g}}{1 \text{ dm}^3} \times 0.0505 \text{ dm}^3 = 0.068 \text{ g}$$

Example 4

The solubility of potassium iodide, KI, is 130 g/100 g H_2O at 0°C and 250 g/100 g H_2O at 70°C. If 55 cm³ of a saturated solution of the salt at 70°C is cooled to 0°C, determine the mass of the salt that would crystallise out of the solution.

Solution

The first step is to calculate the mass, m, of KI that will saturate 55 g of water at each temperature.
At 0°C, we have

$$\frac{100 \text{ g}}{130 \text{ g}} = \frac{55 \text{ g}}{m}$$

So

$$100 \text{ g} \times m = 55 \text{ g} \times 130 \text{ g}$$

Properties of Solutions

$$\therefore m = \frac{55 \text{ g} \times 130 \text{ g}}{100 \text{ g}} = 71.5 \text{ g}$$

At 70°C, we have

$$\frac{100 \text{ g}}{250 \text{ g}} = \frac{55 \text{ g}}{m}$$

So

$$100 \text{ g} \times m = 55 \text{ g} \times 250 \text{ g}$$

$$\therefore m = \frac{55 \text{ g} \times 250 \text{ g}}{100 \text{ g}} = 137.5 \text{ g}$$

The mass, m, of KI that would crystallise out of solution is the difference between the two masses, i.e.

$$m = (137.5 - 71.5) \text{ g} = 66 \text{ g}$$

Example 5

The mass of potassium nitrate, KNO_3, in a saturated solution is 106 g/100 g of solution at 60°C. When the saturated solution is cooled to 20°C, the mass of KNO_3 in the solution decreases to 32 g/100g of solution the salt crystals. Determine the mass of KNO_3 crystals that would form when 250 g of a saturated solution prepared at 60°C is cooled to 20°C?

Solution

The very first step is to obtain the mass, m, of crystals that would form when 100 g of the saturated solution is cooled from 60°C to 20°C. This is the difference in the masses of KNO_3 in the saturated solution at 60°C and 20°C.

$$m = (106 - 32) \text{ g} = 74 \text{ g}$$

Finally, we need to scale up the solution as follows.

$$\frac{74 \text{ g}}{100 \text{ g}} = \frac{m}{250 \text{ g}}$$

So

$$m \times 100 \text{ g} = 74 \text{ g} \times 250 \text{ g}$$

$$\therefore m = \frac{74 \text{ g} \times 250 \text{ g}}{100 \text{ g}} = 190 \text{ g}$$

Example 6

A 2500 g solution at 60°C contains anhydrous copper(II) sulfate and potassium nitrate in a 4:1 ratio. The solution is cooled to 10°C, filtered and reweighed. Calculate the mass of potassium nitrate in the crystals obtained, given that the solution weighs 2240 g after reweighing.

Solution

The mass, m, of the crystals formed is obtained by deduction.

$$m = (2500 - 2240) \text{ g} = 260 \text{ g}$$

Since the ratio of $CuSO_4$ to KNO_3 in the crystals is 4:1, the mass of KNO_3 in the crystals is:

$$m = \frac{1}{5} \times 260 \text{ g} = 52 \text{ g}$$

Example 7

The solubility curve of potassium nitrate, KNO_3, is shown in Figure 9.2. From the graph, determine the mass of the salt that will crystallise out of solution if 100 g of a saturated solution of the salt at 80°C is cooled to 10°C.

Solution

From the curve, the solubilities of KNO_3 at 80°C and 0°C are 170 g/100 g H_2O and 20 g/100 g H_2O, respectively. Thus, the mass, m, of the crystals that would form on cooling is the difference between the two solubility values.

$$m = (170 - 20) \text{ g}/100 \text{ g } H_2O = 150 \text{ g}/100 \text{ g } H_2O$$

PRACTICE PROBLEMS

1. A 25 g saturated solution of sodium chloride contains 1.504 g of the salt at 70°C. Determine the solubility of the salt at this temperature in:
   (a) g/100 g H_2O;
   (b) g dm^{-3}.
2. Potassium sulfate is dissolved in distilled water at 50°C until its crystals begin to form. Given that 150 cm³ of the solution contains 12.5 g of the salt, determine:
   (a) the mass solubility of the salt;
   (b) the molar solubility of the salt.

$$(O = 16, S = 32, K = 39)$$

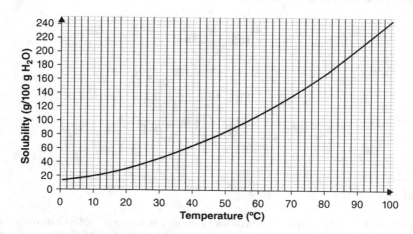

FIGURE 9.2 Solubility curve of KNO_3.

Properties of Solutions

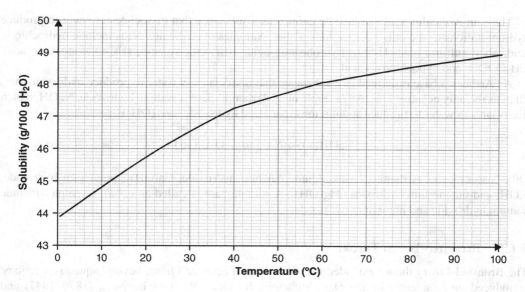

FIGURE 9.3 Solubility curve of $AlCl_3$.

3. The solubility of potassium nitrate is 4.2 mol dm⁻³ at 60°C. Determine the mass of the salt in 25 cm³ of a saturated solution of the salt at the same temperature.

$$(N = 14, \; S = 32, \; K = 39)$$

4. A 100.0 g saturated solution of sodium chloride contains 35.8 g of the salt at 30°C. Another saturated solution contains 38.1 g of the salt in 100.0 g of solution at 80°C. Determine the mass of crystals that would form when 50.0 cm³ of the saturated solution of sodium chloride is cooled from 80°C to 30°C.

5. The solubility curve of aluminium chloride, $AlCl_3$, is shown in Figure 9.3. From the graph, calculate the mass that will crystallise out of solution if 100 g of a saturated solution at 100°C is cooled to 0°C.

9.6 SOLUTIONS OF ACIDS AND BASES

Acids and bases are groups of compounds classified under one of three theories: Arrhenius Theory, Brønsted-Lowry Theory and Lewis Theory. Each of these theories is discussed below.

9.6.1 Arrhenius Theory

The acid-base concept was initially limited to substances in aqueous solutions. Developed by the Swedish scientist Svente Arrhenius (1859–1927) in 1884, the Arrhenius theory defines an acid as a substance that dissociates in water to produce hydrogen ions (protons) as the only positive ions.

A typical example of an Arrhenius acid is nitric acid HNO_3, which dissociates in water to produce hydrogen and nitrate ions:

$$HNO_3(aq) \rightarrow H^+(aq) + NO_3^-(aq)$$

However, it is now known that the hydrogen ion does not exist independently in water but instead combines with water molecule to form a hydronium ion, H_3O^+:

$$H^+(aq) + H_2O(aq) \rightarrow H_3O^+(aq)$$

Thus, an Arrhenius acid can also be defined as a substance that dissociates in water to produce hydronium ions as the only positive ions. Other examples of Arrhenius acids include hydrochloric acid, HCl; sulfuric acid, H_2SO_4; hydrobromic acid, HBr; chloric acid, ($HClO_3$); ethanoic acid, CH_3COOH; and hypochlorous acid, HClO.

An Arrhenius base is defined as a substance that dissociates in water to produce hydroxide ions, OH^-, as the only negative ions. An example of an Arrhenius base is sodium hydroxide, NaOH, which dissociates in water to produce sodium ions (Na^+) and hydroxide ions (OH^-):

$$NaOH(s) \rightarrow Na^+(aq) + OH^-(aq)$$

Other examples of Arrhenius bases include calcium hydroxide, $Ca(OH)_2$; potassium hydroxide, KOH; and magnesium hydroxide, $Mg(OH)_2$. A soluble base is called an *alkali*. Examples include sodium hydroxide and potassium hydroxide.

9.6.2 Brønsted-Lowry Theory

The Brønsted-Lowry theory extended the definitions of acids and bases beyond aqueous reactions. Introduced independently by the Danish chemist Johannes Nicolaus Brønsted (1879–1947) and English chemist Thomas Martin Lowry (1874–1936) in 1923, the theory defines an acid as a proton donor, while a base is a proton acceptor. Examples of Brønsted-Lowry acids and bases are illustrated in the following reactions:

$$HCl(g) + NH_3(g) \rightarrow NH_4^+(s) + Cl^-(s)$$
$$\text{Acid} \quad\quad \text{Base} \quad\quad\quad \text{Acid} \quad\quad \text{Base}$$

$$NH_3(aq) + H_2O(g) \rightarrow NH_4^+(aq) + OH^-(aq)$$
$$\text{Acid} \quad\quad \text{Base} \quad\quad\quad \text{Acid} \quad\quad \text{Base}$$

In these reactions, ammonium, NH_4^+, forms ammonia, NH_3, by losing a proton; thus, ammonium is *the conjugate acid of ammonia*. Conversely, ammonia accepts a proton to become ammonium, making, ammonia the *conjugate base of ammonium*.

An acid-base combination such as ammonia and ammonium, which are related by the gain and loss of a proton, is called a *conjugate acid-base pair*. A list of some conjugate acid-base pairs is given in Table 9.1. From the table, it can be observed from that strong bases have weak conjugate acids and vice versa. For example, ammonia is a strong base, whereas its conjugate acid, ammonium, is a weak acid.

TABLE 9.1
Conjugate Acid-base Pairs

Acid	Base
NH_4^+	NH_3
HCl	Cl^-
H_2SO_4	HSO_4^-
HSO_4^-	SO_4^{2-}
H_3O^+	H_2O
CH_3COOH	CH_3COO^-
$HClO_4$	ClO_4^-
HNO_3	NO_3^-
OH^-	O^{2-}

9.6.3 Lewis Theory

In 1923, American chemist Gilbert Lewis (1875–1946) extended the definitions of acid and base to certain reactions that do not involve proton transfer. According to the Lewis Theory, an acid is an electron-pair acceptor, while a base is an electron-pair donor. The reaction between a Lewis acid and Lewis base always results in the formation of a complex. An example is the reaction between nickel and carbon monoxide:

$$\text{Ni(s)} + 4\text{CO(g)} \rightarrow \text{Ni(CO)}_4$$
$$\text{Acid} \quad\quad \text{Base}$$

9.6.4 pH and Hydrogen Ion Concentration

Pure water dissociates slightly to produce equal concentrations—specifically, 10^{-7} mol dm^{-3}—of hydrogen ions and hydroxide ions:

$$\text{H}_2\text{O(l)} \rightleftharpoons \text{H}^+(\text{aq}) + \text{OH}^-(\text{aq})$$

$$[\text{H}^+] = [\text{OH}^-] = 10^{-7} \text{ mol dm}^{-3}$$

where [H$^+$] and [OH$^-$] represents the molar concentrations of hydrogen and hydroxide ions, respectively. For pure water, the product of these molar concentrations is called the *ionic product* or *autoionisation constant* of water, K_w. In other words,

$$K_w = [\text{H}^+] \times [\text{OH}^-] = 10^{-7} \text{ mol dm}^{-3} \times 10^{-7} \text{ mol dm}^{-3} = 10^{-14} \text{ mol}^2 \text{ dm}^{-6}$$

The acidity or basicity of a solution is determined by the relative amounts of hydrogen ions and hydroxide ions in solution. Pure water is neutral since it contains equal concentrations of both ions. Solutions with excess hydrogen ions are acidic and turn blue litmus paper red, while solutions with excess hydroxide ions are basic (alkaline) and turn red litmus paper blue. However, the product of hydroxide and hydrogen ion concentrations of any solution always remains constant at 10^{-14} mol^2 dm^{-6}.

In 1909, the Danish chemist Søren Peter Lauritz Sørensen (1868–1939) introduced the pH scale (from Latin *pondus Hydrogenii*) as a convenient way to indicate the acidity or alkalinity of a solution. He defined pH as the negative logarithm (base 10) of the hydrogen ion (or hydronium ion) concentration of in a solution, i.e.,

$$\text{pH} = -\log[\text{H}^+] = -\log[\text{H}_3\text{O}^+]$$

Similarly, the hydroxide ion index, pOH, of a solution is given by

$$\text{pOH} = -\log[\text{OH}^-]$$

The full range of pH values is represented by the *pH scale* (see Figure 9.4), which consists of numbers from 0 to 14. A pH of 7 indicates a neutral solution. A pH below 7 indicates acidity, while a pH above 7 indicates alkalinity. Acidity increases as the pH move from 7 to 0, while alkalinity increases as the pH move from 7 to 14.

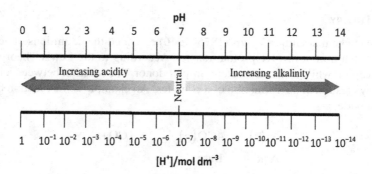

FIGURE 9.4 The pH scale.

The relationship between pH and pOH is given by

$$pOH + pH = 14 = pK_w$$

Example 1

The concentration of hydrogen ions in a solution is 6.5×10^{-5} mol dm^{-3}. Determine:

(a) the pH of the solution;
(b) the pOH of the solution.

Solution

(a) We to apply the relation:

$$pH = -\log[H^+]$$
$$[H^+] = 6.5 \times 10^{-5} \text{ mol dm}^{-3}$$
$$pH = ?$$

Substituting, we have

$$pH = -\log 6.5 \times 10^{-5}$$
$$= -(-4.2) = 4.2$$

(b) Here, we apply the relation:

$$pOH + pH = 14$$
$$\therefore pOH = 14 - pH$$
$$pH = 4.2$$
$$pOH = ?$$

Substituting, we have

$$pOH = 14 - 4.2 = 9.8$$

Properties of Solutions

Example 2

Determine the pH of a 0.10 mol dm^{-3} solution of sulfuric acid.

Solution

We will apply the relation:

$$pH = -\log[H^+]$$

Sulfuric acid is a strong acid that dissociates completely in solution as follows:

$$H_2SO_4(aq) \rightarrow 2H^+(aq) + SO_4^{2-}(aq)$$

From the equation, 1 mol of sulfuric acid dissociates to produce 2 mol of hydrogen or hydronium ions. Thus Therefore, a , 0.10 mol dm^{-3} solution of sulfuric acid dissociates completely to produce a hydrogen or hydronium ion concentration of 2 × 0.10 mol dm^{-3}.
So

$$[H^+] = 0.20 \text{ mol dm}^{-3}$$
$$pH = ?$$

Substituting, we have

$$pH = -\log 0.20$$
$$= -(-0.70) = 0.70$$

Example 3

Determine the pH of a 0.025 mol dm^{-3} solution of sodium hydroxide.

Solution

The first step is to calculate the pOH of the solution.

$$pOH = -\log[OH^-]$$

Sodium hydroxide, NaOH, dissociates completely in solution as follows.

$$NaOH(s) \rightarrow Na(aq) + OH^-(aq)$$

From the equation, 1 mol of NaOH produces 1 mol of hydroxide ions. Therefore, a 0.025 mol dm^{-3} solution of NaOH would produce the hydroxide ion concentration of 0.025 mol dm^{-3}.

$$[OH^-] = 0.025 \text{ mol dm}^{-3}$$
$$pOH = ?$$

Substituting the value, we obtain

$$pOH = -\log 0.025$$
$$= -(-1.6) = 1.6$$

Finally, the pH is determined as follows.

$$pOH + pH = 14$$
$$\therefore pH = 14 - pOH$$
$$pOH = 1.6$$
$$pH = ?$$

Substituting, we obtain

$$pH = 14 - 1.6 = 12.4$$

Example 4

The pH of a solution is 5.8. Determine:

(a) the concentration of hydronium ions in solution;
(b) the concentration of hydroxide ions in solution.

Solution

We apply the relation:

$$pH = -\log[H_3O^+]$$
$$pH = 5.8$$
$$[H_3O^+] = ?$$

Substituting, we have

$$5.8 = -\log[H_3O^+]$$

So

$$\log[H_3O^+] = -5.8$$

Finally, we take the common antilogarithm of both sides of the equation to obtain

$$[H_3O^+] = 10^{-5.8} = 1.6 \times 10^{-6} \text{ mol dm}^{-3}$$

(b) We use the relation:

$$[H_3O^+] \times [OH^-] = 10^{-14} \text{ mol}^2 \text{ dm}^{-6}$$
$$\therefore [OH^-] = \frac{10^{-14} \text{ mol}^2 \text{ dm}^{-6}}{[H_3O^+]}$$
$$[H_3O^+] = 1.6 \times 10^{-6} \text{ mol dm}^{-3}$$
$$[OH^-] = ?$$

Properties of Solutions

Substituting, we have

$$[OH^-] = \frac{10^{-14} \text{ mol}^2 \text{ dm}^{-6}}{1.6 \times 10^{-6} \text{ mol dm}^{-3}} = 6.3 \times 10^{-9} \text{ mol dm}^{-3}$$

PRACTICE PROBLEMS

1. Determine the pH of a solution whose hydrogen ion concentration is 0.0025 mol dm^{-3}.
2. Determine the pOH of a 0.0150 mol dm^{-3} solution of hydrochloric acid.
3. Determine the pH of a 0.0011 mol dm^{-3} solution of potassium hydroxide.
4. Determine the concentration of hydroxide ions in a solution with a pH of 8.5.

9.7 WATER OF CRYSTALLISATION

The molecules of water that are loosely associated with a salt are called *water of crystallisation*. Salts containing water of crystallisation are called *hydrated salts*. Some examples of hydrated salts were given in Chapter 6. The molecules of water of crystallisation of a hydrated salt can be partially or completely removed by heating. The number of molecules of water of crystallisation lost in the process is determined using the following relation:

$$\frac{18x}{M_r} = \frac{m_w}{m_s}$$

where

x = number of molecules of water of crystallisation driven off
M_r = relative molecular mass of the hydrated salt
m_w = mass of water driven off
m_s = mass of hydrated salt sample

The percentage of water of crystallisation in a salt is given by

$$\% \text{ water of crystallisation} = \frac{m_w}{m_s} \times 100\%$$

If the formula of a hydrated salt is known, then the percentage of water of crystallisation in the salt is given by

$$\% \text{ water of crystallisation} = \frac{18x}{M_r} \times 100\%$$

where

x = number of molecules of water of crystallisation in the salt
M_r = relative formula mass of the hydrated salt

Example 1

0.020 mol of the hydrated salt CuSO$_4$.xH$_2$O weighs 5.0 g. Determine:

(a) the number of molecules of water of crystallisation in the salt;
(b) the full formula of the salt.

$$(H = 1, O = 16, S = 32, Cu = 63.5)$$

Solution

(a) The first step is to determine the molar mass of the salt.

$$n = \frac{m}{M}$$

$$\therefore M = \frac{m}{n}$$

$$m = 5.0 \text{ g}$$

$$n = 0.020 \text{ mol}$$

$$M = ?$$

Substituting, we have

$$M = \frac{5.0 \text{ g}}{0.020 \text{ mol}} = 250 \text{ g mol}^{-1}$$

Since the formula of the salt is $CuSO_4.xH_2O$, it follows that

$$63.5 + 32 + (16 \times 4) + x\{(1 \times 2) + 16\} = 250$$

$$159.5 + 18x = 250 \text{ g mol}^{-1}$$

So

$$18x = 250 - 159.5 = 90.5$$

$$\therefore x = \frac{90.5}{18} = 5$$

(b) The full formula of the salt is $CuSO_4.5H_2O$.

Example 2

Determine the percentage of water of crystallisation in the salt $Na_2CO_3.10H_2O$.

$$(H = 1, C = 12, O = 16, Na = 23)$$

Solution

The applicable relation is:

$$\% \text{ water of crystallisation} = \frac{18x}{M_r} \times 100\%$$

$$M_r = (23 \times 2) + 12 + (16 \times 3) + 10(\{1 \times 2\} + 16) = 286$$

$$x = 10$$

$$\% \text{ water of crystallisation} = ?$$

Substituting, we have

$$\% \text{ water of crystallisation} = \frac{18 \times 10}{286} \times 100\% = 63\%$$

Example 3

5.5 g of iron(II) sulfate heptahydrate, $FeSO_4.7H_2O$, was heated to a mass of 3.7 g. Determine the number of molecules of water of crystallisation driven off, and hence, write the formula of the resulting salt.

$$(H = 1, S = 32, O = 16, Fe = 56)$$

Solution

We apply the relation:

$$\frac{18x}{M_r} = \frac{m_w}{m_s}$$

$$M_r = 56 + 32 + (16 \times 4) + 7(\{1 \times 2\} + 16) = 278$$

$$m_w = (5.5 - 3.7) \text{ g} = 1.8 \text{ g}$$

$$m_s = 5.5 \text{ g}$$

$$x = ?$$

Substituting the values, we obtain

$$\frac{18x}{278} = \frac{1.8 \text{ g}}{5.5 \text{ g}}$$

$$18x \times 5.5 = 278 \times 1.8$$

So

$$99x = 500.4$$

$$\therefore x = \frac{500.4}{99} = 5$$

Thus, the remaining molecules of water of crystallisation are $7 - 5 = 2$. Hence, the formula of the resulting salt is $FeSO_4.2H_2O$.

Example 4

4.5 g of a hydrated salt, $Na_2CO_3.xH_2O$, was heated to a constant mass of 3.8 g. Determine:

(a) the percentage composition of water in the sample;
(b) the formula of the salt.

$$(H = 1, C = 12, O = 16, Na = 23)$$

Solution

(a) Since the salt was heated to a constant mass, all the water of crystallisation has been expelled. The number of molecules of water of crystallisation in the salt is calculated as follows.

$$\% \text{ water of crystallisation} = \frac{m_w}{m_s} \times 100\%$$

$$m_w = (4.5 - 3.8)\,g = 0.70\,g$$

$$m_s = 4.5\,g$$

% water of crystallisation = ?

Substituting, we have

$$\% \text{ water of crystallisation} = \frac{0.70\,g}{4.5\,g} \times 100\% = 16\%$$

(b) The number of molecules of water of crystallisation, x, in the salt is calculated as follows.

$$\frac{18x}{M_r} = \frac{m_w}{m_s}$$

$$M_r = (23 \times 2) + 12 + (16 \times 3) + x(\{1 \times 2\} + 16) = 106 + 18x$$

$$m_w = (4.5 - 3.8)\,g = 0.70\,g$$

$$m_s = 4.5\,g$$

$$x = ?$$

Substituting the values, we obtain

$$\frac{18x}{106 + 18x} = \frac{0.70\,g}{4.5\,g}$$

So

$$18x \times 4.5 = 0.70(106 + 18x)$$

$$18x \times 4.5 = 74.2 + 12.6x$$

$$81x = 74.2 + 12.6x$$

Then

$$68.4x = 69.7$$

$$\therefore x = \frac{69.7}{68.4} = 1$$

The number of molecules of water of crystallisation driven off is equal to the number of molecules of water of crystallisation originally present in the sample; hence, the formula of the salt is $Na_2CO_3 \cdot H_2O$.

PRACTICE PROBLEMS

1. 0.0126 mol of a hydrated salt, $CuSO_4 \cdot xH_2O$, weighs 3.15 g. Determine:
    (a) the number of molecules of water of crystallisation in the salt;
    (b) the formula of the salt;
    (c) the percentage composition of water in the salt.

$$(H = 1,\ O = 16,\ S = 32,\ Cu = 63.5)$$

Properties of Solutions

2. 6.50 g of a hydrated salt, CoCl$_2$.xH$_2$O, was heated to a constant mass of 3.55 g. Determine:
   (a) the percentage water of crystallisation in the salt;
   (b) the full formula of the salt.

$$(H = 1,\ O = 16,\ Cl = 35.5,\ Co = 59)$$

FURTHER PRACTICE PROBLEMS

1. A solution of copper(II) sulfate pentahydrate, CuSO$_4$.5H$_2$O, contains 1.5 g of the salt in 250 cm^3 of solution. Determine:
   (a) the mass concentration of the solution;
   (b) the molar concentration of the solution.

$$(H = 1,\ O = 16,\ S = 32,\ Cu = 63.5)$$

2. The concentration of a sodium sulfate solution is 0.015 mol dm^{-3}. What volume of the solution would contain 2.5 g of the salt?

$$(O = 16,\ Na = 23,\ S = 32)$$

3. The molar concentration and mass concentration of a solution of a hydrated salt solution are 0.520 mol dm^{-3} and 128 g dm^{-3}, respectively. Determine the molar mass of the solution.

4. The concentration of a sodium chloride solution is 0.25 mol dm^{-3}. Determine the mass of sodium chloride in 25 cm^3 of the solution.

$$(Na = 23,\ Cl = 35.5)$$

5. A sodium sulfate solution, Na$_2$SO$_4$, has a concentration of 0.150 mol dm^{-3}. Determine the mass concentration of the solution obtained by diluting 25.0 cm^3 of the solution to 50.1 cm^3.

$$(O = 16,\ Na = 23,\ S = 32)$$

6. Calculate the volume of a 0.021 mol dm^{-3} solution of a salt required to prepare 35 cm^3 of a 0.015 mol dm^{-3} solution of the salt.

7. A student intends to prepare a 0.50 mol dm^{-3} solution of sulfuric acid from 15 cm^3 of a 5.5 mol dm^{-3} solution of the acid. What volume of water is required for the dilution?

8. What mass of anhydrous copper(II) sulfate, CuSO$_4$, is required to produce 1.5 dm^3 of a 0.055 mol dm^{-3} solution of the salt?

$$(O = 16,\ S = 32,\ Cu = 63.5)$$

9. Determine the volume of a stock solution of nitric acid required to prepare a 0.25 mol dm^{-3} solution of the acid, given that the mass composition of the acid in the stock solution is 70% and its density is 1.52 g cm^{-3}.

$$(H = 1,\ N = 14,\ O = 16)$$

10. Calculate the molar concentration of a phosphoric acid solution obtained by diluting 25 cm^3 of a stock solution of the acid to 250 cm^3, given that the composition of the acid in the stock solution is 85% and its density is 1.68 g cm^{-3}.

$$(H = 1,\ O = 16,\ P = 31)$$

11. The solubility of anhydrous copper(II) sulfate, CuSO$_4$, is 28.7 g/100 g H$_2$O at 40°C.

Determine the amount of the salt, in moles, in 25 cm³ of a saturated solution at 40°C.

$$(O = 16, \ S = 32, \ Cu = 63.5)$$

12. A saturated solution of potassium nitrate was prepared at 20°C. If the solution contains 7.90 g of the salt per 25.0 g of solution, determine:
    (a) the solubility of the salt in g dm⁻³ at 20°C;
    (b) the molar solubility of the salt at 20°C.

$$(N = 14, \ O = 16, \ K = 39)$$

13. Sodium chloride is recovered from a solution of the salt by evaporating the solution to dryness. Determine the number of moles of sodium chloride that would be obtained from 45 cm³ of a saturated solution at a temperature at which the solubility of the salt is 8.78 g/100 g H₂O.

$$(Na = 23, \ Cl = 35.5)$$

14. The solubility of potassium sulfate, K₂SO₄, is 111 g dm⁻³ at 20°C and 214 g dm⁻³ at 80°C. Determine the amount, in moles, of the salt crystals that would form when 150 cm³ of a saturated solution is cooled from 80°C to 20°C.

$$(O = 16, \ S = 32, \ K = 39)$$

15. Figure 9.5 shows the solubility curves of copper(II) sulfate and sodium chloride. Determine:
    (a) the solubility of copper(II) sulfate in g per 100 g H₂O at 55°C to two significant figures;
    (b) the molar solubility of sodium chloride at 75°C to two significant figures.

$$(Na = 23, \ Cl = 35.5)$$

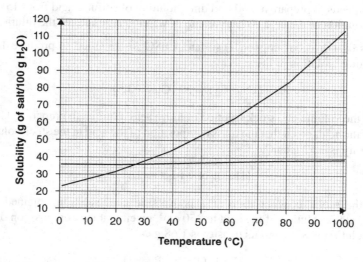

FIGURE 9.5 Solubility curves of CuSO₄ and NaCl.

Properties of Solutions

16. Determine the pH of a 0.0015 mol dm⁻³ solution of chloric acid.

17. The concentration of hydronium ions in a solution is 0.00035 mol dm⁻³. Determine the pOH of the solution.

18. The concentration of hydroxide ions in contaminated river water is 2.5×10^{-5} mol dm⁻³. Giving valid reasons for your answer, predict whether the solution would turn blue litmus paper red.

19. Determine the pH of a solution of anhydrous sodium carbonate with a hydroxide ion concentration of 0.00105 mol dm⁻³.

20. Determine the concentration of hydrogen ions in a solution with a pH of 8.0.

21. Calculate the concentration of hydroxide ions in a solution with a pH of 2.7.

$$\left(K_w = 1 \times 10^{-14} \text{mol}^2 \text{dm}^{-6}\right)$$

22. The concentration of hydroxide ions in a solution is 8.5×10^{-5} mol dm⁻³. Calculate the concentration of hydrogen ions in the solution.

$$\left(K_w = 1 \times 10^{-14} \text{mol}^2 \text{dm}^{-6}\right)$$

23. 10.5 g of magnesium sulfate heptahydrate, $MgSO_4 \cdot 7H_2O$, was heated to a mass of 6.66 g. determine:
    (a) the number of molecules of water of crystallisation expelled during heating;
    (b) the formula of the resulting salt.

$$\left(H = 1,\ O = 16,\ S = 32,\ Mg = 24\right)$$

24. A 2.5 g sample of hydrated calcium sulfate, $CaSO_4 \cdot xH_2O$, contains 0.0132 mol of the salt. Determine the full formula of the salt.

$$\left(H = 1,\ O = 16,\ S = 32,\ Ca = 40\right)$$

25. A 5.0 g sample of hydrated copper(II) chloride, $CuCl_2 \cdot xH_2O$, was heated to a constant mass of 3.9 g. Determine:
    (a) the percentage composition of water in the salt;
    (b) the formula of the salt.

$$\left(H = 1,\ O = 16,\ Cl = 35.5,\ Cu = 63.5\right)$$

10 Chemical Energetics

10.1 INTRODUCTION

Energy is the ability to do work. There are different forms of energy, including heat energy, electrical energy, light energy, mechanical energy, chemical energy nuclear energy and sound energy. The unit of energy is the joule (J).

The energy change associated with a chemical or physical change is measured in terms of the heat evolved or absorbed during the process. In chemical reactions, this heat is called the *heat of reaction*, while for physical changes, the heat change is referred to as the heat of that particular process. For example, the heat change accompanying melting is called the heat of fusion.

A reaction or physical process that absorbs heat from its surroundings is said to be *endothermic*. Examples of endothermic processes include the decomposition of calcium carbonate, electrolysis, dissolution of potassium chloride, the dissolution of ammonium nitrate, melting and evaporation. An endothermic reaction or process is indicated by a positive heat change.

A chemical reaction or physical process that releases heat to its surroundings is said to be *exothermic*. Examples of exothermic processes include combustion of fuels, dissolution of sodium hydroxide pellets, formation of water from hydrogen and oxygen, neutralisation reactions and condensation. The heat of an exothermic reaction or process is denoted by a negative heat change. The heat contents of exothermic and endothermic reactions are illustrated in Figure 10.1.

When a reaction or physical process occurs at constant pressure, the heat of reaction or process is called *enthalpy change*, ΔH. Since most processes occur at constant pressure, this term is often used interchangeably with the heat of reaction or process.

Theoretically, the enthalpy change of a reaction is the difference between the enthalpies, H, also called heat contents or internal energies, of the reactants and products. In other words,

$$\Delta H = H_{products} - H_{reactants}$$

The symbol Δ is the Greek (Greek uppercase letter delta) is used to denote a change in a given quantity. In an endothermic reaction, the heat content of the products is higher than that of the reactants. Conversely, in an exothermic reaction, the heat content of the reactants is higher than that of the products.

Enthalpy change is measured in joules per mole, $J\ mol^{-1}$, of a particular reactant or product. For example, 286 kJ of heat is evolved when 1 mol of liquid water is formed in the reaction between hydrogen and oxygen:

$$H_2(g) + \frac{1}{2}O_2(g) \rightarrow H_2O(l) \qquad \Delta H = -286\ kJ\ mol^{-1}$$

As mentioned earlier, the negative enthalpy change indicates that the formation of water is an exothermic reaction. If the reaction is reversed, the sign of the heat of reaction is also reversed:

$$H_2O(l) \rightarrow H_2(g) + \frac{1}{2}O_2(g) \qquad \Delta H = +286\ kJ\ mol^{-1}$$

This equation shows that 286 kJ of heat must be absorbed by 1 mol of water for it to break down into its constituent elements. As previously explained, reactions with a positive heat of reaction are called endothermic reactions.

Chemical Energetics

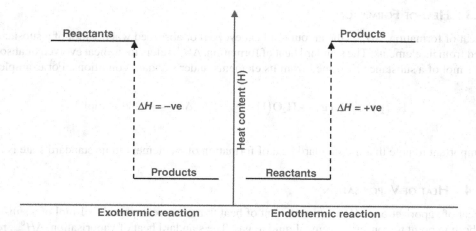

FIGURE 10.1 Explanation of exothermic and endothermic reactions in terms of heat content.

The heat associated with chemical and physical changes depends on the conditions under which it is measured. Therefore, these changes are typically measured under the following standard conditions.

1. A temperature of 25°C (298 K).
2. Normal atmospheric pressure of 1 atm (101325 Pa).
3. A concentration of 1 mol dm^{-3} for substances reacting in aqueous solutions.

The heat associated with a chemical or physical change, when measured under standard conditions, is called the *standard heat* or *standard enthalpy change*, ΔH^θ, of that change.

10.2 TYPES OF HEAT CHANGE

There are different types heat changes associated with chemical reaction and physical processes. Some of these are discussed below.

10.2.1 Heat of Combustion

The heat of combustion, ΔH_c, is the amount of heat evolved when 1 mol of a substance is burnt completely in oxygen. The standard heat of combustion, ΔH_c^θ, refers the heat evolved when 1 mol of a substance undergoes combustion in oxygen under standard conditions. For example,

$$CH_4(g) + 2O_2(g) \rightarrow CO_2(g) + H_2O(g) \qquad \Delta H_c^\theta = -890 \text{ kJ mol}^{-1}$$

10.2.2 Heat of Neutralisation

The heat of neutralisation, ΔH_n, is the amount of heat evolved when 1 mol of hydrogen ions (H$^+$) reacts with 1 mol of hydroxide ions (OH$^-$) to form 1 mol of water. The standard heat of neutralisation, ΔH_c^θ, is the heat evolved under standard conditions. For example,

$$H^+(aq) + OH^-(aq) \rightarrow H_2O(l) \qquad \Delta H_n^\theta = -57.5 \text{ kJ mol}^{-1}$$

Note: The standard heat of neutralisation for any reaction between a strong acid and a strong base is always −57.5 kJ mol^{-1}.

10.2.3 HEAT OF FORMATION

The heat of formation, ΔH_f, is the amount of heat evolved or absorbed when 1 mol of a substance is formed from its elements. The standard heat of formation, ΔH_f^θ, refers to the heat evolved or absolved when 1 mol of a substance is formed from its elements under standard conditions. For example,

$$H_2(g) + \frac{1}{2}O_2(g) \rightarrow H_2O(l) \qquad \Delta H_f^\theta = -286 \text{ kJ mol}^{-1}$$

It is important to note that the standard heat of formation of an element in its standard state is zero.

10.2.4 HEAT OF VAPORISATION

The heat of vaporisation, ΔH_{vap}, is the amount of heat that must be supplied to 1 mol of a substance at its boiling point to convert it from a liquid to gas. The standard heat of vaporisation, ΔH_{vap}^θ, refers to the heat required to vaporise 1 mol of a pure substance at its normal boiling point (at 1 atm). For example,

$$H_2O(l) \rightarrow H_2O(g) \qquad \Delta H_{vap}^\theta = 40.7 \text{ kJ mol}^{-1}$$

10.2.5 HEAT OF CONDENSATION

The heat of condensation, ΔH_{cond}, is the amount of heat that must be removed from 1 mol of a substance to change it from a gas to a liquid at its boiling point. The standard heat of condensation, ΔH_{cond}^θ, refers to the heat that must be removed from 1 mol of a pure substance to convert it from a gas to a liquid at its normal boiling point (at 1 atm). The heat of condensation is numerically equal to the heat of vaporisation but carries a negative sign. For example,

$$H_2O(g) \rightarrow H_2O(l) \qquad \Delta H_{vap}^\theta = -40.7 \text{ kJ mol}^{-1}$$

10.2.6 HEAT OF FUSION

The heat of fusion, ΔH_{fus}, is the amount of heat that must be supplied to 1 mol of a substance at its freezing point to convert it from a solid to a liquid. The standard heat of fusion, ΔH_{fus}^θ, refers to the heat required to melt 1 mol of a pure substance at its normal freezing point (at 1 atm), For example,

$$H_2O(s) \rightarrow H_2O(l) \qquad \Delta H_{fus}^\theta = 6.0 \text{ kJ mol}^{-1}$$

10.2.7 HEAT OF FREEZING

The heat of freezing, ΔH_{fr}, also called the heat of solidification, is the amount of heat that must be removed from 1 mol of a substance in order to convert it from a liquid to a solid its melting or freezing point. The standard heat of freezing, ΔH_{fr}^θ, refers to the heat that must be removed from 1 mol of a pure substance to freeze at its normal freezing or melting point. The heat of freezing is numerically equal to the heat of fusion but carries a negative sign. For example,

$$H_2O(l) \rightarrow H_2O(s) \qquad \Delta H_{fus}^\theta = -6.0 \text{ kJ mol}^{-1}$$

Chemical Energetics

10.2.8 Heat of Sublimation

The heat of sublimation, ΔH_{sub}, is the amount of heat that must be supplied to 1 mol of a substance to change it directly from a solid to a gas. The standard heat of sublimation, ΔH^θ_{fus}, is the heat required to sublime 1 mol of a pure substance at 1 atm of pressure. For example,

$$I_2(s) \rightarrow I_2(g) \qquad \Delta H^\theta_{sub} = 62.4 \text{ kJ mol}^{-1}$$

10.2.9 Heat of Deposition

The heat of deposition, ΔH_{dep}, is the amount of heat that must be removed from 1 mol of a substance to change it directly from a gas to a solid. The standard heat of deposition, ΔH^θ_{dep}, is the heat that must be removed from 1 mol of a pure substance to deposit it at 1 atm of pressure. The heat of deposition is numerically equal to the heat of sublimation but carries a negative sign. For example:

$$I_2(g) \rightarrow I_2(s) \qquad \Delta H^\theta_{dep} = -62.4 \text{ kJ mol}^{-1}$$

10.2.10 Lattice Energy

Lattice energy, $\Delta H_{lattice}$, can be defined in two ways. In terms of bond formation, it is the *energy released* when free gaseous ions combine to form 1 mol of an ionic solid. For example, the lattice energy of solid sodium chloride is -786 kJ mol^{-1}:

$$Na^+(g) + Cl^-(g) \rightarrow NaCl \qquad \Delta H_{lattice} = -787 \text{ kJ mol}^{-1}$$

In terms of bond dissociation energy, lattice energy is the *energy required* to break 1 mol of a solid ionic compound into its free gaseous ions. Based on this definition, the lattice energy of solid sodium chloride is 786 kJ mol^{-1}:

$$NaCl \rightarrow Na^+(g) + Cl^-(g) \qquad \Delta H_{lattice} = 787 \text{ kJ mol}^{-1}$$

Lattice energy measured under standard conditions is called standard lattice energy. It cannot be measured experimentally but can be calculated using Hess's Law and the Born-Haber Cycle, which will be discussed later.

10.2.11 Solvation Energy

Solvation energy, $\Delta H_{solvation}$, is the *energy released* when 1 mol of solute ions become associated with or surrounded by a solvent. When the solvent is water, solvation energy energy is referred to as *hydration energy*, $\Delta H_{hydration}$. For example, the hydration energy of solid sodium chloride is -771 kJ mol^{-1}:

$$Na^+(g) + Cl^-(g) \rightarrow Na^+(aq) + Cl^-(aq) \qquad \Delta H_{hydration} = -771 \text{ kJ mol}^{-1}$$

10.2.12 HEAT OF SOLUTION

The heat of solution, ΔH_{soln}, is the heat absorbed or evolved when 1 mol of a substance dissolves in water. The standard heat of solution, ΔH^{θ}_{soln}, is the heat absorbed or evolved when 1 mol of a substance is dissolved in a large amount of water such that further dilution causes no detectable heat change. For example,

$$NaOH(s) \rightarrow NaOH(aq) \qquad \Delta H^{\theta}_{soln} = -44.5 \text{ kJ mol}^{-1}$$

The heat of solution of a solute is the sum of its lattice dissociation energy and solvation energy. Since lattice (dissociation) energy is positive and solvation energy is negative, dissolution is endothermic if the lattice energy of dissociation exceeds the solvation energy. Conversely, dissolution is exothermic when the solvation energy is greater than the lattice energy of dissociation.

10.3 EXPERIMENTAL DETERMINATION OF HEAT OF REACTION

The heat of reaction is determined in the laboratory using a process called *calorimetry*. This involves measuring the temperature change of the reaction mixture or the extent to which a reaction raises the temperature of a known mass of water. The heat evolved or absorbed in a reaction is given by

$$Q = mc\Delta T$$

where

Q = quantity of heat absorbed or evolved in joules
c = specific heat capacity of water = 4200 J kg^{-1} K^{-1} = 4200 J kg^{-1} °C^{-1}
m = mass of water or reaction mixture
ΔT = change in temperature of water or reaction mixture

The enthalpy change of a process in joules per mole (J mol-1)$^{-1}$ is given by

$$\Delta H = \frac{Q}{n}$$

where n is the number of moles of the substance undergoing the reaction.
In calorimetry, the following assumptions are made:

1. The density of an aqueous solution is assumed to be equal to the density of water.
2. The specific heat capacity of an aqueous solution is assumed to be equal to the specific heat capacity of water.

Example 1

The dissolution of a substance increases the temperature of 25 cm³ of water by 15°C. Calculate the heat energy transferred to the water by the substance.

$$\left(c = 4200 \text{ J kg}^{-1} \text{ K}^{-1}\right)$$

Chemical Energetics

Solution

We use the relation:

$$Q = mc\Delta T$$
$$m = 25\,g = 0.025\,kg$$
$$c = 4200\,J\,kg^{-1}\,K^{-1}$$
$$\Delta T = 15°C = 15\,K$$
$$Q = ?$$

Substituting, we have

$$Q = 0.025\,kg \times \frac{4200\,J}{1\,kg \times 1\,K} \times 15\,K = 1.6\,kJ$$

Note: The temperature change in kelvin (K) is the same as that in °C.

Example 2

The temperature of a 55 cm³ reaction mixture initially at 30°C decreases to 5°C after the reaction. Determine the heat energy released by the mixture.

$$\left(c = 4200\,J\,kg^{-1}\,K^{-1}\right)$$

Solution

We apply the relation:

$$Q = mc\Delta T$$
$$m = 55\,g = 0.055\,kg$$
$$c = 4200\,J\,kg^{-1}\,K^{-1}$$
$$\Delta T = (30 - 5)°C = 25°C = 25\,K$$
$$Q = ?$$

Substituting, we have

$$Q = 0.055\,kg \times \frac{4200\,J}{1\,kg \times 1\,K} \times 25\,K = 5.8\,kJ$$

Example 3

In an experiment to determine the heat of solution of ammonium nitrate, NH_4NO_3, 2.5 g of the salt was dissolved in 25 cm³ of water, initially at 25°C. Given that the temperature of water dropped to 18°C after dissolution, determine the heat of solution of ammonium nitrate.

$$\left(H = 1,\ N = 14,\ O = 16,\ c = 4200\,J\,kg^{-1}\,K^{-1}\right)$$

Solution

The first step is to determine the amount of heat absorbed by water as follows.

$$Q = mc\Delta\theta$$
$$m = 25 \text{ g} = 0.025 \text{ kg}$$
$$c = 4200 \text{ J kg}^{-1} \text{ K}^{-1}$$
$$\Delta\theta = (25-18)°C = 7°C = 7 \text{ K}$$
$$Q = ?$$

Substituting the values, we obtain

$$Q = 0.025 \text{ kg} \times \frac{4200 \text{ J}}{1 \text{ kg} \times 1 \text{ K}} \times 7 \text{ K} = 735 \text{ J}$$

Now, we determine the heat of solution as follows.

$$\Delta H_{sol} = \frac{Q}{n}$$
$$n = \frac{m}{M}$$
$$m = 2.5 \text{ g}$$
$$M = \left(14 + \{1 \times 4\} + 14 + \{16 \times 3\}\right) \text{ g mol}^{-1} = 80 \text{ g mol}^{-1}$$
$$n = ?$$

Substituting, we have

$$n = \frac{2.5 \text{ g} \times 1 \text{ mol}}{80 \text{ g}} = 0.03125 \text{ mol}$$
$$\Delta H_{sol} = ?$$

Substituting, we obtain

$$\Delta H_{sol} = \frac{735 \text{ J}}{0.03125 \text{ mol}} = 24 \text{ kJ mol}^{-1}$$

The data book value is 26 kJ mol^{-1}.

Example 4

In an experiment to determine the heat of neutralisation of hydrochloric acid and sodium hydroxide, 125 cm³ of 1.2 mol dm⁻³ solution of hydrochloric acid was mixed with the same volume and concentration of sodium hydroxide. The initial temperature of each reaction mixture was 20°C, while the final steady temperature recorded after mixing was 28°C. Calculate the heat of neutralisation.

$$\left(c = 4200 \text{ J kg}^{-1}\text{K}^{-1}\right)$$

Chemical Energetics

Solution

We begin by calculating the amount of heat evolved.

$$Q = mc\Delta T$$

$$m = (125 + 125)\ \text{g} = 250\ \text{g} = 0.25\ \text{kg}$$

$$c = 4200\ \text{J kg}^{-1}\ \text{K}^{-1}$$

$$\Delta T = (28 - 20)\ °C = 8\ °C = 8\ \text{K}$$

$$Q = ?$$

Substituting, we have

$$Q = 0.25\ \text{kg} \times \frac{4200\ \text{J}}{1\ \text{kg} \times 1\ \text{K}} \times 8\ \text{K} = 8400\ \text{J}$$

Finally, the heat of neutralisation is calculated as follows.

$$\Delta H_n = \frac{Q}{n}$$

where n is the number of moles of water produced by the reaction, which is obtained from stoichiometry as follows.

$$HCl(aq) + NaOH(aq) \rightarrow NaCl(aq) + H_2O(l)$$

$$n = C \times V$$

For each reactant,

$$C = 1.2\ \text{mol dm}^{-1}$$

$$V = 125\ \text{cm}^3 = 0.125\ \text{dm}^3$$

$$n = ?$$

Substituting, we have

$$n = \frac{1.2\ \text{mol}}{1\ \text{dm}^3} \times 0.125\ \text{dm}^3 = 0.15\ \text{mol}$$

From the equation, 1 mol of NaOH produces 1 mol of H_2O; hence, 0.15 mol of NaOH produces 0.15 mol of water. Similarly, 1 mol of HCl produces 1 mol of H_2O; hence, 0.15 mol of HCl produces 0.15 mol of H_2O. This implies that just the right amounts of NaOH and HCl are mixed.

$$Q = 8400\ \text{J}$$

$$\Delta H_n = ?$$

Substituting, we have

$$\Delta H_n = \frac{8400\ \text{J}}{0.15\ \text{mol}} = -56\ \text{kJ mol}^{-1}$$

The data book value is -57 kJ mol^{-1}. The negative sign indicates that the reaction is exothermic.

PRACTICE PROBLEMS

1. The heat produced from the burning of 50.0 cm³ of a hydrocarbon raises the temperature of 150.0 cm³ of water from 15°C to 85°C. Determine the amount of heat energy gained by the water.

$$\left(c = 4200 \text{ J kg}^{-1} \text{ K}^{-1}\right)$$

2. In an experiment to determine the heat of solution of sodium hydroxide, 5.0 g of sodium hydroxide pellets was dissolved in 150 cm³ of water, initially at 25°C. Determine the heat of solution of the base if the highest temperature recorded after dissolution was 36.3°C.

$$\left(H = 1, \ O = 16, \ Na = 23, \ c = 4200 \text{ J kg}^{-1} \text{ K}^{-1}\right)$$

3. In a laboratory experiment to determine the heat of combustion of ethanol, C_2H_5OH, it was found that the burning of 1.5 g of the compound raised the temperature of 250 cm³ of water from 25°C to 65.4°C. Determine the heat of combustion of the compound.

$$\left(H = 1, \ C = 12, \ O = 16, \ c = 4200 \text{ J kg}^{-1} \text{ K}^{-1}\right)$$

10.4 HESS'S LAW AND HEAT OF REACTION

Hess's Law of constant heat summation states that the total enthalpy change of a chemical reaction remains constant, regardless of the route taken by the reaction. Thus, the heat of reaction for any process can be calculated by dividing the reaction into multiple steps and summing their individual heats of reaction of reaction as illustrated in the energy cycle shown in Figure 10.2.

The energy cycle demonstrates that the heat of reaction remains the same whether A reacts with B to form E and F directly, or whether intermediate products C and D are formed before the final products E and F. The sum of the intermediate reactions A + B → C + D and C + D → E + F yields the overall reaction A + B → E + F, after cancelling substances that appear on both sides of the equation:

$$A + B \rightarrow C + D \qquad \Delta H_1$$
$$+ \ C + D \rightarrow E + F \qquad \Delta H_2$$

So
$$\overline{A + B + \cancel{C} + \cancel{D} \rightarrow \cancel{C} + \cancel{D} + E + F \quad \Delta H_1 + \Delta H_2}$$
$$\Delta H = \Delta H_1 + \Delta H_2$$

When an energy cycle diagram is represented on an energy level diagram, the resulting energy cycle is called *Born-Haber cycle*. In a Born-Haber cycle, the energy of each reaction step or route is represented by a horizontal line, on which the products of the step are written. The energy change for each step is indicated by an arrow connecting two horizontal lines, with the corresponding energy values written alongside it. An exothermic reaction is represented by a downward arrow, while an endothermic reaction is represented by a upward arrow. The Born-Haber cycle for the energy cycle depicted in Figure 10.2 is shown in Figure 10.3.

An alternative application of Hess's Law in the theoretical calculation of heats of reaction is based on the heats of formation of reactants and products. In this method, the standard heat of reaction is calculated as the difference between the sum of the standard heats of formation of the products and the sum of the standard heats of formation of the reactants. This is expressed as:

$$\Delta H^\theta = \Sigma n \Delta H_f^\theta \left(\text{products}\right) - \Sigma n \Delta H_f^\theta \left(\text{reactants}\right)$$

where n is the stoichiometric coefficient of a substance in the chemical equation.

Chemical Energetics

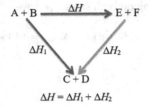

$$\Delta H = \Delta H_1 + \Delta H_2$$

FIGURE 10.2 Using an energy cycle to represent Hess's Law.

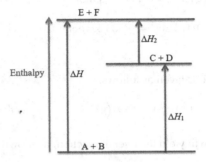

FIGURE 10.3 Using Born-Haber cycle to represent Hess's Law. It can be observed from the diagram that $\Delta H = \Delta H_1 + \Delta H_2$.

The symbol Σ is the Greek uppercase letter Sigma, which means *sum of*.

Example 1

Calculate the standard heat of vaporisation of water, given the following information.

1. $H_2(g) + \frac{1}{2}O_2(g) \rightarrow H_2O(l) \quad \Delta H^\ominus = -286 \text{ kJ mol}^{-1}$
2. $H_2(g) + \frac{1}{2}O_2(g) \rightarrow H_2O(g) \quad \Delta H^\ominus = -242 \text{ kJ mol}^{-1}$

Solution

The conversion for which the standard heat of vaporisation is required is given as follows:

$$H_2O(l) \rightarrow H_2O(g) \qquad \Delta H^\ominus_{vap} = ?$$

We cannot obtain this conversion by merely adding the two equations unless the first equation is reversed, as shown below.

$$H_2O(l) \rightarrow H_2(g) + \frac{1}{2}O_2(g) \qquad \Delta H^\ominus = +286 \text{ kJ mol}^{-1}$$

Now, we add the two equations to obtain the required conversion and its standard heat of vaporisation.

$$H_2(g) + \frac{1}{2}O_2(g) \rightarrow H_2O(g) \quad \Delta H^\ominus = -242 \text{ kJ mol}^{-1}$$
$$+ H_2O(l) \rightarrow H_2(g) + \frac{1}{2}O_2(g) \quad \Delta H^\ominus = +286 \text{ kJ mol}^{-1}$$
$$\overline{H_2(g) + H_2O(l) + \frac{1}{2}O_2(g) \rightarrow H_2O(g) + H_2(g) + \frac{1}{2}O_2(g) \quad \Delta H^\ominus_{vap} = +44 \text{ kJ mol}^{-1}}$$

Cancelling the species appearing on both sides, we have

$$H_2O(l) \rightarrow H_2O(g) \quad \Delta H^\theta_{vap} = +44 \text{ kJ mol}^{-1}$$

Example 2

Calculate the heat of formation of carbon dioxide from the information given below.

1. $C(s) + \frac{1}{2}O_2(g) \rightarrow CO(g) \quad \Delta H^\theta_c = -110.5 \text{ kJ mol}^{-1}$
2. $CO(g) + \frac{1}{2}O_2(g) \rightarrow CO_2(g) \quad \Delta H^\theta_c = -283 \text{ kJ mol}^{-1}$

Solution

The required overall equation is given as follows:

$$C(s) + O_2(g) \rightarrow CO_2(g) \quad \Delta H^\theta_f = ?$$

Adding the given equations will yield the overall reaction, as shown below.

$$C(s) + \frac{1}{2}O_2(g) \rightarrow CO(g) \quad \Delta H^\theta_c = -110.5 \text{ kJ mol}^{-1}$$

$$+ CO(g) + \frac{1}{2}O_2(g) \rightarrow CO_2(g) \quad \Delta H^\theta_c = -283 \text{ kJ mol}^{-1}$$

$$\overline{C(s) + \frac{1}{2}O_2(g) + CO(g) + \frac{1}{2}O_2(g) \rightarrow CO(g) + CO_2(g) \quad \Delta H^\theta_f = -393.5 \text{ kJ mol}^{-1}}$$

Cancelling the species appearing on both sides, we have

$$C(s) + O_2(g) \rightarrow CO_2(g) \quad \Delta H^\theta_f = -393.5 \text{ kJ mol}^{-1}$$

Example 3

Calculate the standard heat of formation of butane from the following information.

1. $C_4H_{10}(g) + \frac{13}{2}O_2(g) \rightarrow 4CO_2(g) + 5H_2O(g) \quad \Delta H^\theta_c = -2878 \text{ kJ mol}^{-1}$
2. $C(s) + O_2(g) \rightarrow CO_2(g) \quad \Delta H^\theta_c = -393.5 \text{ kJ mol}^{-1}$
3. $H_2(g) + \frac{1}{2}O_2(g) \rightarrow H_2O(l) \quad \Delta H^\theta_c = -286 \text{ kJ mol}^{-1}$

Solution

The overall equation is given as follows:

$$4C(s) + 5H_2(g) \rightarrow C_4H_{10}(g) \quad \Delta H^\theta_f = ?$$

Adding the given equations will not result in the overall equation because the direction of the first equation and the stoichiometry of the last two equations do not match the overall equation. To solve this problem, we will reverse the first equation and multiply the second and third equations by 4 and 5, respectively. Now, we can add the equations as follows.

Chemical Energetics

$$4CO_2(g) + 5H_2O(g) \rightarrow C_4H_{10}(g) + \frac{13}{2}O_2(g) \quad \Delta H_c^\theta = +2878$$

$$4C(s) + 4O_2(g) \rightarrow 4CO_2(g) \quad \Delta H_c^\theta = -1574 \text{ kJ}$$

$$+5H_2(g) + \frac{5}{2}O_2(g) \rightarrow 5H_2O(l) \quad \Delta H_c^\theta = -1430 \text{ kJ}$$

$$4CO_2(g) + 5H_2O(g) + 4C(s) + \frac{13}{2}O_2(g) + 5H_2(g) \rightarrow C_4H_{10}(g) + \frac{13}{2}O_2(g) + 4CO_2(g)$$
$$+ 5H_2O(l) \quad \Delta H_f^\theta = -126 \text{ kJ mol}^{-1}$$

Cancelling the common species on both sides, we obtain

$$4C(s) + 5H_2(g) \rightarrow C_4H_{10}(g) \quad \Delta H_f^\theta = -126 \text{ kJ mol}^{-1}$$

Example 4

Consider the thermodynamic data given below.

1. Heat of sublimation of sodium.

   $$Na(s) \rightarrow Na(g) \quad \Delta H_f^\theta = +108 \text{ kJ mol}^{-1}$$

2. Dissociation energy of a chlorine molecule to chlorine atoms.

   $$\frac{1}{2}Cl_2(g) \rightarrow Cl(g) \quad \Delta H_f^\theta = +122 \text{ kJ mol}^{-1}$$

3. First ionisation energy of sodium.

   $$Na(g) \rightarrow Na^+(g) + e^- \quad \Delta H^\theta = +496 \text{ kJ mol}^{-1}$$

4. First electron affinity of chlorine.

   $$Cl(g) + e^- \rightarrow Cl^-(g) \quad \Delta H^\theta = -348 \text{ kJ mol}^{-1}$$

5. The heat of formation of sodium chloride.

   $$Na(s) + \frac{1}{2}Cl_2(g) \rightarrow NaCl(s) \quad \Delta H_f^\theta = -404 \text{ kJ mol}^{-1}$$

Represent the information on a Born-Haber cycle and determine the lattice energy (formation) of sodium chloride.

Solution

The required Born-Haber cycle is shown in Figure 10.4. The diagram clearly shows two routes leading to the formation of solid sodium chloride. By applying Hess's Law, we can sum the energies of the reactions in both routes to determine the heat of formation of sodium chloride.

$$-404 \text{ kJ mol}^{-1} = 108 \text{ kJ mol}^{-1} + 496 \text{ kJ mol}^{-1} + 122 \text{ kJ mol}^{-1} - 348 \text{ kJ mol}^{-1} + \Delta H_{lattice}^\theta$$

$$\therefore \Delta H_{lattice}^\theta = -404 \text{ kJ mol}^{-1} - 377 \text{ kJ mol}^{-1}$$

$$= -782 \text{ kJ mol}^{-1}$$

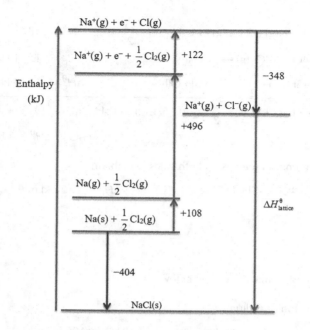

FIGURE 10.4 The Born-Haber cycle of NaCl.

Example 5

Determine the standard heats of vaporisation and condensation of water from the standard heats of formation given below.

$$\Delta H_f^\theta \{H_2O(g)\} = -286 \text{ kJ mol}^{-1}$$

$$\Delta H_f^\theta \{H_2O(g)\} = -242 \text{ kJ mol}^{-1}$$

Solution

Vaporisation and condensation are opposite physical changes and can be represented as follows.

$$H_2O(l) \rightleftharpoons H_2O(g)$$

We can treat this like a chemical reaction and apply Hess's Law.

$$\Delta H^\theta = \Sigma n \Delta H_f^\theta \text{ (products)} - \Sigma n \Delta H_f^\theta \text{ (reactants)}$$

The standard heat of the forward reaction is the standard heat of vaporisation of water.

$$\Delta H_{vap}^\theta = \Delta H_f^\theta \{H_2O(g)\} - \Delta H_f^\theta \{H_2O(l)\}$$

$$\Delta H_f^\theta \{H_2O(g)\} = -242 \text{ kJ mol}^{-1}$$

$$\Delta H_f^\theta \{H_2O(l)\} = -286 \text{ kJ mol}^{-1}$$

$$\Delta H_{vap}^\theta = ?$$

Chemical Energetics

Substituting, we have

$$\Delta H^\theta_{vap} = -242 \text{ kJ mol}^{-1} - \left(-289 \text{ kJ mol}^{-1}\right)$$

So

$$\Delta H^\theta_{vap} = -242 \text{ kJ mol}^{-1} + 286 \text{ kJ mol}^{-1}$$

$$= 44 \text{ kJ mol}^{-1}$$

You will notice that this agrees with the result obtained in Example 1.

The standard heat of condensation, ΔH^θ_{cond}, is the standard heat of reaction for the reverse process; hence, we simply add a negative sign to the standard heat of vaporisation to obtain:

$$\Delta H^\theta_{cond} = -44 \text{ kJ mol}^{-1}$$

Example 6

Carbon monoxide combines with oxygen to form carbon dioxide as follows:

$$2CO(g) + O_2(g) \rightarrow 2CO_2(g)$$

Determine the standard heat of reaction.

$$\left(CO(g) = -110.5 \text{ kJ mol}^{-1}, \; CO_2(g) = -393.5 \text{ kJ mol}^{-1}\right)$$

Solution

As usual, we apply the relation:

$$\Delta H^\theta = \Sigma n H^\theta_f \text{ (products)} - \Sigma n H^\theta_f \text{ (reactants)}$$

For $CO_2(g)$, we have

$$\Delta H^\theta_f = -393.5 \text{ kJ mol}^{-1}$$

$$n = 2 \text{ mol}$$

So

$$\Sigma n \Delta H^\theta_f \text{ (products)} = 2 \text{ mol} \times \left(-\frac{393.5 \text{ kJ}}{1 \text{ mol}}\right) = -787 \text{ kJ}$$

For $CO(g)$, we have

$$\Delta H^\theta_f = -110.5 \text{ kJ mol}^{-1}$$

$$n = 2 \text{ mol}$$

For $O_2(g)$, we have

$$\Delta H^\theta_f = 0$$

So

$$\Sigma n\Delta H_f^\theta \text{ (reactants)} = 2\text{ mol} \times \left(-\frac{110.5\text{ kJ}}{1\text{ mol}}\right) + 0 = -221\text{ kJ}$$

$$\Delta H^\theta = ?$$

Substituting, we have

$$\Delta H^\theta = -787\text{ kJ} - (-221\text{ kJ})$$
$$= -566\text{ kJ}$$

Since 2 mol of CO_2 were produced, it follows that

$$\Delta H^\theta = -\frac{566\text{ kJ}}{2\text{ mol}} = -283\text{ kJ mol}^{-1}$$

Example 7

Methane undergoes combustion to form carbon dioxide and water vapour as follows:

$$CH_4\,(g) + 2O_2\,(g) \rightarrow CO_2\,(g) + 2H_2O\,(g)$$

Determine the standard heat of combustion of methane.

$$\left(CH_4\,(g) = -75\text{ kJ mol}^{-1},\ CO_2\,(g) = -393.5\text{ kJ mol}^{-1},\ H_2O\,(g) = -241.8\text{ kJ mol}^{-1}\right)$$

Solution

The heat of combustion of methane is the standard heat of the reaction.

$$\Delta H_c^\theta = \Sigma n H_f^\theta \text{ (products)} - \Sigma n H_f^\theta \text{ (reactants)}$$

For $CO_2(g)$, we have

$$\Delta H_f^\theta = -393.5\text{ kJ mol}^{-1}$$
$$n = 1\text{ mol}$$

For $H_2O(g)$, we have

$$\Delta H_f^\theta = -241.8\text{ kJ mol}^{-1}$$
$$n = 2\text{ mol}$$

So

$$\Sigma n\Delta H_f^\theta \text{ (products)} = 1\text{ mol} \times \left(-\frac{393.5\text{ kJ}}{1\text{ mol}}\right) + 2\text{ mol} \times \left(-\frac{241.8\text{ kJ}}{1\text{ mol}}\right) = -877.1\text{ kJ}$$

Chemical Energetics

For $CH_4(g)$, we have

$$\Delta H_f^\ominus = -75 \text{ kJ mol}^{-1}$$

$$n = 1 \text{ mol}$$

For $O_2(g)$, we have

$$\Delta H_f^\ominus = 0$$

So

$$\Sigma n\Delta H_f^\ominus \text{ (reactants)} = 1 \text{ mol} \times \left(-\frac{75 \text{ kJ}}{1 \text{ mol}}\right) + 0 = -75 \text{ kJ}$$

$$\Delta H_c^\ominus = ?$$

Substituting the values, we obtain

$$\Delta H_c^\ominus = -877.1 \text{ kJ} - (-75 \text{ kJ})$$

$$= -802 \text{ kJ}$$

1 mol of $CH_4(g)$ was burnt; hence, the standard heat of combustion of the compound is -802 kJ mol^{-1}.

Note: This is the standard heat of combustion of methane when water vapour is produced. The value is -890 kJ mol^{-1} when liquid water is produced.

Example 8

Calculate the enthalpy change of the following reaction.

$$8Al(s) + 3Fe_3O_4(s) \rightarrow 4Al_2O_3(s) + 9Fe(s)$$

$$\left(Fe_3O_4(s) = -1118.4 \text{ kJ mol}^{-1}, Al_2O_3(s) = -1675.7 \text{ kJ mol}^{-1}\right)$$

Solution

As usual, we have

$$\Delta H^\ominus = \Sigma n\Delta H_f^\ominus \text{ (products)} - \Sigma n\Delta H_f^\ominus \text{ (reactants)}$$

For $Fe(s)$, we have

$$\Delta H_f^\ominus = 0$$

For $Al_2O_3(s)$, we have

$$\Delta H_f^\ominus = -1675.7 \text{ kJ mol}^{-1}$$

$$n = 4 \text{ mol}$$

So

$$\Sigma n\Delta H_f^\ominus \text{ (products)} = 4 \text{ mol} \times \left(-\frac{1675.7 \text{ kJ}}{1 \text{ mol}}\right) + 0 = -6702.8 \text{ kJ}$$

For Al(s), we have

$$\Delta H_f^\theta = 0$$

For Fe$_3$O$_4$(s), we have

$$\Delta H_f^\theta = -1118.4 \text{ kJ mol}^{-1}$$

$$n = 3 \text{ mol}$$

So

$$\Sigma n \Delta H_f^\theta \text{ (reactants)} = 3 \text{ mol} \times \left(-\frac{1118.4 \text{ kJ}}{1 \text{ mol}}\right) + 0 = -3355.2 \text{ kJ}$$

$$\Delta H^\theta = ?$$

Finally, we substitute to obtain

$$\Delta H^\theta = -6702.8 \text{ kJ} - (-3355.2 \text{ kJ})$$
$$= -3347.6 \text{ kJ}$$

The heat of reaction is not defined in terms of any specific reactant or product; hence, it is simply expressed in kJ.

Example 9

The standard heat of combustion of benzene, C$_6$H$_6$, is −3267 kJ mol^{-1}. How much heat would be evolved by the complete burning of 5.0 g of benzene under standard condition?

$$(H = 1, C = 12)$$

Solution

The first step is to calculate the number of moles of benzene.

$$n = \frac{m}{M}$$

$$m = 5.0 \text{ g}$$

$$M = (\{12 \times 6\} + \{1 \times 6\}) \text{ g mol}^{-1} = 78 \text{ g mol}^{-1}$$

$$n = ?$$

Substituting, we have

$$n = \frac{5.0 \text{ g} \times 1 \text{ mol}}{78 \text{ g}} = 0.0641 \text{ mol}$$

Let the heat evolved by this amount of benzene under standard conditions be Q. Since 1 mol of benzene liberates 3267 kJ of heat under standard conditions, we have

$$Q = 0.0641 \times 3267 = 210 \text{ kJ}$$

Note: A negative sign is not added to the answer because the phrase *heat evolved* already indicates that the reaction is exothermic.

Chemical Energetics

PRACTICE PROBLEMS

1. Determine the standard enthalpy change of the forward and reverse reactions in the following equation.

$$N_2(g) + 3H_2(g) \rightleftharpoons 2NH_3(g)$$

$$(NH_3(g) = -46 \text{ kJ mol}^{-1})$$

2. Determine the standard enthalpy change of the following reaction.

$$CaCO_3(s) \rightarrow CaO(s) + CO_2(g)$$

$$(CaCO_3(s) = -1207 \text{ kJ mol}^{-1}, \quad CaO(s) = -635 \text{ kJ mol}^{-1}, \quad CO_2(g) = -393.5 \text{ kJ mol}^{-1})$$

3. Consider the following neutralisation reaction.

$$H_2SO_4(aq) + 2KOH(aq) \rightarrow K_2SO_4(aq) + 2H_2O(l) \qquad \Delta H^\theta = -57 \text{ kJ mol}^{-1}$$

Determine the amount of heat that would be evolved when 25 cm³ of a 0.10-mol dm⁻¹ solution of sulfuric acid is completely neutralised by potassium hydroxide.

4. Iodine is a solid in its standard state. Determine the standard heats of sublimation and deposition of the element, given that the standard enthalpy of formation of iodine vapour is 62.4 kJ mol⁻¹.

5. Draw the Born-Haber cycle for the following cycle of reactions and determine the standard heat of formation of solid lithium fluoride.

1. $Li(s) \rightarrow Li(g) \qquad \Delta H_f^\theta = 155 \text{ kJ mol}^{-1}$
2. $\frac{1}{2} F_2(g) \rightarrow F(g) \qquad \Delta H_f^\theta = +75 \text{ kJ mol}^{-1}$
3. $Li(g) \rightarrow Li^+(g) + e^- \quad \Delta H^\theta = +520 \text{ kJ mol}^{-1}$
4. $F(g) + e^- \rightarrow F^-(g) \; \Delta H^\theta = -333 \text{ kJ mol}^{-1}$
5. $Li^+(g) + Cl^-(g) \rightarrow LiF(s) \Delta H_f^\theta = -1012 \text{ kJ mol}^{-1}$

The equation for the formation of LiF(s) from its standard elements is as follows.

$$Li(s) + \frac{1}{2} F_2(g) \rightarrow LiF(s)$$

10.5 CALCULATING HEAT OF REACTION FROM BOND ENERGIES

The bonds between particles of reactants must be broken for chemical reaction to occur. The energy required to break the bonds between the particles of a reactant is called bond energy or bond enthalpy or bond-dissociation energy. Bond breaking is an endothermic process because energy must be supplied to break the bond. In contrast, energy is given off during the formation of new bonds or products. Some standard bond energies are given in Table 10.1.

The standard heat of reaction is estimated from the standard bond energies as follows.

$$\Delta H^\theta = \Sigma \text{ bond breaking energies} - \Sigma \text{ bond forming energies}$$
$$= \Sigma \text{ bond energies of reactants} - \Sigma \text{ bond energies of products}$$

TABLE 10.1
Some Standard Bond Energies

Bond	Standard Bond Energy/kJ mol^{-1}
C–C	347
C=C	612
C≡C	838
Br–Br	193
Cl–Cl	242
H–Cl	431
O=O	497
H–H	436
C–H	412
C=O	1,076
H–Br	366
H–O	463
N≡N	942
S=S	418
H–S	349
N=O	607
S=O	323
C≡O	1,072

Thus, a reaction is exothermic if the sum of bond energies of the products exceeds that of the reactants. In contrast, a reaction is endothermic if the sum of the bond energies of the reactants exceeds that of the products.

Example 1

Consider the following combination reaction.

$$H_2(g) + Br_2(g) \rightarrow 2HBr(g)$$

Calculate the standard heat of reaction from the bond energies given.

$$\left(H\text{–}H = 436 \text{ kJ mol}^{-1}, \ Br\text{–}Br = 193 \text{ kJ mol}^{-1}, \ H\text{–}Br = 66 \text{ kJ mol}^{-1}\right)$$

Solution

The standard heat of reaction is calculated as follows.

$$\Delta H^\theta = \Sigma \text{ bond energies of reactants} - \Sigma \text{ bond energies of products}$$

Now, we rewrite the equation to show all the bonds in the reactants and products.

$$H\text{–}H + Br\text{–}Br \rightarrow 2H\text{–}Br$$

Thus, the bonds in the reactants are as follows:

$$H\text{–}H = 436 \text{ kJ mol}^{-1}$$

$$Br\text{–}Br = 193 \text{ kJ mol}^{-1}$$

Chemical Energetics

So

$$\Sigma \text{bond energies of reactants} = (436 + 193) \text{ kJ} = 629 \text{ kJ}$$

The only bond in the product is H–Br = 366 kJ mol^{-1}.
So

$$\Sigma \text{bond energies of products} = 2 \times 366 \text{ kJ} = 732 \text{ kJ}$$

$$\Delta H^\ominus = ?$$

Substituting, we have

$$\Delta H^\ominus = 629 \text{ kJ} - 732 \text{ kJ} = -103 \text{ kJ}$$

Since our calculation is based on the formation of 2 mol of HBr(g), it follows that

$$\Delta H^\ominus = -\frac{103 \text{ kJ}}{2 \text{ mol}} = -51.5 \text{ kJ mol}^{-1}$$

Example 2

Use bond energies to estimate the standard heat of formation of water vapour. The equation of reaction is as follows:

$$2H_2(g) + O_2(g) \rightarrow 2H_2O(g)$$

$\left(H-H = 436 \text{ kJ mol}^{-1}, O=O = 497 \text{ kJ mol}^{-1}, H=O = 463 \text{ kJ mol}^{-1}\right)$

Solution

The standard heat of reaction is the standard heat of formation of water, which is determined as follows.

$$\Delta H_f^\ominus = \Sigma \text{bond energies of reactants} - \Sigma \text{bond energies of products}$$

$$2H-H + O=O \rightarrow 2H-O-H$$

The bonds in the reactants are

$$H-H = 436 \text{ kJ mol}^{-1}$$

$$O=O = 497 \text{ kJ mol}^{-1}$$

So

$$\Sigma \text{bond energies of reactants} = \left(\{2 \times 436\} + 497\right) \text{ kJ} = 1369 \text{ kJ}$$

The bonds in the product are

$$H-O = 463 \text{ kJ mol}^{-1}$$

$$O-H = 463 \text{ kJ mol}^{-1}$$

So

$$\Sigma \text{bond energies of products} = (\{2\times 463\} + \{2\times 463\})\text{ kJ} = 1712 \text{ kJ}$$

$$\Delta H_f^\theta = ?$$

We now substitute to obtain

$$\Delta H_f^\theta = 1369 \text{ kJ} - 1852 \text{ kJ} = -483 \text{ kJ}$$

2 mol of water were formed; hence,

$$\Delta H_f^\theta = -\frac{483 \text{ kJ}}{2 \text{ mol}} = -242 \text{ kJ mol}^{-1}$$

PRACTICE PROBLEMS

1. Use bond energies to calculate the standard heat of the following reaction:

$$H_2(g) + Cl_2(g) \rightarrow 2HCl(g)$$

$$(H-H = 436 \text{ kJ mol}^{-1},\ Cl-Cl = 242 \text{ kJ mol}^{-1},\ H-Cl = 431 \text{ kJ mol}^{-1})$$

2. Consider the following reaction:

$$2CO(g) + O_2(g) \rightarrow 2CO_2(g)$$

Calculate the standard heat of reaction from bond energies.

$$(C=O = 1076 \text{ kJ mol}^{-1},\ O=O = 497 \text{ kJ mol}^{-1},\ C\equiv O = 1072 \text{ kJ mol}^{-1})$$

Hint: The structures of carbon dioxide and carbon monoxide are O=C=O and C≡O, respectively.

10.6 ENTROPY

Entropy, S, is a measure of the degree of disorder in a system. The unit of entropy is J K^{-1}. The entropy of 1 mol of a substance is called *molar entropy*, which has a unit of J K^{-1} mol^{-1}. When measured at 1 atm, the molar entropy of a substance is referred to as the *standard molar entropy*, S^0.

Unlike the heat of reaction, entropy change, ΔS, can be used to predict the spontaneity of a process. A process will occur spontaneously (without an external supply of energy) only if the total entropy change of the system and its surroundings increases. An increase in entropy is indicated by a positive entropy change.

If the entropy of a substance changes from S_1 to S_2, then the entropy change is given by

$$\Delta S = S_2 - S_1$$

Applying Hess's law, the standard entropy change of a process is calculated from the standard molar entropies of formation of reactants and products as follows:

$$\Delta S^\theta = \Sigma nS^\theta(\text{products}) - \Sigma nS^\theta(\text{reactants})$$

where n is the stoichiometric coefficient of a specie in the chemical reaction.

Chemical Energetics

At constant pressure, the entropy change of a reversible process, such as a change of state, is given by

$$\Delta S = \frac{\Delta H}{T}$$

where T is the absolute transition temperature.

Example 1

Using the values of standard entropy of formation given below, determine the standard entropy change when 1 mol of liquid water is converted into water vapour.

$$\left(H_2O(l) = 70 \text{ JK}^{-1} \text{ mol}^{-1},\ H_2O(g) = 189 \text{ JK}^{-1} \text{ mol}^{-1}\right)$$

Solution

We apply the relation:

$$\Delta S^\theta = S_2^\theta - S_1^\theta$$

The process under consideration is:

$$H_2O(l) \rightarrow H_2O(g)$$

$$S_2^\theta = 189 \text{ JK}^{-1} \text{ mol}^{-1}$$

$$S_1^\theta = 70 \text{ JK}^{-1} \text{ mol}^{-1}$$

$$\Delta S^\theta = ?$$

Substituting the values, we obtain

$$\Delta S^\theta = 189 \text{ JK}^{-1} \text{ mol}^{-1} - 70 \text{ JK}^{-1} \text{ mol}^{-1}$$
$$= 119 \text{ JK}^{-1} \text{ mol}^{-1}$$

Since the entropy change is positive, this indicates an increase in the entropy as liquid water converts to water vapour.

Example 2

Calculate the standard entropy change for the following reaction.

$$2H_2(g) + O_2(g) \rightarrow 2H_2O(g)$$

$$\left(H_2(g) = 131 \text{ JK}^{-1} \text{ mol}^{-1},\ O_2(g) = 205 \text{ JK}^{-1} \text{ mol}^{-1},\ H_2O(g) = 189 \text{ JK}^{-1} \text{ mol}^{-1}\right)$$

Solution

The standard entropy change is determined from the standard molar entropies of formation, as follows.

$$\Delta S^\theta = \Sigma n S^\theta (\text{products}) - \Sigma n S^\theta (\text{reactants})$$

For the product, we have

$$S^\ominus = 189 \text{ JK}^{-1} \text{ mol}^{-1}$$

$$n = 2 \text{ mol}$$

So

$$\sum nS^\ominus (\text{products}) = 2 \text{ mol} \times \frac{189 \text{ J}}{1 \text{ K} \times 1 \text{ mol}} = 378 \text{ JK}^{-1}$$

For $H_2(g)$, we have

$$S^\ominus = 131 \text{ JK}^{-1} \text{ mol}^{-1}$$

$$n = 2 \text{ mol}$$

For $O_2(g)$, we have

$$S^\ominus = 205 \text{ JK}^{-1} \text{ mol}^{-1}$$

$$n = 1 \text{ mol}$$

So

$$\sum nS^\ominus (\text{reactants}) = \left(1 \text{ mol} \times \frac{205 \text{ J}}{1 \text{ K} \times 1 \text{ mol}}\right) + \left(2 \text{ mol} \times \frac{131 \text{ J}}{1 \text{ K} \times 1 \text{ mol}}\right) = 467 \text{ JK}^{-1}$$

$$\Delta S^\ominus = ?$$

Substituting we have

$$\Delta S^\ominus = 378 \text{ JK}^{-1} - 467 \text{ JK}^{-1}$$

$$= -89 \text{ JK}^{-1}$$

The negative sign indicates a decrease in entropy.

Example 3

Determine the standard entropy change of the following reaction.

$$CaCO_3(s) \rightarrow CaO(s) + CO_2(g)$$

$$\left(CaCO_3(s) = 93 \text{ JK}^{-1} \text{ mol}^{-1},\ CaO(s) = 40 \text{ JK}^{-1} \text{ mol}^{-1},\ CO_2(g) = 214 \text{ JK}^{-1} \text{ mol}^{-1}\right)$$

Solution

As usual,

$$\Delta S^\ominus = \sum nS^\ominus (\text{products}) - \sum nS^\ominus (\text{reactants})$$

For the CaO(s), we have

$$S^\ominus = 40 \text{ JK}^{-1} \text{ mol}^{-1}$$

$$n = 1 \text{ mol}$$

Chemical Energetics

For $CO_2(g)$, we have

$$S^\theta = 214 \text{ JK}^{-1} \text{ mol}^{-1}$$

$$n = 1 \text{ mol}$$

So

$$\sum nS^\theta \text{ (products)} = \left(1 \text{ mol} \times \frac{40 \text{ J}}{1 \text{ K} \times 1 \text{ mol}}\right) + \left(1 \text{ mol} \times \frac{214 \text{ J}}{1 \text{ K} \times 1 \text{ mol}}\right) = 254 \text{ JK}^{-1}$$

For $CaCO_3(s)$, we have

$$S^\theta = 93 \text{ JK}^{-1} \text{ mol}^{-1}$$

$$n = 1 \text{ mol}$$

So

$$\sum nS^\theta \text{ (reactants)} = 1 \text{ mol} \times \frac{93 \text{ J}}{1 \text{ K} \times 1 \text{ mol}} = 93 \text{ JK}^{-1}$$

$$\Delta S^\theta = ?$$

Finally, we now substitute to obtain

$$\Delta S^\theta = 254 \text{ JK}^{-1} - 93 \text{ JK}^{-1}$$
$$= 161 \text{ JK}^{-1}$$

Thus, the entropy of the reacting system increases.

Example 4

Determine the entropy change when 1 mol of water is converted to water vapour at 100°C. The standard heat of vaporisation of water is 44 kJ mol^{-1}.

Solution

This is a reversible process; the standard entropy change is calculated as follows.

$$\Delta S^\theta = \frac{\Delta H^\theta}{T}$$

$$\Delta H^\theta = 44 \text{ kJ mol}^{-1} = 44000 \text{ J mol}^{-1}$$

$$T = (100 + 273) \text{ K} = 373 \text{ K}$$

$$\Delta S^\theta = ?$$

Substituting we have

$$\Delta S^\theta = \frac{44000 \text{ J mol}^{-1}}{373 \text{ K}} = 118 \text{ J mol}^{-1} \text{ K}^{-1}$$

This is approximately equal to the result we obtained in Example 1. The slight difference is due to rounding of the values of the standard entropies of formation used.

Example 5

The entropy change for the conversion of 5.0 g of solid iodine to liquid at its melting point of 113.5°C is 0.795 J K⁻¹. Calculate the heat required to melt 5.0 g of iodine completely at constant pressure.

Solution

The heat required is obtained as follows.

$$\Delta S = \frac{\Delta H}{T}$$

$$\therefore \Delta H = T \times \Delta S$$

$$T = (113.5 + 273) \text{ K} = 386.5 \text{ K}$$

$$\Delta S = 0.795 \text{ JK}^{-1}$$

$$\Delta H = ?$$

Substituting, we have

$$\Delta H = 386.5 \text{ K} \times \frac{0.795 \text{ J}}{1 \text{ K}} = 307 \text{ J}$$

PRACTICE PROBLEMS

1. Calculate the standard entropy change when 1 mol of bromine is completely converted to liquid.

   $$\left(Br(l) = 152.2 \text{ J K}^{-1} \text{ mol}^{-1}, \ Br(g) = 245.4 \text{ J K}^{-1} \text{ mol}^{-1} \right)$$

2. Determine the standard entropy change of the following reaction.

   $$2CO(s) + O_2(g) \rightarrow 2CO_2(g)$$

   $$\left(CO(g) = 197.6 \text{ J K}^{-1} \text{ mol}^{-1}, \ O_2(g) = 205 \text{ J K}^{-1} \text{ mol}^{-1}, \ CO_2(g) = 213.6 \text{ J K}^{-1} \text{mol}^{-1} \right)$$

3. The heat of vaporisation of methanol is 35.3 kJ mol⁻¹. Calculate the entropy change when 1 mol of methanol condenses at its boiling point of 64.7°C.
4. How much heat is required to melt 54 g of aluminium completely if the entropy change during the process is 22.9 J K⁻¹. The melting point of aluminium is 660°C.

10.7 FREE ENERGY

Earlier, we discussed the measurement of heat energy accompanying chemical and physical processes. In 1875, American scientist Josiah Willard Gibbs (1839–1903) introduced a much more informative type of energy measurement called *free energy*, G. The free energy of a system is defined as *the energy available for doing work*. The standard free energy of formation of a substance, ΔG_f^θ, is the standard free energy change—or simply free energy—accompanying the formation of 1 mol of the substance from its elements under standard conditions. Like the standard heat of formation, the standard free energy of formation of elements in their standard states is zero.

Chemical Energetics

Just like heat of reaction, Hess's Law is applied in the calculation of the standard free energy of any process using the standard free energies of formation of reactants and products as follows.

$$\Delta G^\theta = \sum n \Delta G_f^\theta (\text{products}) - \sum n \Delta G_f^\theta (\text{reactants})$$

where n is the stoichiometric coefficient of a substance in a chemical reaction.

For any process, the relationship between free energy, absolute temperature, enthalpy change and entropy change is given by

$$\Delta G = \Delta H - T \Delta S$$

At standard conditions, we have

$$\Delta G^\theta = \Delta H^\theta - T \Delta S^\theta$$

Free energy is used to predict the spontaneity of a process as follows.

- If free energy is negative—i.e., $\Delta G < 0$—the reaction or process is spontaneous.
- If the free energy is positive—i.e., $\Delta G > 0$—the process nonspontaneous.
- If the free energy is zero—i.e., $\Delta G = 0$—the process is at equilibrium.

Example 1

Determine the standard free energy for the conversion of water to water vapour.

$$\left(H_2O(g) = -228.6 \text{ kJ mol}^{-1}, \; H_2O(l) = -237.1 \text{ kJ mol}^{-1} \right)$$

Solution

The standard free energy is obtained as follows.

$$\Delta G_f^\theta = \sum n \Delta G_f^\theta (\text{products}) - \sum n \Delta G_f^\theta (\text{reactants})$$

The physical process under consideration is

$$H_2O(l) \rightarrow H_2O(g)$$

For $H_2O(g)$, we have

$$\Delta G_f^\theta = -228.6 \text{ kJ mol}^{-1}$$

$$n = 1 \text{ mol}$$

So

$$\sum n \Delta G_f^\theta (\text{products}) = 1 \text{ mol} \times \left(-\frac{228.6 \text{ kJ}}{1 \text{ mol}} \right) = -228.6 \text{ kJ}$$

For $H_2O(l)$, we have

$$\Delta G_f^\theta = -237.1 \text{ kJ mol}^{-1}$$

$$n = 1 \text{ mol}$$

So

$$\Sigma n\Delta G_f^\theta \text{ (reactants)} = 1 \text{ mol} \times \left(-\frac{237.1 \text{ kJ}}{1 \text{ mol}}\right) = -237.1 \text{ kJ}$$

$$\Delta G^\theta = ?$$

Substituting the values, we have

$$\Delta G^\theta = \left(-228.6 \text{ kJ} - \{-237.1 \text{ kJ}\}\right)$$
$$= 8.5 \text{ kJ}$$

The positive value of the free energy indicates that the conversion of liquid water to water vapour is nonspontaneous. However, the reverse process—the conversion of water vapour to liquid water—is spontaneous, with a free energy of −8.5 kJ.

Example 2

Calculate the free energy of the reaction:

$$2CO(s) + O_2(g) \rightarrow 2CO_2(g)$$

$$\left(CO(g) = -137.2 \text{ kJ mol}^{-1}, \ CO_2(g) = -394.4 \text{ kJ mol}^{-1}\right)$$

Solution

We apply the relation:

$$\Delta G^\theta = \Sigma n\Delta G_f^\theta \text{ (products)} - \Sigma n\Delta G_f^\theta \text{ (reactants)}$$

For $CO_2(g)$, we have

$$\Delta G_f^\theta = -394.4 \text{ kJ mol}^{-1}$$
$$n = 2 \text{ mol}$$

So

$$\Sigma n\Delta G_f^\theta \text{ (products)} = 2 \text{ mol} \times \left(-\frac{394.4 \text{ kJ}}{1 \text{ mol}}\right) + 0 = -788.8 \text{ kJ}$$

For CO(g), we have

$$\Delta G_f^\theta = -137.2 \text{ kJ mol}^{-1}$$
$$n = 2 \text{ mol}$$

For $O_2(g)$, we have

$$\Delta G_f^\theta = 0$$

Chemical Energetics

So

$$\Sigma n \Delta G_f^\theta \text{(reactants)} = 2\text{ mol} \times \left(-\frac{137.2 \text{ kJ}}{1 \text{ mol}}\right) + 0 = -274.4 \text{ kJ}$$

$$\Delta G^\theta = ?$$

Substituting, we have

$$\Delta G^\theta = (-788.8 \text{ kJ}) - \{-274.4 \text{ kJ}\}$$
$$= -514.4 \text{ kJ}$$

The negative value of free energy shows that the reaction is spontaneous, meaning it requires no external energy input to occur.

Example 3

Calculate the free energy of a reaction at 27°C when $\Delta H = 4.35$ kJ and $\Delta S = 14.5$ J K^{-1}.

Solution

The free energy is calculated as follows.

$$\Delta G = \Delta H - T\Delta S$$
$$\Delta H = 4.35 \text{ kJ} = 4350 \text{ J}$$
$$T = (273 + 27) \text{ K} = 300 \text{ K}$$
$$\Delta S = 14.5 \text{ JK}^{-1}$$
$$\Delta G = ?$$

Substituting, we have

$$\Delta G = 4350 \text{ kJ} - \left(300 \text{ K} \times \frac{14.5 \text{ J}}{1 \text{ K}}\right)$$

So

$$\Delta G = 4350 \text{ J} - 4350 \text{ J} = 0$$

The result indicates that the reaction is in equilibrium.

Example 4

The ΔH and ΔS of a sample of water are 20.5 kJ and 55 J K^{-1}, respectively. Determine the temperature at which the liquid water will be in equilibrium with its vapour.

Solution

We begin with the relation:

$$\Delta G = \Delta H - T\Delta S$$

The free energy of any process in equilibrium is zero; hence, $\Delta G = 0$.
So

$$\Delta H - T\Delta S = 0$$

$$\therefore T = \frac{\Delta H}{\Delta S}$$

$$\Delta H = 20.5 \text{ kJ} = 20500 \text{ J}$$

$$\Delta S = 55 \text{ J K}^{-1}$$

$$T = ?$$

Substituting, we have

$$T = \frac{20500 \text{ J} \times 1 \text{ K}}{55 \text{ J}} = 373 \text{ K} = 100°C$$

PRACTICE PROBLEMS

1. Determine the standard free energy for the solidification of 1 mol of liquid aluminium.

$$\left(\text{Al}(l) = 7.2 \text{ kJ mol}^{-1}\right)$$

Hint: Aluminium is a solid in its standard state.

2. Consider the following equation:

$$\text{CaCO}_3(s) \rightarrow \text{CaO}(s) + \text{CO}_2(g)$$

State whether the process requires energy input to occur at standard conditions, providing a reason for your answer.

$$\left(\text{CaCO}_3(s) = -1128.8 \text{ kJ mol}^{-1}, \text{ CaO}(s) = -604 \text{ kJ mol}^{-1}, \text{ CO}_2(g) = -394.4 \text{ kJ mol}^{-1}\right)$$

3. Determine the free energy of a physical process at 78°C when $\Delta H = 35.1$ kJ and $\Delta S = 100.0$ J K^{-1}.
4. The free energy of a reaction is −1.1 kJ at 110°C. Determine the enthalpy change of the reaction if the entropy change is 25 J K^{-1}.

10.8 THE FIRST LAW OF THERMODYNAMICS

Thermodynamics is the branch of science that deals with the conversion of energy from one form to another. The first law of thermodynamics states that energy can neither be created nor destroyed but can be converted from one form to the other. This statement is also called *the law of conservation of energy*. Mathematically, it is expressed as:

$$\Delta Q = \Delta U + W$$

where

ΔQ = the heat absorbed or evolved by a system
ΔU = the change in the internal energy of a system
W = the work done on or by the system

Chemical Energetics

As stated earlier, $\Delta Q = \Delta H$ at constant pressure; hence, at constant pressure the first law is given by

$$\Delta H = \Delta U + W$$

The following sign conventions are followed when performing calculations with the first law of thermodynamics.

- ΔQ is positive if a system absorbs heat, while it is negative if it evolves heat.
- W is positive if the system does work on its surroundings, while it is negative if the system has work done on it by its surroundings.

Example 1

The internal energy of a gas changes by 550 J when it does 2500 J of work on the surroundings. What amount of heat energy is absorbed or evolved by the system?

Solution

According to the first law of thermodynamics,

$$\Delta Q = \Delta U + W$$
$$\Delta U = 550 \text{ J}$$
$$W = 2500 \text{ J}$$
$$\Delta Q = ?$$

Substituting we have

$$\Delta Q = 550 \text{ J} + 2500 \text{ J}$$
$$= 3050 \text{ J}$$

The positive value of the heat change indicates that the system absorbs heat from its surroundings.

Example 2

The internal energy of a system changes by −2.5 kJ when the system has 5.0 kJ of work done on it by its surroundings. Calculate the amount of heat energy absorbed or evolved by the system.

Solution

As usual, we use

$$\Delta Q = \Delta U + W$$
$$\Delta U = -2.5 \text{ kJ}$$
$$W = -5.0 \text{ kJ}$$
$$\Delta Q = ?$$

Substituting, we have

$$\Delta Q = -2.5 \text{ kJ} + (-5.0 \text{ kJ})$$
$$= -7.5 \text{ kJ}$$

Thus, the system releases heat to the surroundings.

Example 3

A system does 7.5 kJ of heat on its surroundings when it releases 4.5 kJ of heat. Determine the change in the internal energy of the system.

Solution

The change in the internal energy of the system is determined as follows.

$$\Delta Q = \Delta U + W$$
$$\therefore \Delta U = \Delta Q - W$$
$$\Delta Q = -4.5 \text{ kJ}$$
$$W = 7.5 \text{ kJ}$$
$$\Delta U = ?$$

Substituting, we have

$$\Delta U = -4.5 \text{ kJ} - 7.5 \text{ kJ}$$
$$= -12 \text{ kJ}$$

The negative sign indicates that the internal energy of the system decreases by 12 kJ.

Example 4

The internal energy of a system changes by 750 J when it absorbs 1500 J of heat from its surroundings at constant pressure. Determine the work done in the process.

Solution

For a constant pressure process, we have

$$\Delta H = \Delta U + W$$
$$\therefore W = \Delta H - \Delta U$$
$$\Delta H = 1500 \text{ J}$$
$$\Delta U = 750 \text{ J}$$
$$W = ?$$

Substituting the values, we have

$$W = 1500 \text{ J} - 750 \text{ J}$$
$$= 750 \text{ J}$$

PRACTICE PROBLEMS

1. Calculate the heat absorbed or evolved by a system doing 2500 J of work on its surroundings if its internal energy changes by −1500 J.
2. The surroundings perform 650 J of work on a system when it releases 1300 J of heat. Calculate the change in the internal energy of the system.
3. The internal energy of a system changes by −250 J when it absorbs 350 J of heat from its surroundings. Calculate the work done in the process.

Chemical Energetics

FURTHER PRACTICE PROBLEMS

1. 250 cm³ of water initially at 10°C was heated to 100°C. Calculate the heat energy absorbed by the water.

$$(c = 4200 \text{ J kg}^{-1} \text{ K}^{-1})$$

2. The dissolution of 5.0 g of potassium chlorate, $KClO_3$, reduces the temperature of 50 cm³ of water from 30°C to 22°C. Calculate the heat of solution of the salt.

$$(O = 16, \; Cl = 35.5, \; K = 39, \; c = 4200 \text{ J kg}^{-1} \text{ K}^{-1})$$

3. Determine the heat of vaporisation of ethanol, C_2H_5OH, if it requires 3.86 kJ of heat to vaporise 4.6 g of the liquid completely at its boiling point.

$$(H = 1, \; C = 12, \; O = 16)$$

4. The complete combustion of 1.0 g of propane, C_3H_8, increases the temperature of 250 cm³ of water from 25°C to 73°C. Determine the heat of combustion of the compound.

$$(H = 1, \; C = 12, \; c = 4200 \text{ J kg}^{-1} \text{ K}^{-1})$$

5. Consider the following neutralisation reaction.

$$NaOH(aq) + HF(aq) \rightarrow NaF(aq) + H_2O(l) \quad \Delta H = -68.6 \text{ kJ mol}^{-1}$$

Calculate the amount of heat released when sodium hydroxide reacts with hydrofluoric acid to produce 4.5 g of water.

$$(H = 1, \; O = 16)$$

6. Consider the following combustion reaction:

$$2C_2H_6(g) + 7O_2(g) \rightarrow 4CO_2(g) + 6H_2O(l) \quad \Delta H^\ominus = -1560 \text{ kJ mol}^{-1}$$

Calculate the amount of heat that would be generated by the complete combustion of 5.0 g of ethane.

$$(H = 1, \; C = 12)$$

7. Calculate the standard heat of combustion of benzene from the information given below.

1. $6C(s) + 3H_2(g) \rightarrow C_6H_6(l) \quad \Delta H_f^\ominus = 49.2 \text{ kJ mol}^{-1}$
2. $C(s) + O_2(g) \rightarrow CO_2(g) \quad \Delta H_f^\ominus = -393.5 \text{ kJ mol}^{-1}$
3. $H_2(g) + \frac{1}{2}O_2(g) \rightarrow H_2O(g) \quad \Delta H_f^\ominus = -242 \text{ kJ mol}^{-1}$

8. Calculate the standard heat of the reaction $2KBr(s) + F_2(g) \rightarrow 2KF(s) + Br_2(g)$ from the following information.

1. $2KBr(s) + Cl_2(g) \rightarrow 2KCl(s) + Br_2(g)$ $\Delta H_f^\theta = -84.4\,kJ$
2. $2KCl(s) + F_2(g) \rightarrow 2KF(s) + Cl_2(g)$ $\Delta H_f^\theta = -261.0\,kJ$

9. Show the following reactions in a Born-Haber cycle and determine the lattice enthalpy of formation of magnesium chloride.

   1. Heat of atomisation of magnesium,

   $$Mg(s) \rightarrow Mg(g) \qquad \Delta H_f^\theta = +148\,kJ\,mol^{-1}$$

   2. First ionisation energy of magnesium,

   $$Mg(g) \rightarrow Mg^+(g) + e^- \qquad \Delta H^\theta = +738\,kJ\,mol^{-1}$$

   3. Second ionisation energy of magnesium,

   $$Mg^+(g) + e^- \rightarrow Mg^{2+}(g) + 2e^- \qquad \Delta H^\theta = +1451\,kJ\,mol^{-1}$$

   4. Dissociation energy of chlorine molecule,

   $$Cl_2(g) \rightarrow 2Cl(g) \qquad \Delta H_f^\theta = +244\,kJ$$

   5. First electron affinity of chlorine,

   $$Cl(g) + e^- \rightarrow Cl^-(g) \qquad \Delta H^\theta = -348\,kJ\,mol^{-1}$$

   6. Heat of formation of $MgCl_2(g)$,

   $$Mg(s) + Cl_2(g) \rightarrow MgCl_2(s) \qquad \Delta H_f^\theta = -643\,kJ\,mol^{-1}$$

10. Determine the standard heat of combustion of methanol from the standard heats of formation given below.

$$2CH_3OH(l) + 3O_2(g) \rightarrow 2CO_2(g) + 2H_2O(l)$$

$$\left(CH_3OH(l) = -238.4\,kJ\,mol^{-1},\; CO_2(g) = -393.5\,kJ\,mol^{-1},\; H_2O(l) = 286\,kJ\,mol^{-1}\right)$$

11. Determine the standard heats of sublimation and deposition of calcium from the standard heat of formation given below.

$$\left(Ca(g) = 178\,kJ\,mol^{-1}\right)$$

12. The heat of fusion of aluminium is 10.7 kJ. How much heat is released when 2.7 g of aluminium solidifies?

$$(of\ Al = 27)$$

13. Use the bond energies given below to determine the heat of formation of hydrogen fluoride. The equation of reaction is given by

$$H_2(g) + F_2(g) \rightarrow 2HF(g)$$

$$\left(H\text{--}H = 436 \text{ kJ mol}^{-1},\ F\text{--}F = 36.6 \text{ kJ mol}^{-1},\ H\text{--}F = 135 \text{ kJ mol}^{-1}\right)$$

14. Determine the standard heat of combustion of methane from the bond energies given below. The equation of reaction is as follows.

$$CH_4(g) + 2O_2(g) \rightarrow CO_2(g) + 2H_2O(g)$$

$$\left(C\text{--}H = 412 \text{ kJ mol}^{-1},\ O=O = 497 \text{ kJ mol}^{-1},\ C=O = 1076 \text{ kJ mol}^{-1},\ H\text{--}O = 463 \text{ kJ mol}^{-1}\right)$$

15. Determine the standard entropy change of the following reaction.

$$H_2(g) + F_2(g) \rightarrow 2HF(g)$$

$$\left(H_2(g) = 130.7 \text{ J K}^{-1} \text{ mol}^{-1},\ F_2(g) = 202.8 \text{ J K}^{-1} \text{ mol}^{-1},\ HF(g) = 173.8 \text{ J K}^{-1} \text{ mol}^{-1}\right)$$

16. The heat of fusion of methanol, CH_3OH, is 3.2 kJ mol^{-1}. Calculate the entropy change when 25 g of solid methanol changes completely into liquid at its melting point of $-97.8°C$.

$$\left(H = 1,\ C = 12,\ O = 16\right)$$

17. The entropy change when 55 g of ethanol, C_2H_5OH, vapour condenses at its boiling point of 78°C is -148 J K^{-1}. How much heat is released in the process?

18. Calculate the standard free energy of the following reaction.

$$CH_4(g) + 2O_2(g) \rightarrow CO_2(g) + 2H_2O(g)$$

$$\left(CH_4(g) = -50.7 \text{ kJ mol}^{-1},\ CO_2(g) = -394.4 \text{ kJ mol}^{-1},\ H_2O(g) = -228.6 \text{ kJ mol}^{-1}\right)$$

19. Calculate the free energy of a reaction at 50°C when ΔS and ΔH are -215 J K^{-1} and 4.5 kJ, respectively.

20. The free energy of a reaction is 25.0 kJ at 25°C. Determine the heat of reaction if the entropy change is 450.5 J K^{-1}.

21. The values of ΔH and ΔS for a system at equilibrium are 25.5 kJ and 450 J K^{-1}, respectively. Determine the temperature of the system.

22. How much heat must be supplied to a system to increase its internal energy by 650 J while allowing make it to perform 5.5 kJ of work on its surroundings.

23. A system evolves 1.5 kJ of heat while the surroundings perform 2.5 kJ of work on it. Calculate the change in the system's internal energy.

24. A system releases 7.0 kJ of heat while its internal energy changes by -1.5 kJ. How much work is done in the process?

11 Chemical Equilibria

11.1 REVERSIBLE REACTIONS

A reversible reaction is a reaction that occurs in both the forward and backward directions. A reversible reaction is indicated with two-way double half arrowheads ($\rightleftharpoons$) between the reactants and products. The symbol $\rightarrow$ denotes the forward reaction, while $\leftarrow$ denotes the reverse or backward reaction. An example of a reversible reaction is the production of ammonia by the Haber process:

$$N_2(g) + 3H_2(g) \rightleftharpoons 2NH_3(g)$$

Another example is the decomposition of hydrogen iodide:

$$2HI(g) \rightleftharpoons H_2(g) + I_2(g)$$

Chemical equilibrium describes the state of a reversible reaction in which the forward and backward reactions occur at the same rate (see Figure 11.1). Stated differently, chemical equilibrium describes the state of a reversible reaction at which there is no net formation or any observable change in the concentrations of the reactants and products with time (see Figure 11.2).

Chemical equilibrium is a typical example of *dynamic equilibrium*—the type of equilibrium in which there is constant motion within a system without any net change in its properties with time. The position of equilibrium or the favoured direction of reaction when a reversible reaction attains equilibrium depends on concentration, temperature and pressure. A catalyst does not affect the position of equilibrium but speeds up the rate at which equilibrium is attained.

11.2 LE CHÂTELIER'S PRINCIPLE

A reaction at equilibrium will remain at equilibrium unless there is a change to any of the equilibrium conditions of concentration, temperature and pressure. According to Le Châtelier's principle, a reaction at equilibrium will always act so as to annul or counteract any change in equilibrium conditions. In other words, when the conditions of a reaction at equilibrium are altered, the reaction will act to re-establish equilibrium by forming more of the products or reactants. The effects of changes to equilibrium conditions are explained below.

11.2.1 Change in Concentration

When the concentration of a product is reduced by removing some of it from an equilibrium mixture, or when the concentration of a reactant is increased by introducing more of it into the equilibrium mixture, the equilibrium position will shift to the right—i.e., equilibrium will be re-established by the formation of more of the product. Conversely, when the concentration of a product is increased, or when the concentration of a reactant is reduced, the equilibrium position will shift to the left—i.e., equilibrium will be re-established by the formation of more of the reactant.

11.2.2 Change in Temperature

The effect of temperature change on a reaction at equilibrium depends on whether the reaction is endothermic or exothermic. When the temperature of a reaction at equilibrium is increased, the equilibrium will shift to the direction of endothermic change. On the other hand, when the temperature is reduced, the equilibrium position will shift to the direction of exothermic change.

Chemical Equilibria

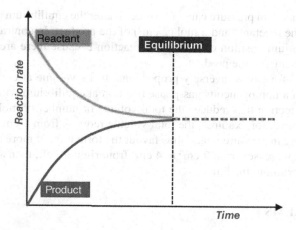

FIGURE 11.1 Equal rates of forward and reverse reactions at equilibrium.

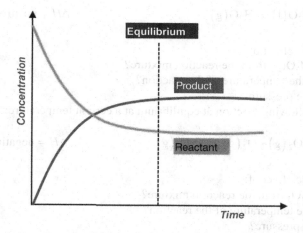

FIGURE 11.2 Constant concentrations of reactants and products at equilibrium.

Consider, for instance, the synthesis of ammonia from nitrogen and hydrogen:

$$N_2(g) + 3H_2(g) \rightleftharpoons 2NH_3(g) \qquad \Delta H = -46 \text{ kJ mol}^{-1}$$

The forward reaction is exothermic, while the backward reaction is endothermic. An increase in temperature will lead to the formation of more hydrogen and nitrogen, while more ammonia will be produced if the temperature is reduced.

11.2.3 Change in Pressure

A change in pressure will only affect a reaction at equilibrium if it contains at least one gaseous substance and if there is a difference between the total number of moles or volumes of gaseous reactants and products. Consider, for instance, the following reactions:

$$2HI(g) \rightleftharpoons H_2(g) + I_2(g)$$

$$N_2(g) + 3H_2(g) \rightleftharpoons 2NH_3(g)$$

In the first reaction, change in pressure cannot be used to alter the equilibrium position because there are 2 mol (2 cm³) of the reactants and 2 mol (2 cm³) of the product. In contrast, change in pressure will distort the equilibrium position of the second reaction because there are 4 mol (4 cm³) of the reactants and 2 mol (2 cm³) of the product.

Since the pressure of a gas is inversely proportional to its volume at constant temperature, an increase in pressure in a homogeneous gas-phase reaction at equilibrium will shift the equilibrium position toward the direction that reduces the total volume or number of moles, and vice versa. In the second reaction above, for example, the total volume reduces from 4 cm³ to 2 cm³ from left to right; hence, an increase in pressure would thus favour the formation of more ammonia. Conversely, since the total volume increases from 2 cm³ to 4 cm³ from right to left, then a reduction in pressure would shift the equilibrium to the left.

PRACTICE PROBLEMS

1. Consider the following reaction at equilibrium at a certain temperature:

$$H_2O(l) \rightleftharpoons H_2O(g) \qquad \Delta H = \text{positive}$$

What will be the effect of:
(a) removing $H_2O(g)$ from the reaction mixture?
(b) increasing the temperature of the reaction?
(c) doubling the pressure?

2. Consider the following reaction at equilibrium at a certain temperature:

$$2SO_2(g) + O_2(g) \rightleftharpoons 2SO_3(g) \qquad \Delta H = \text{negative}$$

What will be the effect of:
(a) removing SO_3 from the reaction mixture?
(b) reducing the temperature of the reaction?
(c) halving the pressure?
(d) adding a catalyst?
(e) adding more SO_2 to the reaction mixture.

11.3 EQUILIBRIUM CONSTANT AND CONCENTRATIONS

The equilibrium constant, K, of a reversible reaction is a number that describes the ratio of the amounts of products to those of reactants at equilibrium. One way of expressing the equilibrium constant is in terms of the equilibrium molar concentrations of reactants and products. The equilibrium constant, when expressed this way, is represented by the symbol K_c.

Consider the following balanced general reversible reaction.

$$aA + bB \rightleftharpoons cC + dD$$

The equilibrium constant, K_c, of the reaction is expressed as follows.

$$K_c = \frac{[C]^c [D]^d}{[A]^a [B]^b}$$

where the square brackets [ ] symbolise the molar concentration of the substance they enclose. The K expression for the reverse equilibrium $cC + dD \rightleftharpoons aA + bB$ is given by

$$K = \frac{1}{K_c} = \frac{[A]^a [B]^b}{[C]^c [D]^d}$$

As we can observe, the K_c expression for a reversible reaction has, as its numerator, the product of the equilibrium molar concentrations of the products of the forward reaction, while the denominator is the product of the molar concentrations of the reactants of the forward reaction. It can also be observed from the same expression that the molar concentration of each substance is raised to its stoichiometric coefficient.

For gas-phase reactions, equilibrium constant is often expressed in terms of the equilibrium partial pressures of the components of the reaction mixture. The equilibrium constant expressed in this way is represented by the symbol K_p. K_p is expressed in much the same way as K_c, except for the fact that equilibrium molar concentrations are replaced by equilibrium partial pressures:

$$K_p = \frac{(P_C)^c (P_D)^d}{(P_A)^a (P_B)^b}$$

Since the concentrations and partial pressures of pure solids and liquids always remain fairly constant, they are usually omitted from the expressions for equilibrium constant by being assumed to be unity. K_c and K_p are interconvertible, as given by the equation

$$K_p = K_c (RT)^{\Delta n}$$

where

R = universal gas constant (0.08206 dm^3 atm K^{-1} mol^{-1})
T = absolute temperature
Δn = the difference in the sum of the number of moles of gaseous products and reactants
Note that for a reaction in which $\Delta n = 0$, K_c is equal to K_p.

The magnitude of the equilibrium constant is very useful for predicting the relative concentrations of products and reactants in an equilibrium mixture. If the value of the equilibrium constant is greater than one ($K_c > 1$), the equilibrium concentrations of the products are larger than those of the reactants. A very high value of K_c indicates that the reactants are completely converted to the products at the prevailing temperature—i.e., the reaction goes to completion before equilibrium is established. On the other hand, if the value of equilibrium constant is less than one ($K_c < 1$), then the equilibrium concentrations of the reactants are larger than those of the products.

Example 1

Write the expression for the K_c of the reaction $CaCO_3(s) \rightleftharpoons CaO(s) + CO_2(g)$.

Solution

The K_c of the reaction is expressed as follows.

$$K_c = \frac{[CaO][CO_2]}{[CaCO_3]}$$

However, $CaCO_3$ and CaO are both solids; hence, their molar concentrations will be assumed to be unity in the expression for the equilibrium constant. $\therefore K_c = [CO_2]$

Example 2

Write the expression for the K_c of the reaction $2H_2(g) + O_2(g) \rightleftharpoons 2H_2O(l)$.

Solution

The K_c of the reaction is expressed as follows:

$$K_c = \frac{[H_2O]^2}{[H_2]^2[O_2]}$$

However, H_2O is a liquid; hence, its molar concentration will be assumed to be unity in the expression for K_c.

$$\therefore K_c = \frac{1}{[H_2]^2[O_2]}$$

Example 3

Write the expressions for the K_c and K_p of the reaction $3Fe(s) + 4H_2O(g) \rightleftharpoons Fe_3O_4(s) + 4H_2(g)$.

Solution

The K_c of the reaction is expressed as follows:

$$K_c = \frac{[Fe_3O_4][H_2]^4}{[Fe]^3[H_2O]^4}$$

However, Fe and Fe_3O_4 are both solids; hence, their molar concentrations are assumed to be unity in the expression for the equilibrium constant. Thus,

$$\therefore K_c = \frac{[H_2]^4}{[H_2O]^4}$$

The K_p is expressed as follows:

$$K_p = \frac{(P_{H_2})^4}{(P_{H_2O})^4}$$

Example 4

Write the expression for the K_p of the reaction $H_2(g) + I_2(g) \rightleftharpoons 2HI(g)$.

Solution

The K_p of the reaction is expressed as follows:

$$K_p = \frac{(P_{HI})^2}{(P_{H_2})(P_{I_2})}$$

Chemical Equilibria

Example 5

In an experiment, 0.200 mol dm³ of H_2 and 0.100 mol dm³ of O_2 are mixed together and allowed to reach equilibrium at a certain temperature. Calculate the K_c of the reaction at the prevailing temperature if the equilibrium molar concentrations of H_2 and O_2 are 0.100 mol dm^{-3} and 0.0500 mol dm^{-3}, respectively. The equation of the reaction is $2H_2(g) + O_2(g) \rightleftharpoons 2H_2O(l)$.

Solution

We will begin by writing the expression for the K_c for the reaction.

$$K_c = \frac{1}{[H_2]^2[O_2]}$$

$$[H_2] = 0.100 \text{ mol dm}^{-3}$$

$$[O_2] = 0.0500 \text{ mol dm}^{-3}$$

$$K_c = ?$$

Substituting, we have

$$K_c = \frac{1}{[0.100 \text{ mol dm}^{-3}]^2[0.0500 \text{ mol dm}^{-3}]} = 2.00 \times 10^3$$

Note that the equilibrium constant is a dimensionless quantity; hence, its values are written without units.

Example 6

Consider the following gas-phase equilibrium.

$$N_2(g) + O_2(g) \rightleftharpoons 2NO(g)$$

At a certain temperature, the amounts of N_2, O_2 and NO in the equilibrium mixture are 0.105 mol, 0.200 mol and 0.250 mol, respectively. Determine the K_c of the reaction at the prevailing temperature, given that the reaction was effected in a 2.00 dm³ reactor. Hence, determine the K_c for the reverse equilibrium:

$$2NO(g) \rightleftharpoons N_2(g) + O_2(g)$$

Solution

We will begin by writing the K_c expression for the equilibrium.

$$K_c = \frac{[NO]^2}{[N_2][O_2]}$$

The equilibrium molar concentration of each substance is obtained as follows.

$$C = \frac{n}{V}$$

Since a gas occupies the entire volume of its container, $V = 2.00$ dm³.

For N_2, we have

$$n = 0.105 \text{ mol}$$
$$[N_2] = ?$$
$$\therefore [N_2] = \frac{0.105 \text{ mol}}{2.00 \text{ dm}^3} = 0.0525 \text{ mol dm}^{-3}$$

For O_2, we have

$$n = 200 \text{ mol}$$
$$[O_2] = ?$$
$$\therefore [N_2] = \frac{0.200 \text{ mol}}{2.00 \text{ dm}^3} = 0.100 \text{ mol dm}^{-3}$$

For NO, we have

$$n = 0.250 \text{ mol}$$
$$[NO] = ?$$
$$\therefore [NO] = \frac{0.250 \text{ mol}}{2.00 \text{ dm}^3} = 0.125 \text{ mol dm}^{-3}$$
$$K_c = ?$$

Finally, substituting the values into the K_c expression, we have

$$K_c = \frac{(0.125)^2}{0.0525 \times 0.100} = 2.98$$

The K_c value for the reverse equilibrium is given by

$$K_c = \frac{1}{2.98} = 0.336$$

Example 7

In an experiment, 0.453 mol dm^{-3} of SO_2 and 0.352 mol dm^{-3} of O_2 are mixed at 973 K and allowed to reach equilibrium. Calculate the K_c and K_p for the reaction at 973 K, given that the equilibrium concentration of SO_3 is 0.303 mol dm^{-3}. The equation for the reaction is $2SO_2(g) + O_2(g) \rightleftharpoons 2SO_3(g)$.

$$(R = 0.08206 \text{ dm}^3 \text{ atm K}^{-1} \text{ mol}^{-1})$$

Solution

The expression for K_c is given by

$$K_c = \frac{[SO_3]^2}{[SO_2]^2 [O_2]}$$

Chemical Equilibria

To determine the equilibrium concentrations of SO_2 and O_2, we have to determine the change in the concentration of each of the reactants from the stoichiometry of the reaction and subtract it from the initial concentration. This information is presented in the table below. Such a table is called an ICE (acronym for **I**nitial concentration or partial pressure, **C**hange in concentration or partial pressure and **E**quilibrium concentration or partial pressure) table.

	SO_2	O_2	SO_3
I (mol dm^{-3})	0.453	0.352	0
C (mol dm^{-3})	−0.303	−0.152	+0.303
E (mol dm^{-3})	0.150	0.200	0.303

Note: the change in the concentration—or partial pressure—of a reactant is indicated with a negative sign, signifying a decrease in value. The change in concentration or partial pressure of a product carries a positive sign, signifying an increase in value.

Finally, we can now substitute the equilibrium concentrations into the expression for K_c to obtain

$$K_c = \frac{(0.303)^2}{(0.150)^2 (0.200)} = 20.4$$

The K_p for the reaction is determined from its K_c as follows.

$$K_p = K_c (RT)^{\Delta n}$$

$$R = 0.08206 \text{ dm}^3 \text{ atm K}^{-1} \text{ mol}^{-1}$$

$$T = 973 \text{ K}$$

$$\Delta n = 2 - 3 = -1$$

$$K_p = ?$$

Substituting, we have

$$K_p = 20.4 \times (0.08206 \times 973)^{-1} = 0.256$$

Example 8

A 1.05 g sample of ammonia, NH_3, at 323 K is added to a 1.00-dm³ reaction vessel and allowed to dissociate until equilibrium is reached. At equilibrium, the partial pressure of nitrogen, N_2, in the vessel is 0.110 atm. Determine the K_p and K_c of the reaction. The equation of reaction is $2NH_3(g) \rightleftharpoons N_2(g) + 3H_2(g)$.

$$\left(N = 14, H = 1, R = 0.08206 \text{ dm}^3 \text{ atm K}^{-1} \text{ mol}^{-1}\right)$$

Solution

The expression for K_p is given by

$$K_p = \frac{P_{N_2} \times (P_{H_2})^3}{(NH_3)^2}$$

The initial partial pressure of NH_3 is required to set up the ICE table. This is determined as follows.

$$n = \frac{m}{M}$$

$$m = 1.05 \text{ g}$$

$$M = \left(14 + \{1 \times 3\}\right) \text{ g mol}^{-1} = 17 \text{ g mol}^{-1}$$

$$n = ?$$

Substituting, we have

$$n = \frac{1.05 \text{ g} \times 1 \text{ mol}}{17 \text{ g}} = 0.06176 \text{ mol}$$

Using the ideal gas law:

$$PV = nRT$$

$$\therefore P = \frac{nRT}{V}$$

$$R = 0.08206 \text{ dm}^3 \text{ atm K}^{-1} \text{ mol}^{-1}$$

$$T = 323 \text{ K}$$

$$V = 1.00 \text{ dm}^3$$

$$P = ?$$

Substituting, we have

$$P = \frac{0.06176 \text{ mol} \times 0.08206 \text{ dm}^3 \text{ atm} \times 323 \text{ K}}{1.00 \text{ dm}^3 \times 1 \text{ mol} \times 1 \text{ K}} = 1.64 \text{ atm}$$

We can now set up the ICE table as follows.

	NH_3	N_2	H_3
I (atm)	1.64	0	0
C (atm)	−0.220	+0.110	+0.330
E (atm)	0.142	0.110	0.330

Finally, we can now substitute the equilibrium partial pressures into the K_p expression to obtain the equilibrium constant.

$$K_p = \frac{0.110 \times (0.330)^3}{(0.142)^2} = 0.196 \text{ atm}$$

The K_c is calculated as follows.

$$K_p = K_c (RT)^{\Delta n}$$

$$\therefore K_c = \frac{K_p}{(RT)^{\Delta n}}$$

$$R = 0.08206 \text{ dm}^3 \text{ atm K}^{-1} \text{ mol}^{-1}$$

Chemical Equilibria

$$T = 323$$
$$\Delta n = 4 - 2 = 2$$
$$K_c = ?$$

Substituting, we have

$$K_c = \frac{0.196}{(0.08206 \times 323)^2} = 2.79 \times 10^{-4}$$

Example 9

Phosphorus(V) chloride, PCl_5, dissociates as follows.

$$PCl_5(g) \rightleftharpoons PCl_3 + Cl_2(g)$$

At a certain temperature, 1.100 atm of PCl_5 was introduced into a reaction vessel and allowed to reach equilibrium. Given that the total pressure is 1.950 atm, determine:

(a) the equilibrium partial pressures of all gases;
(b) the K_p of the reaction at the prevailing temperature.

Solution

(a) We will begin by calculating the conversion of each gas at equilibrium from stoichiometry. From the equation of reaction, we can see that if x atm of PCl_5 is converted, x atm each of PCl_3 and Cl_2 will be formed. With this information, we can set up the ICE table as follows.

	PCl_5	PCl_3	Cl_2
I (atm)	1.100	0	0
C (atm)	$-x$	$+x$	$+x$
E (atm)	$1.100-x$	X	x

Since the total equilibrium pressure is 1.950 atm, it then follows that

$$(1.100 - x) + x + x = 1.950$$

$$\therefore x = 1.950 - 1.100 = 0.850 \text{ atm}$$

Thus, the equilibrium partial pressures are

$$P_{PCl_5} = (1.100 - 0.850) \text{ atm} = 0.250 \text{ atm}$$

$$P_{PCl_3} = P_{Cl_2} = 0.850 \text{ atm}$$

(b) The K_p expression is given by

$$K_P = \frac{P_{PCl_3} \times P_{N_2}}{P_{PCl_5}}$$

$$K_p = ?$$

Substituting, we have

$$K_P = \frac{0.850 \times 0.850}{0.250} = 2.89$$

Example 10

Consider the following gas-phase equilibrium:

$$2HI(g) \rightleftharpoons H_2(g) + I_2(g) \qquad K_c = 0.0207 \ (731K)$$

0.1000 mol dm^{-3} of HI was introduced into a reaction vessel at 731 K and allowed to reach equilibrium. Calculate the concentrations of all gases in the equilibrium mixture.

Solution

The K_c expression is given by

$$K_c = \frac{[I_2][H_2]}{[HI]^2}$$

$$K_c = 0.0207$$

The reaction stoichiometry shows that, if the concentration of HI changes by x mol dm^{-3}, $0.5x$ mol dm^{-3} each of H_2 and I_2 will be formed. We will use this information to set up the ICE table for the equilibrium as follows.

	HI	H_2	I_2
I (mol dm^{-3})	0.1000	0	0
C (mol dm^{-3})	$-2x$	$+x$	$+x$
E (mol dm^{-3})	$0.1000 - 2x$	X	x

Substituting we have

$$\frac{(x)^2}{(0.1000 - 2x)^2} = 0.0207$$

The left side of the equation is a perfect square; hence, we can take the square root of both sides to obtain

$$\frac{x}{(0.1000 - 2x)} = \pm 0.14387$$

So

$$0.014387 - 0.28774x = x \quad \text{or} \quad -0.014387 + 0.28774x = x$$

$$\therefore x = \frac{0.014387}{1.28774} = 0.0112 \text{ mol dm}^{-3} \quad \text{or} \quad x = -\frac{0.014387}{0.71226} = -0.0202 \text{ mol dm}^{-3}$$

The second value of x is not physically meaningful, as concentration cannot be negative. Thus, the admissible value of x is 0.0112 mol dm^{-3}. Finally, we can now substitute the value of x into the equilibrium concentrations to obtain the final values.

$$[HI] = (0.1000 - \{2 \times 0.0112\}) \text{ mol dm}^{-3} = 0.0776 \text{ mol dm}^{-3}$$

$$[H_2] = [I_2] = 0.0112 \text{ mol dm}^{-3}$$

Chemical Equilibria

Example 11

Consider the following gaseous equilibrium:

$$H_2(g) + CO_2(g) \rightleftharpoons H_2O(g) + CO(g) \qquad K_c = 1.60 \quad (1259K)$$

At 1259 K, the equilibrium concentrations of H_2, CO_2, H_2O and CO are 0.150 mol dm^{-3}, 0.0417 mol dm^{-3}, 0.100 mol dm^{-3} and 0.100 mol dm^{-3}, respectively. Determine the new equilibrium concentrations of all components if 0.100 mol dm^{-3} of CO is introduced into the equilibrium mixture.

Solution

The K_c expression for the reaction is as follows:

$$K_c = \frac{[H_2O][CO]}{[H_2][CO_2]}$$

$$K_c = 1.60$$

As we mentioned earlier, the equilibrium constant is independent of concentration. Thus, the equilibrium constant of the reaction will remain unchanged. However, the addition of CO to the equilibrium mixture will shift the position of equilibrium to the left, favouring the conversion of CO and H_2O into H_2 and CO_2.

According to the equation of reaction, if x mol dm^{-3} of CO is converted when its concentration increases by 0.100 mol dm^{-3}, then x mol dm^{-3} each of H_2 and CO_2 will be formed, while x mol dm^{-3} of H_2O will be consumed. This information will be used to set up the ICE table for the reaction as follows.

	H_2	CO_2	H_2O	CO
I (mol dm^{-3})	0.100	0.100	0.150	0.0417 + 0.100
C (mol dm^{-3})	+x	+x	−x	−x
E (mol dm^{-3})	0.100 + x	0.100 + x	0.150 − x	0.1417 − x

Substituting the K_c value and equilibrium concentrations into the K_c expression, we have

$$\frac{(0.150 - x)(0.1417 - x)}{(0.100 + x)(0.100 + x)} = 1.60$$

So

$$\frac{x^2 - 0.2917x + 0.021255}{x^2 + 0.2x + 0.01} = 1.60$$

This is a quadratic equation which when simplified gives the standard form $0.6x^2 + 0.6117x - 0.005255 = 0$. We can solve this equation, using the quadratic formula

$$x = \frac{-b \pm \sqrt{b^2 - 4ac}}{2a}$$

For the equation, $a = 0.6$, $b = 0.6117$, $c = -0.005255$.
Substituting, we have

$$x = \frac{-0.6117 \pm \sqrt{(0.6117)^2 - 4(0.6)(-0.005255)}}{2(0.6)}$$

$$\therefore x = 0.00850 \text{ mol dm}^{-3} \text{ or } x = -1.03 \text{ mol dm}^{-3}$$

The negative value has no physical significance, as conversion cannot be negative. Thus, $x = 0.00850$ mol dm^{-3}. Finally, we will substitute this value into the equilibrium concentrations in the ICE table to obtain the final values.

$$[H_2] = [CO_2] = (0.100 + 0.00850) \text{ mol dm}^{-3} = 0.1085 \text{ mol dm}^{-3}$$

$$[H_2O] = (0.150 - 0.0850) \text{ mol dm}^{-3} = 0.0650 \text{ mol dm}^{-3}$$

$$[CO] = (0.1417 - 0.0850) \text{ mol dm}^{-3} = 0.0579 \text{ mol dm}^{-3}$$

PRACTICE PROBLEMS

1. Write the expression for the K_c of the reaction Zn(s) + Pb^{2+}(aq) ⇌ Zn^{2+}(aq) + Pb(s).
2. Write the expression for the K_p of the reaction H$_2$(g) + I$_2$(g) ⇌ 2HI(g).
3. Consider the following gas-phase equilibrium system in a 1.50 dm^3 reaction vessel.

$$2NO(g) \rightleftharpoons N_2(g) + O_2(g)$$

   Calculate the K_c of the reaction at a temperature at which the equilibrium mixture contains 0.150 mol, 0.110 mol and 0.110 mol of NO, N$_2$ and O$_2$, respectively.
4. Consider the following gas-phase equilibrium system:

$$2SO_3(g) \rightleftharpoons 2SO_2(g) + O_2(g)$$

   2.00 atm of SO$_3$ was added to a reaction vessel at 973 K and allowed to reach equilibrium at a total pressure of 2.41 atm. Calculate:
   (a) the K_p for the reaction;
   (b) the K_c for the reaction.

$$\left(R = 0.08206 \text{ dm}^3 \text{ atm K}^{-1} \text{ mol}^{-1}\right)$$

11.4 REACTION QUOTIENT

Like we said earlier, equilibrium constant, K, provides information on the composition of an equilibrium mixture. However, to determine the direction of a reversible reaction before equilibrium is reached, the reaction quotient, Q, is used as follows.

If $Q = K$, reaction is at equilibrium.
If $Q < K$, the formation of products—i.e., the forward reaction—is favoured.
If $Q < K$, the formation of reactants—i.e., the reverse reaction—is favoured.

The reaction quotient is expressed in just the same way as equilibrium constant. Consider the following balanced general reversible reaction.

$$aA + bB \rightleftharpoons cC + dD$$

$$Q_c = \frac{[C]^c[D]^d}{[A]^a[B]^b}$$

$$Q_p = \frac{(P_C)^c(P_D)^d}{(P_A)^a(P_B)^b}$$

Chemical Equilibria

Example 1

The equilibrium constant, K_c, for the reaction $2PH_3(g) \rightleftharpoons P_2(g) + 3H_2(g)$ is 6.18×10^{-10} at 500 K. As the reaction proceeds to equilibrium, the concentrations of PH_3, P_2 and H_2 are measured to be 0.150 mol dm^{-3}, 0.0205 mol dm^{-3} and 0.0615 mol dm^{-3}, respectively. State the direction of the reaction.

Solution

To determine the direction in which the reaction will proceed, we need to calculate the reaction quotient, Q_c, and compare it to K_c. The expression for Q_c is given as follows:

$$Q_c = \frac{[P_2][H_2]^3}{[PH_3]^2}$$

$[PH_3] = 0.150$ mol dm^{-3}

$[P_2] = 0.0205$ mol dm^{-3}

$[H_2] = 0.0615$ mol dm^{-3}

$Q_c = ?$

Substituting, we have

$$Q_c = \frac{0.150 \times (0.0205)^3}{(0.0615)^2} = 3.42 \times 10^{-4}$$

The Q_c is larger than the K_c value of 6.18×10^{-10}; hence, the direction reaction is from right to left—i.e., the formation of PH_3 is favoured.

Example 2

Consider the following gas-phase equilibrium system:

$$2COCl_2(g) \rightleftharpoons CO(g) + Cl_2(g) \quad K_c = 0.201 \quad (600K)$$

Determine the equilibrium concentrations of all components if 2.90 mol dm^{-3}, 0.150 mol dm^{-3} and 0.100 mol dm^{-3} of $COCl_2$, CO and Cl_2, respectively, are introduced into a reaction vessel at 600 K.

Solution

Since all the components are initially present in the vessel, we cannot determine the direction from which the system will approach equilibrium without knowing the value of Q_c. The reaction quotient, Q_c, is expressed as follows:

$$Q_c = \frac{[CO][Cl_2]}{[COCl_2]^2}$$

$[COCl_2] = 2.90$ mol dm^{-3}

$[CO] = 0.150$ mol dm^{-3}

$[Cl_2] = 0.100$ mol dm^{-3}

$Q_c = ?$

Substituting, we have

$$Q_c = \frac{0.150 \times 0.100}{2.90} = 5.17 \times 10^{-3}$$

The value of Q_c is less than the K_c value of 0.201; hence, the reaction will approach equilibrium from left to right—i.e., $COCl_2$ will dissociate as the system approaches equilibrium.

According to the equation of reaction, x mol dm^{-3} each of CO and Cl_2 is formed for every x mol dm^{-3} of $COCl_2$ that dissociates. We will use this information to set up the ICE table for the system as follows.

	$COCl_2$	CO	Cl_2
I (mol dm^{-3})	2.90	0.150	0.100
C (mol dm^{-3})	$-x$	$+x$	$+x$
E (mol dm^{-3})	$2.90 - x$	$0.150 + x$	$0.100 + x$

The K_c for the reaction is expressed as follows:

$$K_c = \frac{[CO][Cl_2]}{[COCl_2]}$$

$$K_c = 0.201$$

Substituting the values into the K_c expression, we have

$$\frac{(0.150 + x)(0.100 + x)}{(2.90 - x)} = 0.201$$

So

$$\frac{x^2 + 0.25x + 0.015}{0.5829 - 0.201x} = 0.201$$

Simplifying, we have

$$x^2 + 0.451x - 0.5679 = 0$$

The quadratic formula will be used to solve this equation.

$$x = \frac{-b \pm \sqrt{b^2 - 4ac}}{2a}$$

$a = 1$, $b = 0.451$, $c = -0.5679$

Substituting, we have

$$x = \frac{-0.451 \pm \sqrt{(0.451)^2 - 4(1)(-0.5679)}}{2(1)}$$

$x = 0.561$ mol dm^{-3} or $x = -1.01$ mol dm^{-3}

Since conversion cannot be negative, we will discard the negative value; hence, $x = 0.561$ mol dm^{-3}. Substituting the value of x into the equilibrium concentrations in the table, we obtain

$$[COCl_2] = (2.90 - 0.561) \text{ mol dm}^{-3} = 2.34 \text{ mol dm}^{-3}$$

$$[CO] = (0.150 + 0.561) \text{ mol dm}^{-3} = 0.711 \text{ mol dm}^{-3}$$

$$[Cl_2] = (0.100 + 0.561) \text{ mol dm}^{-3} = 0.661 \text{ mol dm}^{-3}$$

Chemical Equilibria

PRACTICE PROBLEM

The K_p for the reaction $PCl_3(g) + Cl_2(g) \rightleftharpoons PCl_5(g)$ is 0.304 at 550 K. The initial partial pressures of PCl_3, Cl_2 and PCl_5 are 0.155 atm, 0.205 atm and 0.150 atm, respectively. Determine the partial pressures of all components when the reaction reaches equilibrium.

11.5 EQUILIBRIUM CONSTANT AND TEMPERATURE

The equilibrium constant of a reaction always remains unchanged as long as the temperature remains constant, irrespective of the presence of a catalyst or variations in concentration and pressure. The variation of equilibrium constant with temperature is given by

$$\ln \frac{K_2}{K_1} = -\frac{\Delta H^\ominus}{R}\left(\frac{1}{T_2} - \frac{1}{T_1}\right) \text{ or } \log \frac{K_2}{K_1} = -\frac{\Delta H^\ominus}{2.303R}\left(\frac{1}{T_2} - \frac{1}{T_1}\right)$$

where

$\Delta H^\ominus$ = the standard heat of the reaction
R = universal gas constant = 8.314 J mol^{-1} K^{-1}
K_1 = equilibrium constant at absolute temperature T_1
K_2 = equilibrium constant at temperature T_2

This equation is called the *van't Hoff Equation*.

Example 1

Consider the Haber process equilibrium system:

$$N_2(g) + 3H_2(g) \rightleftharpoons 2NH_3(g) \qquad \Delta H^\ominus = -92.0 \text{ kJ mol}^{-1}$$

Determine the K_p for the reaction at 523 K, given that the K_p is 2.20×10^4 at 328 K.

$$\left(R = 8.314 \text{ J mol}^{-1} \text{ K}^{-1}\right)$$

Solution

We will apply the van't Hoff Equation as follows.

$$\ln \frac{K_2}{K_1} = -\frac{\Delta H^\ominus}{R}\left(\frac{1}{T_2} - \frac{1}{T_1}\right)$$

$K_1 = 2.20 \times 10^4$

$\Delta H^\ominus = -92.0 \text{ kJ mol}^{-1} = -92000 \text{ J mol}^{-1}$

$T_1 = 328 \text{ K}$

$T_2 = 523 \text{ K}$

$R = 8.314 \text{ J mol}^{-1} \text{ K}^{-1}$

$K_2 = ?$

Substituting, we have

$$\ln\frac{K_2}{2.20\times 10^4} = -\frac{\left(-92000\ \text{J mol}^{-1}\right)\times 1\text{K}}{8.314\ \text{J mol}^{-1}}\left(\frac{1}{523\ \text{K}} - \frac{1}{328\ \text{K}}\right)$$

So

$$\ln\frac{K_2}{2.20\times 10^4} = -12.5787$$

Taking the natural logarithm of both sides, we have

$$\frac{K_2}{2.20\times 10^4} = e^{-12.5787}$$

$$\therefore K_2 = 2.20\times 10^4 \times 3.44\times 10^{-6} = 0.0758$$

Thus, the equilibrium constant of an exothermic reaction decreases with increasing temperature, favouring the reverse reaction or direction of endothermic change.

Example 2

The values of the equilibrium constant, K_c, for the equilibrium system $2Hg(l) + O_2(g) \rightleftharpoons 2HgO(s)$ are 2.25×10^{11} and 1.94×10^9 at 450 K and 560 K, respectively. Determine the standard enthalpy change, $\Delta H^\ominus$, of the reaction.

$$\left(R = 8.314\ \text{J mol}^{-1}\ \text{K}^{-1}\right)$$

Solution

We will apply the van't Hoff Equation:

$$\ln\frac{K_2}{K_1} = -\frac{\Delta H^\ominus}{R}\left(\frac{1}{T_2} - \frac{1}{T_1}\right)$$

$$K_1 = 2.25 \times 10^{11}$$

$$K_2 = 1.94 \times 10^9$$

$$T_1 = 450\text{K}$$

$$T_2 = 560\text{K}$$

$$R = 8.314\ \text{J mol}^{-1}\ \text{K}^{-1}$$

$$\Delta H^\ominus = ?$$

Substituting, we have

$$\ln\frac{1.94\times 10^9}{2.25\times 10^{11}} = -\frac{\Delta H^\ominus \times 1\text{K}}{8.314\ \text{J mol}^{-1}}\left(\frac{1}{560\ \text{K}} - \frac{1}{450\ \text{K}}\right)$$

So

$$-4.753 = \frac{4.365 \times 10^{-4} \Delta H^\theta}{8.314 \text{ J mol}^{-1}}$$

$$4.365 \times 10^{-4} \Delta H^\theta = -39.516 \text{ J mol}^{-1}$$

$$\therefore \Delta H^\theta = \frac{-39.516 \text{ J mol}^{-1}}{4.365 \times 10^{-4}} = -90.5 \text{ kJ mol}^{-1}$$

PRACTICE PROBLEMS

1. The K_c for the equilibrium system $2BrF_3(g) \rightleftharpoons Br_2(g) + 3F_2(g)$ is 1.10×10^{-19} at 1000 K. Determine K_p for the reaction at 1200 K, given that the standard enthalpy change, ΔH^0, of the reaction is 811.5 kJ mol^{-1}.

$$\left(R = 8.314 \text{ J mol}^{-1} \text{ K}^{-1}\right)$$

2. Consider the equilibrium system for the decomposition of $N_2O_4(g)$.

$$N_2O_4(g) \rightleftharpoons 2NO_2(g)$$

Given that the K_c values for the system are 0.125 and 3.19 at 25°C and 77°C, respectively. Determine the standard enthalpy change, ΔH^0, of the reaction.

$$\left(R = 8.314 \text{ J mol}^{-1} \text{ K}^{-1}\right)$$

11.6 SOLUBILITY PRODUCT

Equilibrium is established between a solute and its ions in a saturated solution. For instance, barium chloride, $BaCl_2$, dissolves in water to yield 1 mol of barium ions, Ba^{2+}, and 2 mol of chloride ions, Cl^-, which establishes the following equilibrium on saturation.

$$BaCl_2(s) \rightleftharpoons Ba^{2+}(aq) + 2Cl^-(aq)$$

Just like reversible reactions at equilibrium, we can determine the equilibrium constant for a solute in its saturated solution. This type of equilibrium constant is called *solubility product*, K_{sp}.

Consider the following equilibrium established between a general solute MN and its ions, M$^+$ and N$^-$, in a saturated solution.

$$MN(s) \rightleftharpoons mM^+(aq) + nN^-(aq)$$

The solubility product, K_{sp}, for the solute is expressed as follows.

$$K_{sp} = \frac{[M^+]^m [N^-]^n}{[MN]}$$

Since equilibrium constant expressions are written without the concentrations of pure solids, it follows that

$$K_{sp} = [M^+]^m [N^-]^n$$

Thus, the solubility product of a substance is expressed as the product of the concentrations of its ions in a saturated solution, each raised to its respective stoichiometric coefficients, as given by the dissolution equation of the substance. Being an equilibrium constant, the solubility product varies with temperature. The standard temperature for reporting solubility product is 25°C.

Example 1

250.0 cm³ of a saturated solution of silver chloride contains 2.50×10^{-7} mol of the salt at a particular temperature. Calculate:

(a) the molar concentration of the saturated solution;
(b) the solubility product of the salt at the prevailing temperature.

Solution

(a) The molar concentration of the solution is given by

$$C = \frac{n}{V}$$

$$n = 2.5 \times 10^{-7} \text{ mol}$$

$$V = 250.0 \text{ cm}^3 = 0.250 \text{ dm}^3$$

$$C = ?$$

Substituting, we have

$$C = \frac{2.50 \times 10^{-7} \text{ mol}}{0.25 \text{ dm}^3} = 1.00 \times 10^{-6} \text{ mol dm}^{-3}$$

(b) The dissolution equilibrium of AgCl:

$$AgCl(s) \rightleftharpoons Ag^+(aq) + Cl^-(aq)$$

$$\therefore K_{sp} = [Ag^+][Cl^-]$$

According to the equation, 1 mol of AgCl produces 1 mol each of Ag^+ and Cl^-; hence,

$$[AgCl] = [Ag^+] = [Cl^-] = 1.00 \times 10^{-6} \text{ mol dm}^{-3}$$

$$K_{sp} = ?$$

Substituting, we have

$$K_{sp} = (1.00 \times 10^{-6} \text{ mol dm}^{-3}) \times (1.00 \times 10^{-6} \text{ mol dm}^{-3}) = 1.00 \times 10^{-12}$$

Chemical Equilibria

Example 2

A saturated solution of silver carbonate, Ag_2CO_3, contains 0.00800 g of the salt per 250.0 cm³ of solution at a certain temperature. Determine:

(a) the molar concentration of the saturated solution;
(b) the solubility product of the salt at the prevailing temperature.

$$(C = 12, \ O = 16, \ Ag = 108)$$

Solution

(a) We will begin by calculating the number of moles of the solute as follows.

$$n = \frac{m}{M}$$

$$m = 0.00800 \text{ g}$$

$$M = (\{108 \times 2\} + 12 + \{16 \times 3\}) \text{ g mol}^{-1} = 276 \text{ g mol}^{-1}$$

$$n = ?$$

Substituting, we have

$$n = \frac{0.00800 \text{ g} \times 1 \text{ mol}}{276 \text{ g}} = 2.90 \times 10^{-5}$$

We can now determine the molar concentration as follows.

$$C = \frac{n}{V}$$

$$V = 250.0 \text{ cm}^3 = 0.250 \text{ dm}^3$$

$$C = ?$$

Substituting, we have

$$C = \frac{2.90 \times 10^{-5} \text{ mol}}{0.250 \text{ dm}^3} = 1.16 \times 10^{-4} \text{ mol dm}^{-3}$$

Alternatively, the molar concentration can be calculated as follows.

$$\rho = \frac{m}{V}$$

$$m = 0.00800 \text{ g}$$

$$V = 250.0 \text{ cm}^3 = 0.250 \text{ dm}^3$$

$$\rho = ?$$

Substituting, we have

$$\rho = \frac{0.00800 \text{ g}}{0.250 \text{ dm}^3} = 0.0320 \text{ g}$$

We now apply the relation:

$$C = \frac{\rho}{M}$$

$$M = 276 \text{ g mol}^{-1}$$

$$C = ?$$

Substituting, we obtain

$$C = \frac{0.320 \text{ g} \times 1 \text{ mol}}{276 \text{ g}} = 1.16 \times 10^{-4} \text{ mol dm}^{-3}$$

(b) The dissolution equilibrium of Ag_2CO_3 is as follows:

$$Ag_2CO_3(s) \rightleftharpoons 2Ag^+(aq) + CO_3^{2-}(aq)$$

$$\therefore K_{sp} = [Ag^+]^2 [CO_3^{2-}]$$

According to the equation, 1 mol of Ag_2CO_3 produces 2 mol and 1 mol of Ag^+ and CO_3^{2-}, respectively. Hence,

$$[Ag^+] = 2[Ag_2CO_3] = 2 \times (1.16 \times 10^{-4} \text{ mol dm}^{-3}) = 2.32 \times 10^{-4} \text{ mol dm}^{-3}$$

$$[CO^{2-}] = [Ag_2CO_3] = 1.16 \times 10^{-4} \text{ mol dm}^{-3}$$

$$K_{sp} = ?$$

Substituting, we have

$$K_{sp} = (2.32 \times 10^{-4} \text{ mol dm}^{-3})^2 \times 1.16 \times 10^{-4} \text{ mol dm}^{-3} = 6.24 \times 10^{-12}$$

Example 3

The K_{sp} of copper(II) bromide, $CuBr_2$, is 4.20×10^{-8} mol³ dm⁻⁹ at 25°C. Determine the solubility of the salt in mol dm⁻³ at this temperature.

Solution

The dissolution equilibrium of $CuBr_2$ is as follows:

$$CuBr_2(s) \rightleftharpoons Cu^+(aq) + 2Br^-(aq)$$

$$\therefore K_{sp} = [Cu^+][Br^-]^2$$

$$K_{sp} = 4.20 \times 10^{-8}$$

Recall that the molar solubility, S, of a substance at a particular temperature is the molar concentration of its saturated solution at that temperature. Hence,

$$[Cu^+] = [CuBr] = S$$

$$[Br^-] = 2[CuBr] = 2S$$

Chemical Equilibria

Substituting the values into the K_{sp} expression, we have

$$4.20 \times 10^{-8} \text{ mol}^3 \text{ dm}^{-9} = S \times (2S)^2$$

So

$$4S^3 = 4.20 \times 10^{-8} \text{ mol}^3 \text{ dm}^{-9}$$

Dividing both sides by 4 and then taking the cube root, we obtain

$$S = \sqrt[3]{\frac{4.20 \times 10^{-8} \text{ mol}^3 \text{ dm}^{-9}}{4}} = 0.00219 \text{ mol dm}^{-3}$$

PRACTICE PROBLEMS

1. 150 cm³ of a saturated solution of magnesium hydroxide, $Mg(OH)_2$, contains 0.0015 g of the base at 25°C. Calculate:
   (a) the molar solubility of the base at 25°C;
   (b) the solubility product of the base at 25°C.

$$(H = 1,\ O = 16,\ Mg = 24)$$

2. Calculate the solubility of iron(II) hydroxide, $Fe(OH)_2$, in 100 g of H_2O at a temperature at which the solubility product, K_{sp}, is 2.0×10^{-39} mol³ dm⁻⁹.

$$(H = 1,\ O = 16,\ Fe = 56)$$

11.7 ION PRODUCT

The *ion product*, Q_{sp}, is a quantity in solubility equilibria that is analogous to the reaction quotient, Q. Consider the dissolution of a general solute, as shown below.

$$MN(s) \rightleftharpoons mM^+(aq) + nN^-(aq)$$

The Q_{sp} of the solution is expressed in just the same way as the K_{sp}:

$$Q_{sp} = [M^+]^m [N^-]^n$$

The Q_{sp} value of a solution provides the following information about the solution:

If $Q_{sp} = K_{sp}$, solution is at equilibrium—i.e., solution is saturated.
If $Q_{sp} < K_{sp}$, solution is unsaturated—i.e., more solute dissolves.
If $Q_{sp} > K_{sp}$, the solute precipitates—i.e., no more solute dissolves.

Example 1

Calculate the ion product, Q_{sp}, of a calcium hydroxide, $Ca(OH)_2$, solution containing 1.50 g of the solute in 2.50 dm³ of solution. Predict whether more of the solute will dissolve in the solution, given that its K_{sp} is 5.02×10^{-6}.

$$(H = 1,\ O = 16,\ Ca = 40)$$

Solution

We will begin by calculating the molar concentration of the solution.

$$C = \frac{n}{V}$$

The number of moles of $Ca(OH)_2$ is obtained as follows.

$$n = \frac{m}{M}$$

$$m = 1.50 \text{ g}$$

$$M = \left(40 + \{16 \times 2\} + \{1 \times 2\}\right) \text{ g mol}^{-1} = 74 \text{ g mol}^{-1}$$

$$n = ?$$

$$\therefore n = \frac{1.50 \text{ g} \times 1 \text{ mol}}{74 \text{ g}} = 0.02027 \text{ mol}$$

$$V = 2.50 \text{ dm}^3$$

$$C = ?$$

Substituting, we have

$$C = \frac{0.02027 \text{ mol}}{2.50 \text{ dm}^3} = 0.00811 \text{ mol dm}^{-3}$$

The dissolution equilibrium of $Ca(OH)_2$ is as follows:

$$Ca(OH)_2 (s) \rightleftharpoons Ca^{2+} (aq) + 2OH^- (aq)$$

$$Q_{sp} = \left[Ca^{2+}\right]\left[OH^-\right]^2$$

According to the equation, 1 mol of $Ca(OH)_2$ produces 1 mol of Ca^{2+} and 2 mol of OH^-. It then follows that

$$\left[Ca(OH)_2\right] = \left[Ca^{2+}\right] = 0.00811 \text{ mol dm}^{-3}$$

$$\left[OH^-\right] = 2\left[Ca(OH)_2\right] = 0.0162 \text{ mol dm}^{-3}$$

$$Q_{sp} = ?$$

Substituting, we have

$$Q_{sp} = 0.00811 \times (0.0162)^2 = 2.13 \times 10^{-6}$$

Since the Q_{sp} (2.13×10^{-6}) < K_{sp} (5.02×10^{-6}), then the solution is unsaturated and more of the solute will dissolve.

Chemical Equilibria

Example 2

Determine if barium carbonate, $BaCO_3$, will precipitate when 20.00 cm³ of a 0.400 mol dm⁻³ barium sulfate, $BaSO_4$, solution is added to 100.00 cm³ of a 0.250 mol dm⁻³ silver carbonate, Ag_2CO_3, solution. The K_{sp} for $BaCO_3$ is 8.1×10^{-9}.

Solution

We will calculate the Q_{sp} of $BaCO_3$ and compare it with the K_{sp}. The equation for the solubility equilibrium of the solute is as follows:

$$BaCO_3(s) \rightleftharpoons Ba^{2+}(aq) + CO_3^{2-}(aq)$$

$$Q_{sp} = [Ba^{2+}][CO_3^{2-}]$$

The amount of Ba^{2+} and CO_3^{2-} in the resulting mixture corresponds to the amount originally present in the $BaSO_4$ and Ag_2CO_3 solutions, respectively. We can determine this as follows.

$$n = C \times V$$

For $BaSO_4$, we have

$$BaSO_4(s) \rightleftharpoons Ba^{2+}(aq) + SO_4^{2-}(aq)$$

$$\therefore [BaSO_4] = [Ba^{2+}] = 0.400 \text{ mol dm}^{-3}$$

$$V = 20.00 \text{ cm}^3 = 0.0200 \text{ dm}^3$$

$$n = ?$$

Substituting, we have

$$n = \frac{0.400 \text{ mol}}{1 \text{ dm}^3} \times 0.0200 \text{ dm}^3 = 0.00800 \text{ mol}$$

For Ag_2CO_3, we have

$$Ag_2CO_3(s) \rightleftharpoons 2Ag^+(aq) + CO_3^{2-}(aq)$$

$$\therefore [Ag_2CO_3] = [CO_3^{2-}] = 0.250 \text{ mol dm}^{-3}$$

$$V = 100.00 \text{ cm}^3 = 0.100 \text{ dm}^3$$

$$n = ?$$

Substituting, we have

$$n = \frac{0.250 \text{ mol}}{1 \text{ dm}^3} \times 0.100 \text{ dm}^3 = 0.0250 \text{ mol}$$

The concentration of each ion in the resulting mixture is determined as follows.

$$C = \frac{n}{V}$$

$$V = 20.00 \text{ cm}^3 + 100.00 \text{ cm}^3 = 120.00 \text{ cm}^3 = 0.120 \text{ dm}^3$$

$$[Ba^{2+}] = ?$$

$$[CO_3^{2-}] = ?$$

For Ba^{2+}, we have

$$[Ba^{2+}] = \frac{0.00800 \text{ mol}}{0.120 \text{ dm}^3} = 0.0667 \text{ mol dm}^{-3}$$

For $[CO_3^{2-}]$, we have

$$[CO_3^{2-}] = \frac{0.0250 \text{ mol}}{0.120 \text{ dm}^3} = 0.208 \text{ mol dm}^{-3}$$

Substituting the concentrations into the Q_{sp} expression, we have

$$Q_{sp} = 0.0667 \times 0.208 = 0.0139$$

Since the Q_{sp} (0.0139) > K_{sp} (8.1 × 10^{-9}), then a precipitate will form when the solutions mix.

PRACTICE PROBLEMS

1. Predict whether more of barium sulfate, $BaSO_4$, will dissolve in a solution containing 4.50 g of the salt per dm³ of solution. The K_{sp} for $BaSO_4$ is 2.40 × 10^{-5}.

$$(O = 16, \ S = 32, \ Ba = 137)$$

2. Will a precipitate form when 25.00 cm³ of a 0.00120 mol dm^{-3} copper(I) iodide, CuI, solution is added to 100.00 cm³ of a 0.200 mol dm^{-3} solution of silver bromide, AgBr? The K_{sp} for copper(I) bromide, CuBr, is 4.2 × 10^{-8}.

11.8 SOLUBILITY AND THE COMMON-ION EFFECT

When a substance is added to a solution containing a common ion, the solubility equilibrium will shifts to the left or the direction of the undissolved solute, decreasing its solubility. This phenomenon is termed the common-ion effect.

Example 1

The solubility product, K_{sp}, of calcium hydroxide, $Ca(OH)_2$, is 5.02 × 10^{-6}. At what pH will $Ca(OH)_2$ just start to precipitate from a 0.00110 mol dm^{-3} calcium chloride, $CaCl_2$, solution.

Solution

The very first step is to determine the molar concentration of hydroxide ions, OH$^-$, in the resulting saturated solution of $Ca(OH)_2$.

The solubility equilibrium of $Ca(OH)_2$ is as follows:

$$Ca(OH)_2(s) \rightleftharpoons Ca^{2+}(aq) + 2OH^-(aq)$$

$$K_{sp} = [Ca^{2+}][OH^-]^2 = 5.02 \times 10^{-6}$$

Chemical Equilibria

The solubility equilibrium of $CaCl_2$ is as follows:

$$CaCl_2(s) \rightleftharpoons Ca^{2+}(aq) + 2Cl^-(aq)$$

According to the equation,

$$[CaCl_2] = [Ca^{2+}] = 0.00110 \text{ mol dm}^{-3}$$
$$[OH^-] = ?$$

Substituting into the K_{sp} expression for $Ca(OH)_2$, we have

$$0.00110 \text{ mol dm}^{-3} \times [OH^-]^2 = 5.02 \times 10^{-6}$$

$$\therefore [OH^{-1}] = \sqrt{\frac{5.02 \times 10^{-6}}{0.00110}} = 0.06755 \text{ mol dm}^{-3}$$

Now, we will determine the pOH using the relation:

$$pOH = -\log[OH^-]$$

Substituting, we have

$$pOH = -\log[0.06755] = 1.17$$

But

$$pH + pOH = 14.00$$
$$\therefore pH = 14.00 - 1.17 = 12.83$$

Example 2

Determine the molar solubility of silver chloride, AgCl, in a 0.150 mol dm^{-3} sodium chloride, NaCl, solution. The K_{sp} for AgCl is 1.60×10^{-10}.

Solution

The solubility equilibrium of $PbCl_2$ is as follows:

$$AgCl(s) \rightleftharpoons Ag^+(aq) + Cl^-(aq)$$

$$K_{sp} = [Ag^+][Cl^-] = 1.60 \times 10^{-10}$$

The solubility equilibrium of NaCl is as follows:

$$NaCl(s) \rightleftharpoons Na^+(aq) + Cl^-(aq)$$

The common ion is Cl$^-$. According to the first equilibrium, if the molar concentration of dissolved Ag$^+$ is x, then the molar concentration of Cl$^-$ from AgCl will also be x. Similarly, from the second equilibrium, the concentration of Cl$^-$ from the 0.150 mol dm^{-3} NaCl solution is 0.150 mol dm^{-3}. We will use this information to set up the ICE table for the system as follows:

	Ag⁺	Cl⁻
I (mol dm⁻³)	0	0.150
C (mol dm⁻³)	+x	+x
E (mol dm⁻³)	x	0.150 + x

Substituting the equilibrium concentrations into the K_{sp} expression, we have

$$x \times (0.150 + x) = 1.60 \times 10^{-10}$$

So

$$x^2 + 0.15x - 1.60 \times 10^{-10} = 0$$

$$x = \frac{-b \pm \sqrt{b^2 - 4ac}}{2a}$$

$$a = 1,\ b = 0.15,\ c = -1.60 \times 10^{-10}$$

Substituting, we have

$$x = \frac{-0.15 \pm \sqrt{(0.15)^2 - 4(1)(-1.60 \times 10^{-10})}}{2(1)}$$

$\therefore x = -0.150$ mol dm⁻³ or $x = 1.07 \times 10^{-9}$ mol dm⁻³

The negative value is inadmissible because concentration cannot be negative. Thus, $x = 1.07 \times 10^{-9}$ mol dm⁻³. Since, according to the equation, [AgCl] = [Ag⁺], the solubility, S, of AgCl is:

$$x = 1.07 \times 10^{-9} \text{ mol dm}^{-3}.$$

PRACTICE PROBLEMS

1. Calculate the number of moles of silver carbonate, Ag_2CO_3, that will dissolve in 2.50 dm³ of a 0.010 mol dm⁻³ calcium carbonate, $CaCO_3$ solution. The K_{sp} for Ag_2CO_3 is 6.20×10^{-12}.
2. The K_{sp} for copper(II) hydroxide, $Cu(OH)_2$, is 2.20×10^{-20}. Calculate the pH at which $Cu(OH)_2$ will just start to precipitate from a solution containing 0.0015 mol dm⁻³ of copper(II) ions, Cu^{2+}.

FURTHER PRACTICE PROBLEMS

1. (a) Explain chemical equilibrium.
   (b) What is a reversible reaction? Give two examples.
2. Explain Le Châtelier's principle.
3. State the effect of increasing the concentration of the substance in bold on each equilibrium.
   (a) **2NO₂(g)** ⇌ $N_2O_4(g)$
   (b) **3Fe(s)** + 4H₂O(g) ⇌ $Fe_3O_4(s) + 4H_2(g)$
4. State the effect of reducing the concentration of the substance in bold on each equilibrium.
   (a) $N_2(g) + $ **O₂(g)** ⇌ 2NO(g)
   (b) 3Fe(s) + **4H₂O(g)** ⇌ $Fe_3O_4(s) + 4H_2(g)$

Chemical Equilibria

5. State the effect of temperature increase on the following equilibrium systems.
   (a) $N_2(g) + O_2(g) \rightleftharpoons 2NO(g)$ $\Delta H = +90.4$ kJ mol^{-1}
   (b) $2SO_2(g) + O_2(g) \rightleftharpoons 2SO_3(g)$ $\Delta H = -395.7$ kJ mol^{-1}

6. How would an increase in pressure affect the following reactions at equilibrium?
   (a) $N_2(g) + O_2(g) \rightleftharpoons 2NO$
   (b) $2H_2(g) + O_2(g) \rightleftharpoons 2H_2O(g)$

7. Write the K_c and K_p expressions for each of the following reactions.
   (a) $2NO(g) \rightleftharpoons N_2O_4(g)$
   (b) $2SO_2(g) + O_2(g) \rightleftharpoons 2SO_3(g)$
   (c) $N_2(g) + O_2(g) \rightleftharpoons 2NO(g)$
   (d) $CH_3COOH(aq) + NaOH(aq) \rightleftharpoons CH_3COONa(aq) + H_2O(l)$.
   (e) $PCl_3(g) + Cl_2(g) \rightleftharpoons PCl_5(g)$
   (f) $HBrO(aq) + H_2O(l) \rightleftharpoons H_3O^+(aq) + BrO^-(aq)$

8. The K_c for the equilibrium system $PCl_5(g) \rightleftharpoons PCl_3(g) + Cl_2(g)$ is 0.800 at 613 K. Determine:
   (a) the K_p for the reaction;
   (b) the K_p for the reverse reaction.

$$\left(R = 0.08206 \text{ dm}^3 \text{ atm K}^{-1} \text{ mol}^{-1}\right)$$

9. Consider the following equilibrium system:

$$3Fe(s) + 4H_2O(g) \rightleftharpoons Fe_3O_4(s) + 4H_2(g)$$

At a certain temperature, a 36.0 g sample of H_2O and an unknown amount of Fe were introduced into a 2.50 dm^3 reaction vessel, and the reaction was allowed to reach equilibrium. Determine the K_c for the reaction if the equilibrium concentration of H_2 is 0.650 mol dm^{-3}.

$$(H = 1, O = 16)$$

10. Consider the following equilibrium.

$$2NH_3(g) \rightleftharpoons N_2(g) + 3H_2(g)$$

Initially, 0.250 atm of NH_3 was introduced into a reaction vessel at a certain temperature. Determine the K_p of the system if 1.00 atm of NH_3 is left in the vessel at equilibrium.

11. Consider the equilibrium system.

$$CH_3OH(g) \rightleftharpoons CO(g) + 2H_2(g)$$

0.0273 g of CH_3OH is added to a 1.75 dm^3 reaction vessel at 700 K, and the dissociation is allowed to reach equilibrium. Determine the K_p for the reaction given that the partial pressure of CH_3OH in the equilibrium mixture is 0.0200 atm.

$$(H = 1, C = 12, O = 16, R = R = 0.08206 \text{ dm}^3 \text{ atm K}^{-1} \text{ mol}^{-1})$$

12. The K_c for the equilibrium $H_2(g) + CO_2(g) \rightleftharpoons H_2O(g) + CO(g)$ is 1.60 at 1260 K. Initially, 0.150 mol dm^3 of H_2 was mixed with 0.250 mol dm^3 of CO_2. Calculate the equilibrium concentrations of all components.

13. The K_c for the reaction $H_2(g) + I_2(g) \rightleftharpoons 2HI(g)$ is 45.9 at 763 K. The initial concentrations of H_2, I_2 and HI in a reaction vessel are 0.150 mol dm^{-3}, 0.200 mol dm^{-3} and 0.250 mol dm^{-3}, respectively. Determine the concentration of all components at equilibrium.

14. Consider the following gas-phase equilibrium.

$$PCl_5(g) \rightleftharpoons PCl_3(g) + Cl_2(g)$$

Calculate the K_c for the reaction at 300 K, given that PCl_5 was found to be 14.32% dissociated when the system reached equilibrium at this temperature.

15. The K_p for the Haber equilibrium $N_2(g) + 3H_2(g) \rightleftharpoons 2NH_3(g)$ is 2.15×10^{-5} at 800 K. Briefly comment on the direction of reaction when 10.0 atm, 15.5 atm and 12.2 atm of nitrogen, hydrogen and ammonia, respectively, are added to a reaction vessel at 800 K.

16. The K_c for the equilibrium system $2NO(g) \rightleftharpoons N_2(g) + O_2(g)$ is 1.00×10^5 at 1500 K. At this temperature, 0.450 mol dm^{-3} of NO was added to a reaction vessel and allowed to equilibrate. Determine the percent dissociation of NO at equilibrium.

17. The K_p for the dissociation $N_2O_4(g) \rightleftharpoons 2NO_2(g)$ is 3.06 at 298 K. 1.05 g of N_2O_4 was added to a 1.75-dm^3 reaction vessel containing 0.764 atm and 1.53 atm of N_2O_4 and NO_2, respectively, at equilibrium. Determine the new equilibrium partial pressures of the components.

$$\left(N = 14,\ H = 1,\ R = 0.08206\ dm^3\ atm\ K^{-1}\ mol^{-1}\right)$$

18. Consider the equilibrium.

$$COCl_2(g) \rightleftharpoons CO(g) + Cl_2(g)$$

At a certain temperature, 10.0 atm of $COCl_2$ was added to a reaction vessel and allowed to reach equilibrium at a total pressure of 18.5 atm. Determine:
(a) the equilibrium partial pressures of all components;
(b) the K_p for the reaction.

19. Consider the equilibrium.

$$2CO(g) \rightleftharpoons C(s) + CO_2(g) \qquad \Delta H^\ominus = -172.5\ kJ\ mol^{-1}$$

Determine the K_c at 1100 K if the K_c is 7.70×10^{-15} at 1120 K.

$$\left(R = 8.314\ J\ mol^{-1}\ K^{-1}\right)$$

20. The K_p values for the reaction $COCl_2(g) \rightleftharpoons CO(g) + Cl_2(g)$ are 7.41×10^{-5} and 4.09×10^{-3} at 550 K and 600 K, respectively. Determine the standard enthalpy change of the reaction.

$$\left(R = 8.314\ J\ mol^{-1} K^{-1}\right)$$

21. A saturated solution of lead(II) iodide, PbI_2, contains 0.54 g of the salt per 100.0 cm^3 of solution at 25°C. Determine the solubility product of the salt at the given temperature.

$$(Pb = 207,\ I = 127)$$

22. The solubility product of calcium fluoride, CaF_2, is 3.70×10^{-11} mol^3 dm^{-9} at 25°C. Determine:
   (a) the molar solubility of the salt at 25°C;
   (b) the solubility of the salt in g dm^{-3} of H_2O at 25°C.

$$(F = 17, \ Ca = 40)$$

23. What volume of a 0.00500 mol dm^{-3} silver nitrate solution, $AgNO_3$, is required to precipitate silver chloride, $AgCl$, from a 0.00550 mol dm^{-3} barium chloride, $BaCl_2$, solution. The K_{sp} for $AgCl$ is 1.60×10^{-10}.

24. Determine the pH at which silver hydroxide, $AgOH$, will start to precipitate from a 0.0500 mol dm^{-3} $AgNO_3$ solution. The K_{sp} for $AgOH$ is 1.1×10^{-4}.

12 Electrochemistry

12.1 METAL ION-METAL SYSTEM

Some metals undergo oxidation when dipped into solutions of their salts. For example, when a zinc plate is dipped into a solution of zinc sulfate, $ZnSO_4$, zinc atoms from the plate lose two electrons each to the plate and move into the solution as positively charged zinc ions, Zn^{2+} (see Figure 12.1):

$$Zn(s) \rightarrow Zn^{2+}(aq) + 2e^-$$

In other words, the zinc plate has been oxidised. In time, the number of negative charge on the plate and the number of positive charges in solution remain constant. At this point, we say equilibrium has been established between the zinc plate and its ions in solution. This equilibrium is denoted as follows.

$$Zn^{2+}(aq) + 2e^- \rightleftharpoons Zn(s)$$

Note: Metal-metal ion equilibrium is written as a reduction process—i.e., electrons are written on the left side of the equation.

In contrast, some metals such as copper will undergo reduction when dipped into solutions of their salts. For instance, when a copper plate is inserted into copper(II) sulfate solution, copper(II) ions, Cu^{2+}, from the solution will take up two electrons each from the surface of the plate and get deposited on the plate as neutral copper atoms (see Figure 12.2):

$$Cu^{2+}(aq) + 2e^- \rightarrow Cu(s)$$

In other words, the copper(II) ions have been reduced. With time, the following equilibrium is established between the copper plate and its ions in solution.

$$Cu^{2+}(aq) + 2e^- \rightleftharpoons Cu(s)$$

The two systems we have just described are called *metal ion-metal systems* or *half-cells*. The solution in each half-cell is called an *electrolyte*. The metal plates are called *electrodes*.

12.1.1 Electrode Potential

In a half-cell, a potential difference, known as *electrode potential (E)* or *voltage*, is established between the electrode and the electrolyte. The unit of electrode potential is the volt (V).

To measure the electrode potential of a half-cell, the half-cell must be coupled with the standard hydrogen electrode (SHE). A *voltmeter*—an instrument for measuring voltage or potential difference—is then connected to the resulting system as shown in Figure 12.3. Electrode potential varies with the concentration of ions in an electrolyte, temperature, pressure (for gaseous systems) and the nature of the electrode. The standard electrode potential, E^0, of a half-cell is the potential difference measured when the electrode is in contact with 1 mol dm^{-3} solution of its ions at 25°C.

By convention, the electrode potential of a cell is assigned a positive value if reduction occurs in the half-cell. For example, the standard electrode (reduction) potential of the half-cell $Cu^{2+}(aq)/$

Electrochemistry

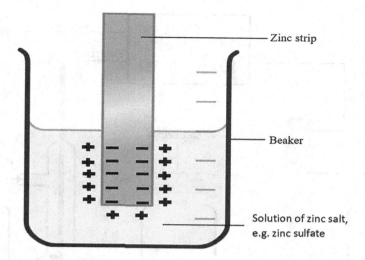

FIGURE 12.1 Zinc–zinc ion half-cell.

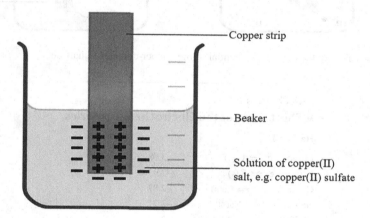

FIGURE 12.2 Copper–copper ion half-cell.

Cu(s) is +0.34 V. The values of the standard electrode potential listed in tables are all reduction potentials. This means that, if oxidation occurs in a half-cell, its electrode potential is assigned a negative sign, signifying that reduction is nonspontaneous. For example, the standard electrode potential of the half-cell Zn^{2+}(aq)/Zn(s) is −0.76 V.

A list of different half-cells and their standard electrode (reduction) potentials arranged in ascending order of magnitude is called *electrochemical series*. In the electrochemical series, half-cells with the most negative standard electrode potentials—i.e., the strong reducing agents—are listed at the top, while those with the most positive standard electrode potentials—i.e., the strong oxidising agent are listed at the bottom. A short form of the electrochemical series is given in Table 12.1. A longer list is given in Appendix F.

12.1.2 Electrochemical Cell

If two half-cells are coupled together, electrons will flow from the cell with the smaller potential to the one with the higher potential, producing an electric current. This device or system in which chemical energy is converted to electrical energy is called an *electrochemical cell* or *voltaic cell* or *galvanic cell*.

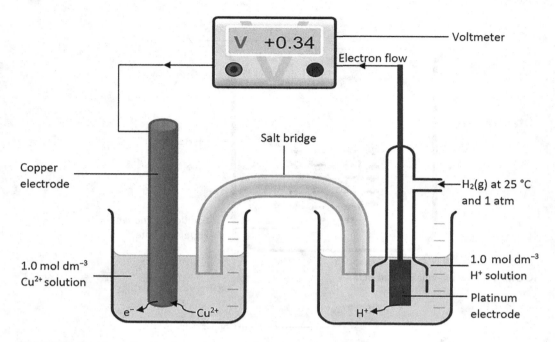

FIGURE 12.3 Measuring the electrode potential of the copper-copper ion half-cell.

TABLE 12.1
A Short Form of the Electrochemical Series

Half-Cell	E^0 (V)
$Li^+(aq) + e^- \rightleftharpoons Li(s)$	−3.08
$K^+(aq) + e^- \rightleftharpoons K(s)$	−2.93
$Ca^{2+}(aq) + 2e^- \rightleftharpoons Ca(s)$	−2.87
$Na^+(aq) + e^- \rightleftharpoons Na(s)$	−2.71
$Mg^{2+}(aq) + 2e^- \rightleftharpoons Mg(s)$	−2.37
$Al^{3+}(aq) + 3e^- \rightleftharpoons Al(s)$	−1.66
$Zn^{2+}(aq) + 2e^- \rightleftharpoons Zn(s)$	−0.76
$Fe^{2+}(aq) + 2e^- \rightleftharpoons Fe(s)$	−0.44
$Pb^{2+}(aq) + 2e^- \rightleftharpoons Pb(s)$	−0.13
$2H^+ + 2e^- \rightleftharpoons H_2(g)$	0.00 (Reference electrode)
$Cu^{2+}(aq) + 2e^- \rightleftharpoons Cu(s)$	+0.34
$Fe^{3+}(aq) + e^- \rightleftharpoons Fe^{2+}(aq)$	+0.80
$Ag^+ + e^- \rightleftharpoons Ag(s)$	+0.77
$Cl_2(g) + 2e^- \rightleftharpoons 2Cl^-(aq)$	+1.36
$Au^+(aq) + e^- \rightleftharpoons Au(s)$	+1.69
$F_2(g) + 2e^- \rightleftharpoons 2F^-(aq)$	+2.87

While the two electrolytes in an electrochemical cell are prevented from mixing, a link is provided between the solutions for exchange of ions. This is necessary to maintain the electrical neutrality of the solutions. A link can be established with a porous partition—if the two half-cells share the same electrolyte compartment—or by means of a salt bridge, if the electrolytes occupy different compartments.

Electrochemistry

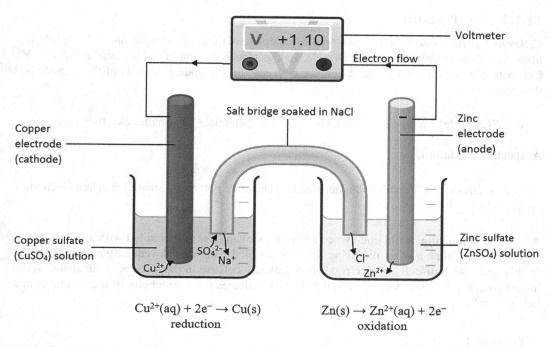

FIGURE 12.4 The Daniell cell.

The electrode from which electrons flow is called anode, while the electrode that receives the electrons is called cathode. Oxidation occurs at the anode, while reduction occurs at the cathode. As indicated earlier, the cathode has a higher electrode potential than the anode. The anode is marked as the negative (−) electrode—because it is the source of electrons—while the cathode is marked as positive (+). The reaction at the anode is called oxidation (anodic) half-reaction, while the reaction at the cathode is called reduction (cathodic) half-reaction.

An example of a galvanic cell is the *Daniell cell* which consists of the $Cu^{2+}(aq)/Cu(s)$ half-cell coupled with the $Zn^{2+}(aq)/Zn(s)$ half-cell (see Figure 12.4). The cell is named after British chemist Professor John Frederic Daniell (1790–1845) who invented it in 1836. The reactions in the Daniell cell are as follows.

Anodic (oxidation) half-reaction: $Zn(s) \rightarrow Zn^{2+}(aq) + 2e^-$
Cathodic (reduction) half-reaction: $Cu^{2+}(aq) + 2e^- \rightarrow Cu(s)$
Overall (net) reaction: $Zn(s) + Cu^{2+}(aq) \rightarrow Zn^{2+}(aq) + Cu(s)$

12.1.3 Cell Diagram

The notation for an electrochemical cell is called a *cell diagram*. The IUPAC rule for writing cell diagram is as follows.

anode | anode electrolyte or ion ‖ cathode electrolyte or ion | cathode

The double vertical lines represents a *salt bridge* or *porous partition*.
The notation for the Daniell cell, for example, is as follows.

$$Zn(s)|ZnSO_4(aq)\|CuSO_4(aq)|Cu(s) \text{ or } Zn(s)|Zn^{2+}(aq)\|Cu^{2+}(aq)|Cu(s)$$

12.1.4 CELL POTENTIAL

Cell potential, E, or *electromotive force (emf)* is a measure of the ability of a galvanic cell to push electrons through an external circuit. Like the electrode potential, the unit of cell potential is the volt (V). Cell potential is expressed as the difference between the electrode potentials of the cathode and the anode.

$$E = E(\text{cathode}) - E(\text{anode}) = \text{the higher electrode potential} - \text{the smaller electrode potential}$$

At standard conditions,

$$E^\theta = E^\theta(\text{cathode}) - E^\theta(\text{anode}) = \text{higher standard electrode potential} - \text{smaller standard electrode potential}$$

A higher cell potential is obtained when a strong oxidising agent is coupled with a strong reducing agent than when a weak oxidising agent is coupled with a weak reducing agent. Thus, the farther apart the half-cells or electrodes of a galvanic cell are in the electrochemical series, the higher its cell potential. A positive cell potential indicates the spontaneity of the reactions in a galvanic cell.

Example 1

Calculate the standard potential of the Daniell cell $Zn(s)|Zn^{2+}(aq)||Cu^{2+}(aq)|Cu(s)$.

$$\left(E^\theta\left(Cu^{2+}(aq)/Cu(s)\right) = +0.34 \text{ V}, E^\theta\left(Zn^{2+}(aq)/Zn(s)\right) = -0.76 \text{ V}\right)$$

Solution

We apply the relation:

$$E^\theta = E^\theta(\text{cathode}) - E^\theta(\text{anode})$$

$$E^\theta(\text{cathode}) = 0.34 \text{ V}$$

$$E^\theta(\text{anode}) = -0.76 \text{ V}$$

$$E^\theta = ?$$

Substituting, we have

$$E^\theta = 0.34 \text{ V} - (-0.76 \text{ V})$$
$$= 1.10 \text{ V}$$

Example 2

Determine the potential of the cell $Ca(s)|Ca^{2+}(aq)||Fe^{3+}(aq)|Fe(s)$.

$$\left(E^\theta\left(Cu^{2+}(aq)/Cu(s)\right) = -2.51 \text{ V}, \ E^\theta\left(Fe^{3+}(aq)/Fe(s)\right) = +0.60 \text{ V}\right)$$

Electrochemistry

Solution

We apply the relation:

$$E = E(\text{cathode}) - E(\text{anode})$$

$$E(\text{cathode}) = +0.60 \text{ V}$$

$$E(\text{anode}) = -2.51 \text{ V}$$

$$E = ?$$

Substituting the values, we obtain

$$E = 0.60 \text{ V} - (-2.51 \text{ V})$$
$$= 3.11 \text{ V}$$

Example 3

The standard potential of the cell Sn(s)|Sn^{2+}(aq)||Ag^+(aq)|Ag(s) is 0.94 V. Determine the standard electrode potential of Ag^+(aq)/Ag(s) if the standard electrode potential of Sn^{2+}(aq)/Sn(s) is −0.14 V.

Solution

The required standard electrode potential is determined as follows.

$$E^\ominus = E^\ominus(\text{cathode}) - E^\ominus(\text{anode})$$

$$E^\ominus = +0.94 \text{ V}$$

$$E^\ominus(\text{anode}) = -0.14 \text{ V}$$

$$E^\ominus(\text{cathode}) = ?$$

Substituting, we have

$$E^\ominus(\text{cathode}) = 0.94 \text{ V} + (-0.14 \text{ V})$$
$$= 0.80 \text{ V}$$

PRACTICE PROBLEMS

1. Determine the standard potential of the cell Zn(s)|Z^{2+}(aq)||Ag^+(aq)|Ag(s).

$$\left(E^\ominus\left(Ag^+(aq)|Ag(s)\right) = +0.77, \ E^\ominus\left(Zn^{2+}(aq)|Zn(s)\right) = -0.76 \text{ V}\right)$$

2. The standard potential of the cell Zn(s)|Z^{2+}(aq)||Pb^{2+}(aq)|Pb(s) is +0.63 V. Determine the standard electrode potential of Pb^{2+}(aq)/Pb(s) if the standard electrode potential of Zn^{2+}(aq)/Zn(s) is −0.76 V.

12.1.5 CELL POTENTIAL AND FREE ENERGY

The free energy of a galvanic cell is given by the relation:

$$\Delta G = -nFE$$

where

n = number of moles of electrons transferred
F = Faraday constant = 96500 C mol^{-1}

At standard conditions, we have

$$\Delta G^\ominus = -nFE^\ominus$$

The faraday or Faraday constant (F) is the quantity of electricity required to liberate or deposit 1 mol of a univalent ion or electrons from a solution. It is measured in coulombs per mol (C mol^{-1}).

Example 1

Calculate the standard free energy of the Daniell cell.

$$(F = 96500\, C\, mol^{-1})$$

Solution

We use the relation:

$$\Delta G^\ominus = -nFE^\ominus$$

For the Daniell cell, we know that

$$n = 2\, mol$$
$$E^\ominus = 1.10\, V$$
$$F = 96500\, C\, mol^{-1}$$
$$\Delta G^\ominus = ?$$

Substituting, we have

$$\Delta G^\ominus = -2\, mol \times \frac{96500\, C}{1\, mol} \times 1.10\, V$$
$$= -212\, kJ$$

You should recall that a negative value of free energy indicates spontaneity.
Note: 1 J = 1 C × 1 V.

Example 2

Determine the free energy of the cell Mg(s)|Mg^{2+}(aq)||Al^{3+}(aq)|Al(s).

$$\left(E^\ominus\left(Mg^{2+}(aq)/Mg(s)\right) = -2.38\, V,\ E^\ominus\left(Al^{3+}(aq)/Al(s)\right) = -1.66\, V,\ F = 96500\, C\, mol^{-1}\right)$$

Electrochemistry

Solution

The free energy of the cell is obtained as follows.

$$\Delta G = -nFE$$

The cell potential is determined using the relation:

$$E = E(\text{cathode}) - E(\text{anode})$$
$$E(\text{cathode}) = -1.66 \text{ V}$$
$$E(\text{anode}) = -2.38 \text{ V}$$
$$E = ?$$

So

$$E = -1.66 \text{ V} - (-2.38 \text{ V})$$
$$= 0.72 \text{ V}$$

To obtain the number of electrons transferred, we need to split the cell reaction into two half-reactions as follows:

Oxidation half-reaction: $Mg(s) \rightarrow Mg^{2+}(aq) + 2e^-$
Reduction half-reaction: $Al^{3+}(aq) + 3e^- \rightarrow Al(s)$

The balanced half-reactions are as follows:

Oxidation half-reaction: $3Mg(s) \rightarrow 3Mg^{2+}(aq) + 6e^-$
Reduction half-reaction: $2Al^{3+}(aq) + 6e^- \rightarrow 2Al(s)$

So

$$n = 6$$
$$F = 96500 \text{ C mol}^{-1}$$
$$\Delta G = ?$$

Finally, we now substitute to obtain

$$\Delta G = -6 \text{ mol} \times \frac{96500 \text{ C}}{1 \text{ mol}} \times 0.72 \text{ V}$$
$$= -420 \text{ kJ}$$

Note: A quick way to determine the number of moles of electrons transferred in a galvanic cell reaction is to multiply the valencies (the charges with the signs dropped) of the electrodes by each other. However, this method works only if the valencies of the electrodes are different. If the valencies of the electrodes are the same, then the number of moles of electrons transferred is equal to the valency of one of the electrodes.

Example 3

The free energy of the cell Fe(s)|Fe^{2+}(aq)||Cu^{2+}(aq)|Cu(s) is -144.8 kJ. Calculate the cell potential.

$$\left(F = 96500 \text{ C mol}^{-1}\right)$$

Now, we apply the relation:

$$\Delta G = -nFE$$

$$\therefore E = -\frac{\Delta G}{nF}$$

Since the charges on the two ions are equal, it follows that $n = 2$ mol.

$$\Delta G = -144.8 \text{ kJ} = -144800 \text{ J} = -144800 \text{ CV}$$

$$F = 96500 \text{ C mol}^{-1}$$

$$E = ?$$

Substituting, we have

$$E = -\frac{(-144800 \text{ CV}) \times 1 \text{ mol}}{2 \text{ mol} \times 96500 \text{ C}} = 0.75 \text{ V}$$

PRACTICE PROBLEMS

1. Calculate the standard free energy of the cell Pb(s)|Pb^{2+}(aq)||Ag$^+$(s)|Ag(s).

   $\left(E^\theta\left(Pb^{2+}(aq)/Pb(s)\right) = -0.13 \text{ V}, \; E^\theta\left(Ag^+(aq)/Ag(s)\right) = +0.77 \text{ V}, \; F = 96500 \text{ C mol}^{-1}\right)$

2. The free energy of the cell Ni(s)|Ni^{2+}(aq)||Cu^{2+}(aq)|Cu(s) is −98.4 kJ. Calculate the cell potential.

   $\left(F = 96500 \text{ C mol}^{-1}\right)$

12.1.6 CELL POTENTIAL AND EQUILIBRIUM CONSTANT

Earlier, we encountered the equation $\Delta G^\theta = -nFE^\theta$ that describes the relationship between free energy and cell potential. You will recall that free energy can also be expressed in terms of the equilibrium constant, K, as follows.

$$\Delta G^\theta = -RT \ln K$$

A relationship can be established between the standard cell potential and the equilibrium constant as follows:

$$-nFE = -RT \ln K$$

So

$$E^\theta = \frac{RT}{nF} \ln K$$

Electrochemistry

Substituting $T = 298$ K, $R = 8.314$ J mol^{-1} K^{-1} and $F = 96500$ C mol^{-1}, we have

$$E^\ominus = \frac{0.02569}{n} \ln K$$

The equation can also be written in terms of the common logarithm as follows:

$$E^\ominus = \frac{0.0592}{n} \log K$$

Note: $\ln K = 2.303 \log K$.

Example 1

The half-cell reactions of the lead-acid accumulator are as follows:

Anodic half-reaction: $Pb(s) + SO_4(aq) \rightarrow PbSO_4(s) + 2e^-$
Cathodic half-reaction: $PbO_2(s) + SO_4^{2-}(aq) + 4H^+(aq) + 2e^- \rightarrow PbSO_4(s) + 2H_2O(l)$

Calculate the standard cell potential of the cell if its equilibrium constant is 9.25×10^{68} at 298 K.

Solution

The standard cell potential is given by

$$E^\ominus = \frac{0.02569}{n} \ln K$$

$$K = 9.25 \times 10^{68}$$

$$n = 2$$

$$E^\ominus = ?$$

Substituting, we have

$$E^\ominus = \frac{0.02569}{2} \ln 9.25 \times 10^{68}$$

So

$$E^\ominus = 0.012845 \times 158.8$$
$$= +2.04 \text{ V}$$

Example 2

Calculate the equilibrium constant of the Daniell cell $Zn(s)|Zn^{2+}(aq)||Cu^{2+}(aq)|Cu(s)$ at 298 K, given that the standard cell potential is +1.10 V.

Solution

The equilibrium constant is calculated as follows:

$$E^\ominus = \frac{0.02569}{n} \ln K$$

$$E^\theta = +1.10 \text{ V}$$
$$n = 2$$
$$K = ?$$

Substituting, we have

$$1.10 = \frac{0.02569}{2} \ln K$$

So

$$\ln K = \frac{2.20}{0.02569} = 85.636$$

Taking the natural antilogarithm of both sides, we have

$$\ln K = e^{85.636}$$
$$= 1.55 \times 10^{37}$$

PRACTICE PROBLEMS

1. The equilibrium constant of the cell Sn(s)|Sn^{2+}(aq)||Ag$^+$(aq)||Ag(s) is 6.05×10^{31} at 298 K. Calculate the standard cell potential.
2. Calculate the equilibrium constant of the cell Zn(s)|Zn^{2+}(aq)||Fe(s)|Fe(s) at 298 K.

$$\left(E^\theta\left(Zn^{2+}(aq)/Zn(s)\right) = -0.76 \text{ V}, \; E^\theta\left(Fe^{3+}(aq)/Fe(s)\right) = +0.04 \text{ V}\right)$$

12.1.7 Cell Potential and Concentration: The Nernst Equation

As stated earlier, the cell potential varies with the concentration of an electrolyte. The relationship between free energy and concentration is given by

$$\Delta G = -RT \ln K + RT \ln Q$$

where Q is the reaction quotient.
Substituting $\Delta G^\theta = -RT\ln Q$ into the equation, we have

$$\Delta G = \Delta G^\theta + RT \ln Q$$

You will recall that ΔG and ΔG^θ are related to cell potential as follows:

$$\Delta G = -nFE$$
$$\Delta G^\theta = -nFE^\theta$$

Substituting the equations into $\Delta G = \Delta G^\theta + RT\ln Q$, we have

$$-nFE = -nFE^\theta + RT \ln Q$$

Electrochemistry

Dividing the equation by $-nF$ gives the Nernst Equation:

$$E = E^\theta - \frac{RT}{nF} \ln Q$$

By substituting $T = 298$ K, $R = 8.314$ J mol^{-1} K^{-1} and $F = 96500$ C mol^{-1} into the equation, we will obtain the simplified version

$$E = E^\theta - \frac{0.02569}{n} \ln Q$$

The Nernst equation can also be written in terms of the common logarithm as follows.

$$E = E^\theta - \frac{0.0592}{n} \log Q$$

Example 1

Predict the electrode potential of the cell Mg(s)|Mg^{2+}(aq)||Cu^{2+}(aq)|Cu(s) when [Mg^{2+}] = 0.75 mol dm^{-3} and [Cu^{2+}] = 0.015 mol dm^{-3}.

$$\left(E^\theta\left(Mg^{2+}(aq)/Mg(s)\right) = -2.37 \text{ V}, \ E^\theta\left(Cu^{2+}(aq)/Cu(s)\right) = +0.34 \text{ V}\right)$$

Solution

We need to apply the Nernst Equation as follows.

$$E = E^\theta - \frac{0.02569}{n} \ln Q$$

The overall cell reaction is Mg(s) + Cu^{2+}(aq) → Cu(s) + Mg^{2+}(aq).

So

$$E = E^\theta - \frac{0.02569}{n} \ln \frac{\left[Mg^{2+}\right]}{\left[Cu^{2+}\right]}$$

The standard cell potential is obtained as follows.

$$E^\theta = E^\theta(\text{cathode}) - E^\theta(\text{anode})$$

$$E^\theta(\text{cathode}) = +0.34 \text{ V}$$

$$E^\theta(\text{anode}) = -2.37 \text{ V}$$

$$E^\theta = ?$$

So

$$E^\theta = 0.34 \text{ V} - (-2.37 \text{ V}) = 2.71 \text{ V}$$

$$\left[Mg^{2+}\right] = 0.75 \text{ mol dm}^{-3}$$

$$[Cu^{2+}] = 0.015 \text{ mol dm}^{-3}$$

$$n = 2$$

$$E = ?$$

Substituting the values, we have

$$E = 2.17 \text{ V} - \frac{0.02569}{2} \ln \frac{0.75 \text{ mol dm}^{-3}}{0.015 \text{ mol dm}^{-3}}$$

So

$$E = 2.17 \text{ V} - 0.050 \text{ V}$$
$$= +2.12 \text{ V}$$

Example 2

A galvanic cell is made by coupling a standard $Co^{2+}(aq)/Co(s)$ half-cell to a $Pt^{2+}(aq)/Pt(s)$ half-cell of an unknown concentration. Determine the concentration of Pt^{2+} in the cell if the cell potential is 1.35 V at 298 K.

$$\left(E^{\theta}\left(Co^{2+}(aq)/Co(s) \right) = -0.28 \text{ V}, \; E^{\theta}\left(Pt^{2+}(aq)/Pt(s) \right) = +1.20 \text{ V} \right)$$

Solution

The required concentration is calculated using the relation:

$$E = E^{\theta} - \frac{0.02569}{n} \ln Q$$

The overall cell reaction is $Co(s) + Pt^{2+}(aq) \rightarrow Pt(s) + Co^{2+}(aq)$.

So

$$E = E^{\theta} - \frac{0.02569}{n} \ln \frac{[Co^{2+}]}{[Pt^{2+}]}$$

The standard cell potential is obtained as follows.

$$E^{\theta} = E^{\theta}(\text{cathode}) - E^{\theta}(\text{anode})$$

$$E^{\theta}(\text{cathode}) = +1.20 \text{ V}$$

$$E^{\theta}(\text{anode}) = -0.28 \text{ V}$$

$$E^{\theta} = ?$$

So

$$E^{\theta} = 1.20 \text{ V} - (-0.28 \text{ V}) = 1.48 \text{ V}$$

$$[Co^{2+}] = 1.00 \text{ mol dm}^{-3} \text{ (standard concentration)}$$

Electrochemistry

$$n = 2$$
$$E = 1.35 \text{ V}$$
$$[Pt^{2+}] = ?$$

Substituting, we have

$$1.35 \text{ V} = 1.48 \text{ V} - \frac{0.02569}{2} \ln \frac{1.00 \text{ mol dm}^{-3}}{[Pt^{2+}]}$$

So

$$-0.13 = -0.012845 \ln \frac{1.00 \text{ mol dm}^{-3}}{[Pt^{2+}]}$$

$$\therefore \ln \frac{1.00 \text{ mol dm}^{-3}}{[Pt^{2+}]} = 10.12$$

Taking the natural antilogarithm of both sides, we have

$$\frac{1.00 \text{ mol dm}^{-3}}{[Pt^{2+}]} = e^{10.12}$$

So

$$[Pt^{2+}] = \frac{1.00 \text{ mol dm}^{-3}}{e^{10.12}} = 4.03 \times 10^{-05} \text{ mol dm}^{-3}$$

PRACTICE PROBLEMS

1. Calculate the cell potential of the cell $Cr(s)|Cr^{3+}(aq)\|Fe^{2+}(aq)|Fe(s)$ when the concentrations of Cr^{3+} and Fe^{2+} are 0.70 mol dm^{-3} and 0.010 mol dm^{-3}, respectively.

$$\left(E^{\ominus}\left(Cr^{3+}(aq)/Cr(s)\right) = -0.74 \text{ V}, \; E^{\ominus}\left(Fe^{2+}(aq)/Fe(s)\right) = -0.44 \text{ V}\right)$$

2. Calculate the concentration of Cu^{2+} in the half-cell $Cu^{2+}(aq)/Cu(s)$ when the electrode potential is +0.30 V.

12.2 ELECTROLYSIS

Electrolysis is the process by which electrolytes are decomposed by the passage of a direct electric current through their solutions or molten state. Unlike the reaction in a galvanic cell, which is spontaneous, an electrolytic cell requires an external supply of energy, in the form of electric current to bring about the decomposition of electrolytes.

The compartment or apparatus in which electrolysis occurs is called an *electrolytic cell*. As shown in Figure 12.5, an electrolytic cell consists of two electrodes inserted into an electrolyte solution. A direct current is supplied to the electrolyte through the electrodes. The electrode at which oxidation occurs is called *anode*; electrons leave the electrolyte through the anode. Another way of saying this is that current enters the electrolyte through the anode. The electrode at which reduction occurs is

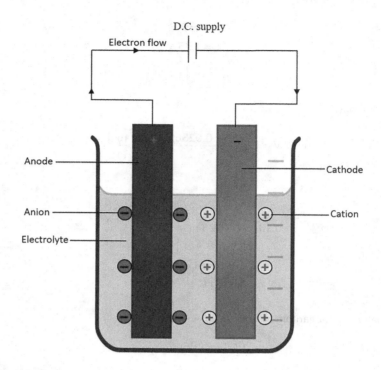

FIGURE 12.5 An electrolytic cell.

called *cathode*; electrons enter the electrolyte through the cathode. Another way of saying this is that current leaves the electrolyte through the cathode. The markings of the electrodes in an electrolytic cell are opposite to those in a galvanic cell: the anode is marked positive (+), while the cathode is marked negative (−).

12.2.1 Mechanism of Electrolysis

In 1887, Swedish chemist Svante Arrhenius put forward the *ionic theory* to explain how electrolysis works. According to this theory, when electrolytes are dissolved in water, they split into positively and negatively charged ions that move randomly in solution. This process is called *ionisation*. Sodium chloride, for example, splits into its ions in solution as follows.

$$NaCl(s) \rightarrow Na^+(aq) + Cl^-(aq)$$

If a direct electric current is passed through the solution, the positively charged ions—called cations—will move to the cathode, while the negatively charged ions—called anions—will move to the anode. At the anode, anions lose their electrons—or get oxidised—and may then be discharged from the solution, depending on a number of factors which we will discuss shortly. At the cathode, cations take up electrons—or get reduced—and may also get discharged from the solution. Since water also ionises, two different types of ions usually move to one or both of the electrodes during electrolysis. However, only one type of ion gets discharged at each electrode. For instance, in the electrolysis of copper(II) sulfate, the ions present at the electrodes are shown below.

Substance	Anode	Cathode
$CuSO_4$	Cu^{2+}	SO_4^{2-}
H_2O	H^+	OH^-

Electrochemistry

In the case of dilute sulfuric acid, only one type of ion is present at the anode.

Substance	Anode	Cathode
H_2SO_4	H^+	SO_4^{2-}
H_2O	H^+	OH^-

One of the factors that influence the preferential discharge of an ion during electrolysis is its position relative to the competing ion in the electrochemical series. Cations lower in the series are discharged in preference to those higher up in the series. Ions such as K^+ and Na^+ are so high up in the series that they never get discharged from solution. In contrast, anions higher in the electrochemical series will be discharged in preference to those lower in the series. For example, in the electrolysis of sodium chloride solution, OH^- will be preferentially discharged over Cl^- because OH^- occupies a higher position in the series. The positions of anions such as SO_4^{2-} and NO_3^- are so low in the series that they never get discharged from solution.

Another factor governing the preferential discharge of ions during electrolysis is the relative concentrations of the competing ions. As long as the two competing ions are very close in the electrochemical series, an increase in the the concentration of one will promote its discharge from solution. For example, in the electrolysis of concentrated sodium chloride, the high concentration of Cl^- will promote its discharge from solution—as opposed to OH^-—which would normally be preferentially discharged due to its favourable position in the series.

Lastly, the nature of the electrodes used may influence the preferential discharge of ions in electrolysis. For instance, using an electrode that has an affinity for any of the competing ions in solution will promote its discharge, regardless of the influence of the other two factors we have discussed, as the ion will readily bind with such an electrode.

12.2.2 Faraday's First Law of Electrolysis

The quantitative relationships between electrical energy and the amount of substances formed at the electrodes during electrolysis was first proposed by the great British experimental scientist Michael Faraday (1791–1867) in 1833. These relationships are known as Faraday's laws of electrolysis.

Faraday's first law of electrolysis states that the mass of an element deposited during electrolysis is directly proportional to the quantity of electricity passed through an electrolyte. This is expressed mathematically as follows:

$$m \: \alpha \: Q$$

So

$$m = ZQ$$

where

m = mass of an element discharged in grams (g)
Q = quantity of electricity (in coulombs, C).
Z = the electrochemical equivalent of an element

The quantity of electricity is the product of current, I, and time, t.

$$Q = It$$

So

$$m = ZIt$$

The electrochemical equivalent of an element is defined as the mass of the element deposited by passing 1 C of electricity through an electrolyte. It has a unit of gram per coulomb (g C^{-1}).

As stated in Section 8.1, it requires 1 faraday (1 F) or 96500 C of electricity to liberate or discharge 1 mol of electrons or a univalent ion from solution. Similarly, it requires $2F$ or 193000 C of electricity to liberate 2 mol of electrons or 1 mol of a divalent ion from solution, and so on.

Example 1

2.5 A of electric current was passed through an electrolyte for 30 min. Calculate the quantity of electricity used.

Solution

We determine the quantity of electricity as follows.

$$Q = It$$
$$I = 2.5 \text{ A}$$
$$t = 30 \text{ min} = (30 \times 60) \text{ s} = 1800 \text{ s}$$
$$Q = ?$$

Substituting, we have

$$Q = 2.5 \text{ A} \times 1800 \text{ s}$$
$$= 4500 \text{ C}$$

Note: 1 C = 1 A × 1 s.

Example 2

18000 C of electricity was passed through an electrolyte for 2.5 h. Determine the amount of current used.

Solution

We apply the relation:

$$Q = It$$

$$\therefore I = \frac{Q}{t}$$

$$Q = 18000 \text{ C} = 18000 \text{ A s}$$
$$t = 2.5 \text{ h} = (2.5 \times 60 \times 60) \text{ s} = 9000 \text{ s}$$
$$I = ?$$

Substituting, we have

$$I = \frac{18000 \text{ A s}}{9000 \text{ s}} = 2 \text{ A}$$

Electrochemistry

Example 3

Calculate the time required for an electric current of 1.5 A to produce 19300 C of electricity.

Solution

As usual, we apply the relation:

$$Q = It$$
$$\therefore t = \frac{Q}{I}$$

$Q = 19300$ C $= 19300$ As
$I = 1.5$ A
$t = ?$

Substituting, we have

$$t = \frac{19300 \text{ As}}{1.5 \text{ A}} = 13000 \text{ s} = 3.6 \text{ h}$$

Example 4

Determine the quantity of electricity required to discharge 1 mol of aluminium ions from solution.

$$\left(F = 96500 \text{ C mol}^{-1}\right)$$

Solution

The reaction at the cathode is $Al^{3+}(aq) + 3e^- \rightarrow Al(s)$.

Since it requires 1 F (96500 C) of electricity to discharge 1 mol of a univalent or singly charged ion, then it requires 3 F or 289500 C of electricity to discharge 1 mol of Al^{3+}, a trivalent ion, from solution.

Example 5

Determine the quantity of electricity required to deposit 5.5 g of silver from a solution of silver nitrate.

$$\left(Ag = 108, F = 96500 \text{ C mol}^{-1}\right)$$

Solution

The first step is to write the cathodic half-reaction:

$$Ag^+(aq) + e^- \rightarrow Ag(s)$$

Let the quantity of electricity required to deposit 5.5 g of silver be Q. Since the stoichiometry of reaction shows that it requires 1 F (96500 C) of electricity to deposit 1 mol (108 g) of silver, it then follows that

$$\frac{96500 \text{ C}}{108 \text{ g}} = \frac{Q}{5.5 \text{ g}}$$

So

$$96500 \text{ C} \times 5.5 \text{ g} = Q \times 108 \text{ g}$$

$$\therefore Q = \frac{96500 \text{ C} \times 5.5 \text{ g}}{108 \text{ g}} = 4900 \text{ C}$$

Example 6

Determine the mass of copper that would be deposited when 2.5 A of electric current is passed through a solution of copper(II) sulfate for 37 min.

$$(Cu = 63.5, F = 96500 \text{ C mol}^{-1})$$

Solution

The reaction at the cathode is $Cu^{2+}(aq) + 2e^- \rightarrow Cu(s)$.
The quantity of electricity is calculated as follows.

$$Q = It$$

$$I = 2.5 \text{ A}$$

$$t = 37 \text{ min} = (37 \times 60) \text{ s} = 2220 \text{ s}$$

$$Q = ?$$

Substituting, we have

$$Q = 2.5 \text{ A} \times 2220 \text{ s} = 5550 \text{ C}$$

Let the mass of copper deposited by this quantity of electricity be m. Since the equation shows that 2 F (193000 C) of electricity is required to deposit 1 mol (63.5 g) of copper, it follows that

$$\frac{193000 \text{ C}}{63.5 \text{ g}} = \frac{5550 \text{ C}}{m}$$

So

$$193000 \text{ C} \times m = 5550 \text{ C} \times 63.5 \text{ g}$$

$$m = \frac{5550 \text{ C} \times 63.5 \text{ g}}{193000 \text{ C}} = 1.8 \text{ g}$$

Example 7

1.5 A of electric current was passed through acidified water for 45 min at STP. Calculate the volume of hydrogen liberated at the cathode at 850 mmHg and 27°C.
(H = 1, molar volume of gases at STP = 22.4 dm³ mol⁻¹, F = 96500 C mol⁻¹)

Solution

The equation of reaction at the cathode is $2H^+(aq) + 2e^- \rightarrow H_2(g)$.
The quantity of electricity used is calculated as follows.

$$Q = It$$

$$I = 1.5 \text{ A}$$

$$t = 45 \text{ min} = 2700 \text{ s}$$

$$Q = ?$$

Electrochemistry

Substituting, we have

$$Q = 1.5 \text{ A} \times 2700 \text{ s} = 4050 \text{ C}$$

Let the volume of hydrogen liberated by 4050 C of electricity be V at STP. Since it requires 2 F (193000 C) of electricity to liberate 1 mol (22.4 dm³ at STP) of hydrogen gas, it follows that

$$\frac{193000 \text{ C}}{22.4 \text{ dm}^3} = \frac{4050 \text{ C}}{V}$$

So

$$193000 \text{ C} \times V = 4050 \text{ C} \times 22.4 \text{ dm}^3$$

$$\therefore V = \frac{4050 \text{ C} \times 22.4 \text{ dm}^3}{193000 \text{ C}} = 0.47 \text{ dm}^3$$

Finally, we have to convert this volume to the volume at 950 mmHg and 27°C, using the general gas equation.

$$\frac{P_1 V_1}{T_1} = \frac{P_2 V_2}{T_2}$$

$$\therefore V_2 = \frac{P_1 V_1 T_2}{P_2 T_1}$$

$$P_1 = 760 \text{ mmHg}$$

$$V_1 = 0.47 \text{ dm}^3$$

$$T_1 = 273 \text{ K}$$

$$P_2 = 950 \text{ mmHg}$$

$$T_2 = 27°C = (273 + 27) \text{ K} = 300 \text{ K}$$

$$V_2 = ?$$

Substituting, we have

$$V_2 = \frac{760 \text{ mmHg} \times 0.47 \text{ dm}^3 \times 300 \text{ K}}{950 \text{ mmHg} \times 273 \text{ K}} = 0.41 \text{ dm}^3 = 410 \text{ cm}^3$$

Example 8

Calculate the mass of copper that would be obtained from the electrolysis of copper(II) nitrate by the same quantity of electricity required to deposit 5.5 g of aluminium.

$$\left(Al = 27, \; Cu = 63.5; \; F = 96500 \text{ C mol}^{-1}\right)$$

Solution

The first step is to determine the quantity of electricity required to deposit 5.5 g of aluminium from its electrolyte.

The cathodic half-reaction is $Al^{3+} + 3e^- \rightarrow Al(s)$. Let the quantity of electricity required to deposit 5.5 g of aluminium be Q. From the equation, 3 F (289500 C) of electricity would deposit 1 mol (27 g) of aluminium. It then follows that

$$\frac{289000 \text{ C}}{27 \text{ g}} = \frac{Q}{5.5 \text{ g}}$$

So

$$289500 \text{ C} \times 5.5 \text{ g} = Q \times 27 \text{ g}$$

$$\therefore Q = \frac{289500 \text{ C} \times 5.5 \text{ g}}{27 \text{ g}} = 58972.22 \text{ C}$$

We must now determine the mass of copper that would be deposited from copper(II) nitrate solution by this quantity of electricity.

$$Cu^{2+}(aq) + 2e^- \rightarrow Cu(s)$$

Let the mass of copper that would be deposited by 58972.22 C of electricity be m. Since the equation shows that it requires 2 F (193000 C) of electricity to deposit 1 mol (63.5 g) of copper, it follows that

$$\frac{193000 \text{ C}}{63.5 \text{ g}} = \frac{58972.22 \text{ C}}{m}$$

So

$$193000 \text{ C} \times m = 58972.22 \text{ C} \times 63.5 \text{ g}$$

$$\therefore m = \frac{58972.22 \text{ C} \times 63.5 \text{ g}}{193000} = 19 \text{ g}$$

Example 9

3700 C of electricity was passed through a solution of a chloride of gold. Determine the charge of the gold ion if the mass of gold deposited is 2.5 g. Hence, state the IUPAC name of the chloride.

$$(Au = 197, F = 96500 \text{ C})$$

Solution

Let the charge on the gold ion be x. We can then write the cathodic half-reaction as

$$Au^{x+}(aq) + xe^- \rightarrow Au(s)$$

From the equation, it requires x F (96500x C) of electricity to deposit 1 mol (197 g) of gold. Since it requires 3700 C to deposit 2.5 g, it then follows that

$$\frac{96500x \text{ C}}{197 \text{ g}} = \frac{3700 \text{ C}}{2.5 \text{ g}}$$

Electrochemistry

So

$$96500x \text{ C} \times 2.5 \text{ g} = 3700 \text{ C} \times 197 \text{ g}$$

$$\therefore x = \frac{3700 \text{ C} \times 197 \text{ g}}{96500 \text{ C} \times 2.5 \text{ g}} = 3$$

Since gold is a metal, then the charge on the ion is 3+; hence, the IUPAC name of the chloride is gold(III) chloride.

Example 10

For how long must 3.5 A of electric current be passed through silver chloride solution to deposit 5.0 g of silver at the cathode?

$$\left(\text{Ag} = 108, \; F = 96500 \text{ C mol}^{-1}\right)$$

Solution

The very first step is to work out the quantity of electricity required to deposit 5.0 g of silver.
The reaction at the cathode is $Ag^+(aq) + e^- \rightarrow Ag(s)$.
Let the quantity of electricity required to deposit 5.0 g of silver be Q. According to the equation, it requires 1 F (96500 C) of electricity to deposit 1 mol (108 g) of silver; hence,

$$\frac{96500 \text{ C}}{108 \text{ g}} = \frac{Q}{5.0 \text{ g}}$$

So

$$Q \times 108 \text{ g} = 96500 \text{ C} \times 5.0 \text{ g}$$

$$\therefore Q = \frac{96500 \text{ C} \times 5.0 \text{ g}}{108 \text{ g}} = 4467.59 \text{ C}$$

Finally, we need to determine the required time for the passage of this quantity of electricity as follows.

$$Q = It$$

$$\therefore t = \frac{Q}{I}$$

$$Q = 4467.59 \text{ C} = 4467.59 \text{ A s}$$

$$I = 3.5 \text{ A}$$

$$t = ?$$

Substituting, we have

$$t = \frac{4467.59 \text{ A s}}{3.5 \text{ A}} = 1300 \text{ s} = 22 \text{ min}$$

PRACTICE PROBLEMS

1. Determine the quantity of electricity produced when 5.50 A of electric current is passed through an electrolyte for 65.5 min.
2. What mass of aluminium would be deposited when 1.5 A of electric current is passed through a solution of aluminium sulfate for 2.5 h.

$$\left(Al = 27,\ F = 96500\,C\,mol^{-1}\right)$$

3. Calculate the quantity of electricity required to liberate 250 cm³ of oxygen at STP. (Molar volume of gases at STP = 22.4 dm³ mol⁻¹, F = 96500 C mol⁻¹)
4. 1.5 g of copper is deposited at the cathode in 35 min in a laboratory experiment on the electrolysis of copper(II) sulfate. Determine the amount of electric current used. (Molar volume of gases at STP = 22.4 dm³ mol⁻¹, F = 96500 C mol⁻¹)

12.2.3 Faraday's Second Law of Electrolysis

Faraday's second law of electrolysis states that if the same quantity of electricity is passed through different electrolytes, the mass of each element discharged is directly proportional to its equivalent mass or mass equivalent. The mass equivalent of an element is its relative atomic mass, if a metal, or its relative molecular mass, if a gas, divided by the number of moles of electrons involved in the deposition or liberation of 1 mol of the element.

Suppose the same quantity of electricity is passed through two electrolytes. According to Faraday's second law of electrolysis, the relative masses of the elements A and B discharged in their respective electrolytic cells are in the ratio of their mass equivalents. This is given by

$$\frac{m_A}{m_B} = \frac{M_{rA}/x_A}{M_{rB}/x_B}$$

where

m_A = mass of A
m_B = mass of B
M_{rA} = relative atomic mass of A, if a metal, or the relative molecular mass, if a gas
M_{rB} = relative atomic mass of B, if a metal, or the relative molecular mass, if a gas
x_A = number of moles of electrons required to liberate or deposit 1 mol of A
x_B = number of moles of electrons required to liberate or deposit 1 mol of B

A simplified form of the above equation is given as:

$$\frac{n_A}{n_B} = \frac{x_B}{x_A}$$

Example 1

Calculate the mass of copper that would be obtained from the electrolysis of copper(II) nitrate by the same quantity of electricity required to deposit 5.5 g of aluminium.

$$\left(Al = 27,\ Cu = 63.5\right)$$

Electrochemistry

Solution

We have solved this problem earlier with the first law in (see Example 8, Section 12.2.2). Here, we will illustrate how the second law can be used to solve the same problem. We will apply the equation

$$\frac{m_{Al}}{m_{Cu}} = \frac{M_{rAl}/x_{Al}}{M_{rCu}/x_{Cu}}$$

The equations of reaction at both cathodes are as follows:

$$Cu^{2+}(aq) + 2e^- \rightarrow Cu(s)$$

$$Al^{3+}(s) + 3e^- \rightarrow Al(s)$$

$$m_{Al} = 5.5 \text{ g}$$

$$M_{rAl} = 27$$

$$M_{rCu} = 63.5$$

$$x_{Al} = 3$$

$$x_{Cu} = 2$$

$$m_{Cu} = ?$$

Substituting, we have

$$\frac{5.5 \text{ g}}{m_{Cu}} = \frac{27/3}{63.5/2}$$

So

$$m_{Cu} \times 9 = 5.0 \text{ g} \times 31.75$$

$$\therefore m_{Cu} = \frac{5.5 \text{ g} \times 31.75}{9} = 19 \text{ g}$$

This agrees with the answer we got with the first law.

Example 2

A quantity of electricity was passed through solutions of a chloride of cobalt and copper(II) sulfate which were connected in series. If 0.83 g of cobalt and 0.89 g of copper were deposited in the first and second cells, respectively, determine the charge of the cobalt ions in the first solution. Hence, state the IUPAC name of the chloride of cobalt.

$$(Co = 59, Cu = 63.5)$$

Solution

We will apply the relation:

$$\frac{m_{Cu}}{m_{Co}} = \frac{M_{rCu}/x_{Cu}}{M_{rCo}/x_{Co}}$$

The cathodic half-reaction of the second cell is $Cu^{2+}(aq) + 2e^- \rightarrow Cu(s)$.

Let the charge on the cobalt ions be x; hence, the cathodic half-reaction of the first cell is $Cu^{x+}(aq) + xe^- \rightarrow Cu(s)$.

$$m_{Cu} = 0.89 \text{ g}$$
$$m_{Co} = 0.83 \text{ g}$$
$$M_{rCu} = 63.5$$
$$M_{rCo} = 59$$
$$x_{Cu} = 2$$
$$x_{Cu} = x$$

Substituting, we have

$$\frac{0.89 \text{ g}}{0.83 \text{ g}} = \frac{63.5/2}{59/x}$$

$$\frac{0.89 \text{ g}}{0.83 \text{ g}} = \frac{31.75x}{59}$$

So

$$0.89 \text{ g} \times 59 = 0.83 \text{ g} \times 31.75x_B$$

$$\therefore x = \frac{0.89 \text{ g} \times 59}{0.83 \text{ g} \times 31.75} = 2$$

Since cobalt is a metal, then its ionic charge in the solution is 2+. Consequently, the IUPAC name of the chloride is cobalt(II) chloride.

Example 3

A quantity of electricity was passed through two electrolytic cells containing solutions of sodium chloride and aluminium sulfate, leading to the deposition of 0.466 g of aluminium in the second cell. If the cells were connected in series, determine:

(a) the amount of chlorine liberated in the first cell at STP;
(b) the volume of chlorine liberated in the first cell at STP.

(Al = 27, molar volume of gases at STP = 22.4 dm³ mol⁻¹)

Solution

(a) We will apply the relation:

$$\frac{n_{Al}}{n_{Cl}} = \frac{x_{Cl}}{x_{Al}}$$

The number of moles of aluminium deposited is determined as follows.

$$n_{Al} = \frac{m_{Al}}{M_{Al}}$$

Electrochemistry

$$m_{Al} = 0.466 \text{ g}$$
$$M_{Al} = 27 \text{ g}$$
$$n_{Al} = ?$$

Substituting, we have

$$n_{Al} = \frac{0.466 \text{ g} \times 1 \text{ mol}}{27 \text{ g}} = 0.01726 \text{ mol}$$

The two equations of reaction are as follows:

$$Al^{3+}(aq) + 3e^- \rightarrow Al(s)$$
$$2Cl^-(aq) \rightarrow Cl_2(g) + 2e^-$$

$$x_{Al} = 3$$
$$x_{Cl} = 2$$
$$n_{Cl} = ?$$

Substituting the values, we have

$$\frac{0.01726 \text{ mol}}{n_{Cl}} = \frac{2}{3}$$

So

$$n_{Cl} \times 2 = 0.01726 \text{ mol} \times 3$$

$$\therefore n_{Cl} = \frac{0.01726 \times 3 \text{ mol}}{2} = 0.02589 \text{ mol at STP}$$

(b) The volume of chlorine produced at STP is calculated as follows.

$$n = \frac{V}{22.4 \text{ dm}^3 \text{ mol}^{-1}}$$

$$\therefore V = n \times 22.4 \text{ dm}^3 \text{ mol}^{-1}$$

$$n = 0.02589 \text{ mol}$$
$$V = ?$$

Substituting, we have

$$V = 0.02589 \text{ mol} \times \frac{22.4 \text{ dm}^3}{1 \text{ mol}} = 0.58 \text{ dm}^3 = 580 \text{ cm}^3$$

PRACTICE PROBLEMS

1. Determine the volume of oxygen that would be liberated at STP by the same quantity of electricity required to liberate 2.5 g of copper.

$$\left(Cu = 63.5, \text{ molar volume of gases at STP} = 22.4 \text{ dm}^3 \text{ mol}^{-1}\right)$$

2. A quantity of electricity was passed through solutions of silver bromide and acidified water. Determine the mass of silver deposited in the first cell, given that 5.5 dm³ of oxygen was liberated in the second cell at STP.

$$\left(O = 16;\ Ag = 108,\ \text{molar volume of gases at STP} = 22.4\,dm^3\,mol^{-1}\right)$$

FURTHER PRACTICE PROBLEMS

1. Determine the standard emf of the cell Al(s)|Al³⁺(aq)||Mn²⁺(aq)|Mn(s).

$$\left(E^\theta\left(Al^{3+}(aq)/Al(s)\right) = -1.66\,V,\ E^\theta\left(Mn^{2+}(aq)/Mn(s)\right) = -1.19\,V\right)$$

2. The potential of the cell Fe(s)|Fe²⁺(aq)||Sn²⁺(aq)|Sn(s) is 0.31 V. Determine the potential of the half-cell Sn²⁺(aq)/Sn(s) if the potential of Fe(s)|Fe²⁺(aq) is –0.42 V.

3. Determine the free energy of the cell Fe(s)|Fe²⁺(aq)||Ag⁺(aq)|Ag(s).

$$\left(E^\theta\left(Fe^{2+}(aq)/Fe(s)\right) = -0.44\,V,\ E^\theta\left(Ag^+(aq)/Ag(s)\right) = 0.72\,V,\ F = 96500\,C\,mol^{-1}\right)$$

4. The free energy of the cell Ag(s)|Ag⁺(aq)||Pb⁴⁺(aq)|Pb²⁺(aq) is –172 kJ. Calculate the emf of the cell.

$$\left(F = 96500\,C\,mol^{-1}\right)$$

5. The standard free energy of the cell Ag(s)|Ag⁺(aq)||Au⁺(aq)|Au(s) is –99.3 kJ. Determine the potential of the half-cell Au⁺(aq)/Au(s).

$$\left(E^\theta\left(Ag^+(aq)/Ag(s)\right) = 0.80\,V,\ F = 96500\,C\,mol^{-1}\right)$$

6. Calculate the equilibrium constant of the reaction Zn²⁺(aq) + 2e⁻ ⇌ Zn(s) at 298 K.

$$\left(E^\theta\left(Zn^{2+}(aq)/Zn(s)\right) = -0.76\right)$$

7. The equilibrium constant of a galvanic cell is 6.05 × 10³¹ at 298 K. Determine the standard cell potential of the cell if 2 mol of electrons was exchanged in the reaction.

8. The reaction in a half-cell is oxidation. Determine the valency of the electrode if its standard cell potential and equilibrium constant at 298 K are –1.66 V and 6.49 × 10⁸⁵, respectively.

9. Calculate the electrode potential of the half-cell whose half-cell reaction is Fe²⁺(aq) + e⁻ → Fe²⁺(aq) when the concentration of Fe²⁺ is 100 times the concentration of Fe²⁺. The standard electrode potential is –0.77 V.

10. The cell potential of the cell K(s)|K⁺(aq)||Pt²⁺(aq)|Pt(s) is 4.00 V when the concentration of Pt²⁺ is 0.15 mol dm⁻³. Determine the concentration of Pt²⁺.

$$\left(E^\theta\left(K^+(aq)/K(s)\right) = -2.93\,V,\ E^\theta\left(Pt^{2+}(aq)/Pt(s)\right) = +1.20\,V\right)$$

Electrochemistry

11. Calculate the time required for 1.5 A of electric current to liberate 25 cm³ of hydrogen from acidified water at STP.

    $$\left(\text{Molar volume of gases at STP} = 22.4\,\text{dm}^3\,\text{mol}^{-1},\ F = 96500\,\text{C}\,\text{mol}^{-1}\right)$$

12. 0.50 A of electric current was passed through a solution of cobalt(II) chloride for 1.5 h. Determine the mass of cobalt deposited at the cathode.

    $$\left(Co = 59,\ F = 96500\,\text{C}\,\text{mol}^{-1}\right)$$

13. 2.2 A of electric current liberated 450 cm³ of chlorine from a solution of sodium chloride at STP. How long did it take to liberate this volume of chlorine?

    $$\left(\text{Molar volume of gases at STP} = 22.4\,\text{dm}^3\,\text{mol}^{-1},\ F = 96500\,\text{C}\,\text{mol}^{-1}\right)$$

14. Determine the amount of current required to deposit 1.2 g of aluminium from a solution of its salt in 75 min.

    $$\left(Al = 27,\ F = 96500\,\text{C}\,\text{mol}^{-1}\right)$$

15. 2.0 A of electricity was passed through acidified water for 45 min at STP. Determine the volume of hydrogen liberated at 55°C and 780 mmHg.

    $$\left(\text{Molar volume of gases at STP} = 22.4\,\text{dm}^3\,\text{mol}^{-1},\ F = 96500\,\text{C}\,\text{mol}^{-1}\right)$$

16. Determine the volume of oxygen gas that would be liberated by the same quantity of electricity required to liberate 460 cm³ of chlorine at STP?

    $$\left(\text{Molar volume of gases at STP} = 22.4\,\text{dm}^3\,\text{mol}^{-1}\right)$$

17. A quantity of electricity was passed through solutions of copper(II) nitrate and sodium chloride which were connected in series. Determine the volume of hydrogen liberated from the second solution at STP if 8.2 g of copper was deposited in the first solution.

    $$\left(Cu = 63.5,\ \text{molar volume of gases at STP} = 22.4\,\text{dm}^3\,\text{mol}^{-1}\right)$$

18. A quantity of electricity was passed through solutions of a nickel salt and aluminium sulfate. Determine the charge of the nickel ions in the first solution if the masses of nickel and aluminium deposited were 19.8 g and 6.1 g, respectively.

    $$\left(Al = 27,\ Ni = 58.7\right)$$

13 Volumetric Analysis

13.1 TITRATION

Volumetric analysis is a branch of analytical chemistry that involves determining the amount of a substance in solution based on the volume of a standard solution required for its complete reaction. The standard solution is called a *titrant*, while the solution with an unknown concentration is known as an *analyte* or *titrand*. The method used in volumetric analysis is called *titration*.

There are four types of titrations: *acid-base titration*, *redox titration*, *complexometric titration* and *precipitation titration*. In this book, we will focus on acid-base and redox titrations. In acid-base titration, an acid is titrated against a base, while an oxidising agent is titrated against a reducing agent in redox titration. Similar calculations are encountered in acid-base and redox titrations.

In titration, a specific volume—usually 20 or 25 cm³—of the titrand or analyte is transferred into a *conical flask* using a *pipette*. A titrant solution from a *burette* is then added to the solution in the conical flask until the reaction is just complete. This point is called the *equivalence point* of the titration.

For better accuracy, the titration is repeated to obtain two to three concordant values, which must lie within 0.10 cm³ of each other. The average of the concordant values is called the *average titre value*. The first value is called the *rough titre* because it usually differs significantly from the other values. The main steps involved in titration are illustrated in Figure 13.1. A sample titration result is given in Table 13.1.

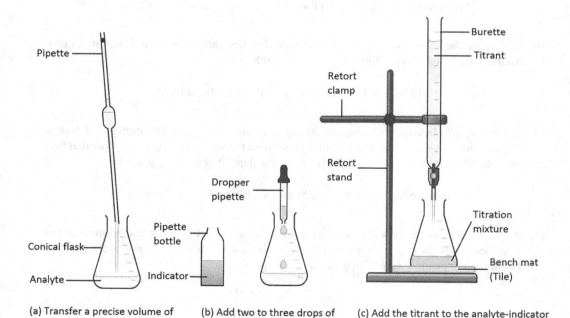

(a) Transfer a precise volume of the analyte into a conical flask.

(b) Add two to three drops of an indicator to the analyte.

(c) Add the titrant to the analyte-indicator until the endpoint is reached.

FIGURE 13.1 The main steps of titration.

TABLE 13.1
A Sample Titration Result

		Rough	1	2	3
Burette reading (cm³)	Final	22.80	44.90	22.10	44.25
	Initial	0.00	22.80	0.00	22.10
Titre value (cm³)		22.80	22.10	22.10	22.15

$$\text{Average titre value} = \frac{(22.10 + 22.10 + 22.15)\ cm^3}{3} = 22.12\ cm^3$$

TABLE 13.2
Some Indicators Used in Acid-Base Titration

Indicator	Natural Colour	pH Range of Colour Change	Colour in Acid	Colour in Base	Suitability
Methyl orange	Orange	3.1–4.4	Red	Yellow	Strong acid with weak base
Phenolphthalein	Colourless	8.2–10.0	Colourless	Pink	Strong acid with strong base/ weak acid with strong base
Methyl red	Red	4.8–6.0	Red	Yellow	Strong acid with weak base
Methyl yellow	Yellow	2.9–4.0	Red	Yellow	Strong acid with weak base
Bromocresol green	Green	4.0–5.6	Yellow	Blue	Strong acid with weak base
Bromocresol purple	Purple	5.2–6.8	Yellow	Purple	Strong acid with weak base
Phenol red	Red	6.8–8.4	Yellow	Red	Strong acid with strong base
Bromothymol blue	Blue	6.2–7.6	Yellow	Blue	Strong acid with strong base
Thymol blue	Blue	8.0–9.6	Red, Yellow	Blue	Weak acid with strong base

In most cases, it is impossible to detect the equivalence point of a titration by merely looking at the reaction mixture. To overcome this problem, an *indicator* is added to the solution in the conical flask, either at the beginning or at some other point during the titration. Indicators are organic dyes that give different colours in acid and base solutions and change colours with the pH of the solution. The *endpoint* of a titration, which occurs just after the equivalence point, is indicated by the permanent colour change of the reaction mixture. The endpoint is taken as an approximation of the equivalence point.

All acid-base titrations require an indicator, as acid and base solutions are usually colourless. The choice of indicator in an acid-base titration depends on the strengths of the acid and base involved. Some examples of common indicators used in acid-base titration are given in Table 13.2.

Some redox titrations do not require an indicator because the reaction itself causes a permanent colour change in the reacting solutions. Such titrations are said to be *self-indicating*. For instance, the endpoint of a redox titration involving acidified potassium permanganate, $KMnO_4$, is signalled by its colour change from purple to persistent pink. Some redox titrations require an indicator. For example, starch is used as an indicator in the titration of iodine solution with a reducing agent. Redox titrations are named according to the titrant, as shown in Table 13.3.

13.1.1 STANDARDISATION

One of the problems most commonly encountered in titration is the determination of the concentration of a solution. This process is called standardisation. The concentration of an acid or base solution is calculated from the relation:

TABLE 13.3
Types of Redox Titration

Redox Titration	Titrant
Permanganometry	Potassium permanganate ($KMnO_4$) (OA)
Dichromatometry	Potassium dichromate ($K_2Cr_2O_7$) (OA)
Iodimetry	Iodine (OA)
Iodometry	Iodine (RA)
Bromatometry	Potassium bromate ($KBrO_3$) (RA)
Cerimetry	Cerium(IV) salts (OA)

$$\frac{C_a V_a}{C_b V_b} = \frac{n_a}{n_b}$$

where

C_a = molar concentration of acid solution
V_a = volume of acid solution
C_b = molar concentration of base solution
V_b = volume of base solution
n_a = stoichiometric coefficient (number of moles) of acid in the balanced equation of the reaction
n_b = stoichiometric coefficient (number of moles) of base in the balanced equation of the reaction

The analogous relation for redox titration is given as follows:

$$\frac{C_{ox} V_{ox}}{C_{red} V_{red}} = \frac{n_{ox}}{n_{red}}$$

where

C_{ox} = molar concentration of oxidising agent solution
V_{ox} = volume of oxidising agent solution
C_{red} = molar concentration of reducing agent solution
V_{red} = volume of reducing agent solution
n_{ox} = stoichiometric coefficient (number of moles) of the oxidising agent in the balanced equation of the reaction
n_{red} = stoichiometric coefficient (number of moles) of the reducing agent in the balanced equation of the reaction

The relations above are based on the principle that, if the correct titre value is obtained in a titration, the ratio with which the analyte and titrant react should correspond with the ratio given by the balanced equation for the reaction. You should recall that the numerator and denominator of the left-hand side each relation represent the number of moles of each reactant involved in the reaction.

Volumetric Analysis

Example 1

In the titration of 25.00 cm³ of a 0.15 mol dm⁻³ sodium hydroxide solution against a 0.10 mol dm⁻³ hydrochloric acid solution, it was found that 18.50 cm³ of the acid solution was required to neutralise the base solution. Write a balanced equation for the reaction and determine the molar concentration of the acid.

Solution

The balanced equation of reaction is as follows:

$$HCl(aq) + NaOH(aq) \rightarrow NaCl(aq) + H_2O(l)$$

For an acid-base titration, we have

$$\frac{C_a V_a}{C_b V_b} = \frac{n_a}{n_b}$$

$$\therefore C_a = \frac{C_b V_b n_a}{V_a n_b}$$

$C_b = 0.15$ mol dm⁻³

$V_b = 25.00$ cm³

$V_a = 18.50$ cm³

$n_a = 1$

$n_b = 1$

$C_a = ?$

Substituting the values, we have

$$C_a = \frac{0.15 \text{ mol} \times 25.00 \text{ cm}^3 \times 1}{18.50 \text{ cm}^3 \times 1 \times 1 \text{ dm}^3} = 0.20 \text{ mol dm}^{-3}$$

Example 2

25.00 cm³ of a 0.20 mol dm⁻³ oxalic (ethanedioic) acid, $H_2C_2O_4$, solution is titrated against a 0.10 mol dm⁻³ acidified potassium permanganate, $KMnO_4$, solution. Determine the volume of $KMnO_4$ used in the titration. The equation of reaction is $2MnO_4^-(aq) + 5C_2O_4^{2-}(aq) + 16H^+(aq) \rightarrow 2Mn^{2+}(aq) + 8H_2O(l) + 10CO_2(g)$.

Solution

This is a redox titration in which the oxidising and reducing agents are $KMnO_4$ and $H_2C_2O_4$, respectively. For a redox titration, we have

$$\frac{C_{ox} V_{ox}}{C_{red} V_{red}} = \frac{n_{ox}}{n_{red}}$$

$$\therefore V_{ox} = \frac{C_{red} V_{red} n_{ox}}{C_{ox} n_{red}}$$

$$C_{red} = 0.20 \text{ mol dm}^{-3}$$

$$V_{red} = 25.00 \text{ cm}^3$$

$$C_{ox} = 0.10 \text{ mol dm}^{-3}$$

$$n_{ox} = 2$$

$$n_{red} = 5$$

$$V_{ox} = ?$$

Substituting, we have

$$V_{ox} = \frac{0.20 \text{ mol dm}^{-3} \times 25.00 \text{ cm}^3 \times 2}{0.10 \text{ mol dm}^{-3} \times 5} = 20.00 \text{ cm}^3$$

Example 3

It requires 21.20 cm³ of a 0.15 mol dm⁻³ solution of sulfuric acid to reach the endpoint when 20.00 cm³ of potassium hydroxide, KOH, was titrated against the solution.

(a) write the balanced equation of reaction;
(b) calculate the concentration of the base solution;
(c) determine the number of moles of the base in the base solution.

Solution

(a) The balanced equation of reaction is as follows:

$$H_2SO_4(aq) + 2KOH(aq) \rightarrow K_2SO_4(aq) + 2H_2O(l)$$

(b) We know that, for an acid-base reaction,

$$\frac{C_a V_a}{C_b V_b} = \frac{n_a}{n_b}$$

$$\therefore C_b = \frac{C_a V_a n_b}{V_b n_a}$$

$$C_a = 0.15 \text{ mol dm}^{-3}$$

$$V_a = 21.20 \text{ cm}^3$$

$$V_b = 20.00 \text{ cm}^3$$

$$n_a = 1$$

$$n_b = 2$$

$$C_b = ?$$

Volumetric Analysis

Substituting, we have

$$C_b = \frac{0.15 \text{ mol} \times 21.20 \text{ cm}^3 \times 2}{20.00 \text{ cm}^3 \times 1 \times 1 \text{ dm}^3} = 0.32 \text{ mol dm}^{-3}$$

(c) We will apply the relation:

$$n = C \times V$$

$$C = 0.32 \text{ mol dm}^{-3}$$

$$V = 20.00 \text{ cm}^3 = 0.0020 \text{ dm}^3$$

$$n = ?$$

Substituting the values, we obtain

$$n = \frac{0.32 \text{ mol}}{1 \text{ dm}^3} \times 0.020 \text{ dm}^3 = 0.0064 \text{ mol}$$

Example 4

It requires 15.50 cm^3 of acidified potassium permanganate, $KMnO_4$, to reach the endpoint in the titration of 20.00 cm^3 of a 10.5-g dm^{-3} solution of anhydrous iron(II) sulfate, $FeSO_4$, against the oxidising agent. The equation of reaction is $MnO_4^-(aq) + 5Fe^{2+}(aq) + 8H^+(aq) \rightarrow Mn^{2+}(aq) + 5Fe^{3+}(aq) + 4H_2O(l)$. Determine the molar concentration of the $KMnO_4$ solution.

$$(O = 16, S = 32, Fe = 56)$$

This is a redox titration in which the oxidising and reducing agents are $KMnO_4$ and $FeSO_4$, respectively. For a redox titration, we have

$$\frac{C_{ox}V_{ox}}{C_{red}V_{red}} = \frac{n_{ox}}{n_{red}}$$

$$\therefore C_{ox} = \frac{C_{red}V_{red}n_{ox}}{V_{ox}n_{red}}$$

The molar concentration of the $FeSO_4$ solution is obtained as follows.

$$C = \frac{\rho}{M}$$

$$\rho = 10.5 \text{ g dm}^{-3}$$

$$M = (56 + 32 + \{16 \times 4\}) \text{ g mol}^{-1} = 152 \text{ g mol}^{-1}$$

$$C = ?$$

Substituting, we have

$$C = \frac{10.5 \text{ g} \times 1 \text{ mol}}{152 \text{ g} \times 1 \text{ dm}^3} = 0.06908 \text{ mol dm}^{-3}$$

So

$$C_{red} = 0.06908 \text{ mol dm}^{-3}$$

$$V_{red} = 20.00 \text{ cm}^3$$

$$V_{ox} = 15.50 \text{ cm}^3$$

$$n_{ox} = 1$$

$$n_{red} = 5$$

$$C_{ox} = ?$$

Substituting, we have

$$C_{ox} = \frac{0.06908 \text{ mol} \times 20.00 \text{ cm}^3 \times 1}{15.50 \text{ cm}^3 \times 5 \times 1 \text{ dm}^3} = 0.018 \text{ mol dm}^{-3}$$

PRACTICE PROBLEMS

1. It requires just 10.50 cm³ of a solution of potassium dichromate, $K_2Cr_2O_7$, for the complete reaction of 25.00 cm³ of a 0.45 mol dm⁻³ solution of acidified iron(II) sulfate, $FeSO_4$. Calculate the concentration of the oxidising agent solution. The equation of reaction is $Cr_2O_7^{2-}(aq) + 6Fe^{2+}(aq) + 14H^+(aq) \rightarrow 2Cr^{3+}(aq) + 6Fe^{3+}(aq) + 7H_2O(l)$.
2. It requires 18.70 cm³ of a 0.15 mol dm⁻³ solution of nitric acid to reach the endpoint in the titration of 20.00 cm³ of a sodium hydroxide, NaOH, solution against the acid. Write the balanced equation for the reaction and determine:
   (a) the concentration of the base solution;
   (b) the number of moles of the base in the base solution;
   (c) the mass of the base in the base solution.

$$(H = 1, O = 16, Na = 23)$$

13.1.2 Solubility

The solubility of a substance at a particular temperature can be determined by dissolving it in water and titrating the resulting solution with a standard solution. If a *saturated* solution of the substance is used for titration, then the solubility of the substance at the temperature at which the solution was prepared corresponds to the *concentration* of the solution. On the other hand, if an unsaturated solution of the substance is used, its solubility must be determined using the procedure explained in Chapter 9.

Volumetric Analysis

Example 1

25.00 cm³ of a saturated solution of sodium hydroxide prepared at 25°C was titrated against a 0.15 mol dm⁻³ solution of nitric acid. Determine the molar solubility of the base at 25°C if the average titre value is 27.00 cm³.

Solution

The balanced equation for the reaction is $HNO_3(aq) + NaOH(aq) \rightarrow NaNO_3(aq) + H_2O(l)$.
The first step is to calculate the molar concentration of the base as follows.

$$\frac{C_a V_a}{C_b V_b} = \frac{n_a}{n_b}$$

$$\therefore C_b = \frac{C_a V_a n_b}{V_b n_a}$$

$$C_a = 0.15 \text{ mol dm}^{-3}$$

$$V_a = 27.00 \text{ cm}^3$$

$$V_b = 25.00 \text{ cm}^3$$

$$n_a = 1$$

$$n_b = 1$$

$$C_b = ?$$

Substituting, we have

$$C_b = \frac{0.15 \text{ mol} \times 27.00 \text{ cm}^3 \times 1}{25.00 \text{ cm}^3 \times 1 \times 1 \text{ dm}^3} = 0.16 \text{ mol dm}^{-3}$$

The base solution is saturated; hence, the molar solubility of the base is 0.16 mol dm⁻³.

Example 2

35.00 cm³ of a saturated iodine solution was prepared by dissolving crystals of the element in potassium iodide solution at 25°C. The solution was then made up to 1.0 dm³ at the same temperature. Determine the solubility of the iodine crystals in g/100 g H_2O at 25°C, given that an average titre value of 12.50 cm³ was obtained when 25.00 cm³ of the iodine solution was titrated against a 0.25 mol dm⁻³ sodium thiosulfate, $Na_2S_2O_3$, solution. The equation for the reaction is $I_2(aq) + 2S_2O_3^{2-}(aq) \rightarrow 2I^-(aq) + S_4O_6^{2-}(aq)$.

$$(I = 127)$$

Solution

This is a redox titration in which the oxidising and reducing agents are iodine and sodium thiosulfate, respectively. To determine the solubility of iodine, we must first calculate the molar concentration of its solution.

$$\frac{C_{ox}V_{ox}}{C_{red}V_{red}} = \frac{n_{ox}}{n_{red}}$$

$$\therefore C_{ox} = \frac{C_{red}V_{red}n_{ox}}{V_{ox}n_{red}}$$

$$C_{red} = 0.25 \text{ mol dm}^{-3}$$

$$V_{ox} = 25.00 \text{ cm}^3$$

$$V_{red} = 12.50 \text{ cm}^3$$

$$n_{ox} = 1$$

$$n_{red} = 2$$

$$C_{ox} = ?$$

Substituting, we have

$$C_{ox} = \frac{0.25 \text{ mol} \times 12.50 \text{ cm}^3 \times 1}{25.00 \text{ cm}^3 \times 2 \times 1 \text{dm}^3} = 0.0625 \text{ mol dm}^{-3}$$

Hence, the 85.00 g saturated solution contains 0.0625 mol of iodine at 25°C.
The next step is to calculate the mass of iodine in this amount of the element, using the relation:

$$n = \frac{m}{M}$$

$$\therefore m = n \times M$$

where $M = (2 \times 127)$ g mol^{-1} = 254 g mol^{-1}
So

$$m = 0.0625 \text{ mol} \times \frac{254 \text{ g}}{1 \text{ mol}} = 15.875 \text{ g}$$

Let the maximum amount of iodine that will dissolve in 100 g of water at the specified temperature be S. It then follows that

$$85.00 \text{ g} = 15.875 \text{ g}$$

$$100 \text{ g} = S$$

So

$$85.00 \text{ g} \times S = 100 \text{ g} \times 15.875 \text{ g}$$

$$\therefore S = \frac{100 \text{ g} \times 15.875 \text{ g}}{85.00 \text{ g}} = 18.68 \text{ g}/100 \text{ g H}_2\text{O}$$

PRACTICE PROBLEMS
1. 25.00 cm³ of a saturated solution of potassium hydroxide, KOH, at 25°C was titrated against a 0.15 mol dm⁻³ solution of sulfuric acid, H₂SO₄. Determine the solubility of the base in g/100 g H₂O at 25°C, given that 17.50 cm³ of the acid is required for the complete reaction of the base solution.

$$(H = 1, O = 16, K = 39)$$

Volumetric Analysis

2. A solution of anhydrous atomic sodium carbonate was prepared by making up 55.00 cm^5 of its saturated solution to 1.0 dm^3 at 20°C. 25.00 cm^3 of the solution was then titrated against a 0.10 mol dm^{-3} sulfuric acid solution. Determine the solubility of the base in mol dm^{-3} at 20°C if the average titre value is 21.50 cm^3.

13.1.3 Mole Ratio

As we have seen in the previous sections, the mole ratio of analyte to titrant can be easily obtained from the balanced equation of the reaction if the two reactants are specified. However, if either of the reactants cannot be identified, the mole ratio must be determined from the amount of each reactant consumed in the reaction, as illustrated in the following examples.

Example 1

25.00 cm^3 of a 0.14 mol dm^{-3} solution of sodium hydroxide was titrated against a 0.10 mol dm^{-3} solution of a strong acid. Determine the mole ratio of acid to base if the average titre value is 17.20 cm^3.

Solution

The very first step is to calculate the number of moles of each reactant as follows.

$$n = C \times V$$

For the acid, we have

$$C = 0.10 \text{ mol dm}^{-3}$$

$$V = 17.20 \text{ cm}^3 = 0.0172 \text{ dm}^3$$

$$n = ?$$

Substituting, we have

$$n = \frac{0.10 \text{ mol}}{1 \text{ dm}^3} \times 0.0172 \text{ dm}^3 = 0.00172 \text{ mol}$$

For the base, we have

$$C = 0.14 \text{ mol dm}^{-3}$$

$$V = 25.00 \text{ cm}^3 = 0.025 \text{ dm}^3$$

$$n = ?$$

Substituting, we have

$$n = \frac{0.14 \text{ mol}}{1 \text{ dm}^3} \times 0.025 \text{ dm}^3 = 0.0035 \text{ mol}$$

The whole-number mole ratio of acid to base is obtained by dividing the number of mole of acid by the number of moles of base.

$$\frac{n_a}{n_b} = \frac{0.00172 \text{ mol}}{0.0035 \text{ mol}} = \frac{1}{2}$$

Thus, the mole ratio is 1 : 2. Only dibasic acids—i.e., acids with two replaceable hydrogen atoms—react with sodium hydroxide in this ratio. The acid in question is definitely sulfuric acid, as it is the only well-known strong dibasic acid. The equation of reaction is as follows:

$$H_2SO_4 (aq) + 2NaOH(aq) \rightarrow Na_2SO_4 (aq) + 2H_2O(l)$$

Example 2

18.87 cm³ of a 0.15 mol dm⁻³ solution of a mineral acid is required for the complete reaction of 25.00 cm³ of a solution containing 6.0 g of anhydrous sodium carbonate, Na_2CO_3, per dm³ of solution. Determine the mole ratio of acid to base.

$$(C = 12, O = 16, Na = 23)$$

Solution

The very first step is to calculate the number of moles of each reactant.

$$n = C \times V$$

For the acid, we have

$$C = 0.15 \text{ mol dm}^{-3}$$

$$V = 18.87 \text{ cm}^3 = 0.01887 \text{ dm}^3$$

$$n = ?$$

Substituting, we have

$$n = \frac{0.15 \text{ mol}}{1 \text{ dm}^3} \times 0.01887 \text{ dm}^3 = 0.0028305 \text{ mol}$$

For the base, we have

$$C = \frac{\rho}{M}$$

$$\rho = 6.0 \text{ g dm}^{-3}$$

$$M = (23 + 12 + \{16 \times 3\}) \text{ g mol}^{-3} = 106 \text{ g mol}^{-1}$$

$$C = ?$$

$$\therefore C = \frac{6.0 \text{ g} \times 1 \text{ mol}}{106 \text{ g} \times 1 \text{ dm}^3} = 0.0566 \text{ mol dm}^{-3}$$

$$V = 25.00 \text{ cm}^3 = 0.025 \text{ dm}^3$$

$$n = ?$$

Substituting, we have

$$n = \frac{0.0566 \text{ mol}}{1 \text{ dm}^3} \times 0.025 \text{ dm}^3 = 0.001415 \text{ mol}$$

So

$$\frac{n_a}{n_b} = \frac{0.0028305 \text{ mol}}{0.001415 \text{ mol}} = \frac{2}{1} = 2:1$$

Volumetric Analysis

PRACTICE PROBLEMS

1. The average titre value when 25.00 cm³ of a 0.10 mol dm⁻³ solution of a base was titrated against a 0.085 mol dm⁻³ solution of sulfuric acid, H_2SO_4, is 14.71 cm³. Determine the mole ratio of acid to base.

2. A student titrated 20.00 cm³ of a 0.15 mol dm⁻³ solution of a base against a solution of hydrochloric acid containing 4.5 g of the acid in 1000 cm³ of solution. Determine the mole ratio of acid to base if the average titre value is 24.33 cm³. Hence, if possible, suggest the base used in the titration.

$$(H = 1, Cl = 35.5)$$

13.1.4 PERCENT PURITY

Purity is very important in certain branches of the chemical industry, such as food and drug manufacturing. Titration is used to determine the purity of a substance as follows:

$$\%\text{purity or mass} = \frac{\text{mass of pure substance}}{\text{mass of impure substance}} \times 100\%$$

$$= \frac{\text{mass concentration of pure sustance}}{\text{mass concentration of impure substance}} \times 100\%$$

Since impurities will not partake in the desired analyte-titrant reaction, it follows that all calculated values will be based solely on the pure substance.

Example 1

18.00 cm³ of a 9.0 g dm⁻³ solution of impure oxalic acid, $H_2C_2O_4$, is required to reach the endpoint when 20.00 cm³ of a 0.15 mol dm⁻³ solution of sodium hydroxide, NaOH, was titrated against it. Determine the percent purity of the acid.

$$(H = 1, C = 12, O = 16)$$

Solution

The balanced equation of reaction is as follows:

$$H_2C_2O_4(aq) + 2NaOH(aq) \rightarrow Na_2C_2O_4(aq) + 2H_2O(l)$$

The first step is to obtain the molar concentration of the acid.

$$\frac{C_a V_a}{C_b V_b} = \frac{n_a}{n_b}$$

$$\therefore C_a = \frac{C_b V_b n_a}{V_a n_b}$$

$$C_b = 0.15 \text{ mol dm}^{-3}$$

$$V_b = 20.00 \text{ cm}^3$$

$$V_a = 18.00 \text{ cm}^3$$

$$n_a = 1$$

$$n_b = 2$$

$$C_a = ?$$

Substituting, we have

$$C_a = \frac{0.15 \text{ mol} \times 20.00 \text{ cm}^3 \times 1}{18.00 \text{ cm}^3 \times 2 \times 1 \text{dm}^3} = 0.0833 \text{ mol dm}^{-3}$$

The next step is to convert the molar concentration of the pure acid to mass concentration.

$$C = \frac{\rho}{M}$$

$$\therefore \rho = C \times M$$

$$C = 0.0833 \text{ mol dm}^{-3}$$

$$M = (\{1 \times 2\} + \{12 \times 2\} + \{16 \times 4\}) \text{ g mol}^{-1} = 90 \text{ g mol}^{-1}$$

$$\rho = ?$$

Substituting, we have

$$\rho = \frac{0.0833 \text{ mol}}{1 \text{dm}^3} \times \frac{90 \text{ g}}{1 \text{mol}} = 7.497 \text{ g dm}^{-3}$$

Finally, we can calculate the percent purity of the acid as follows.

$$\% \text{ purity} = \frac{\text{mass concentration of pure acid}}{\text{mass concentration of impure acid}} \times 100\%$$

Mass concentration of pure acid = 7.497 g dm^{-3}

Mass concentration of impure acid = 9.0 g dm^{-3}

% purity = ?

Substituting, we have

$$\% \text{ purity} = \frac{7.497 \text{ g dm}^{-3}}{9.0 \text{ g dm}^{-3}} \times 100\% = 83\%$$

Volumetric Analysis

Example 2

A 2.5 g sample of iron was completely dissolved in dilute sulfuric acid. The resulting solution requires 27.50 cm³ of a 0.10 mol dm⁻³ solution of potassium permanganate, $KMnO_4$, to reach the endpoint during titration. Calculate the percent mass of iron in the sample. The equation for the reaction is $MnO_4^-(aq) + 5Fe^{2+}(aq) + 8H^+(aq) \rightarrow Mn^{2+}(aq) + 5Fe^{3+}(aq) + 4H_2O(l)$.

$$(Fe = 56)$$

Solution

We will begin by calculating the number of moles of $KMnO_4$ that took part in the reaction.

$$n = C \times V$$
$$C = 0.10 \text{ mol dm}^{-3}$$
$$V = 27.50 \text{ cm}^3 = 0.0275 \text{ dm}^3$$
$$n = ?$$

Substituting we have

$$n = \frac{0.10 \text{ mol}}{1 \text{ dm}^3} \times 0.0275 \text{ dm}^3 = 0.00275 \text{ mol}$$

As shown by the equation of reaction, the mole ratio of $KMnO_4$ to Fe is 1:5. Thus, the number of moles of Fe that took part in the reaction is given by

$$n = 5 \times 0.00275 \text{ mol} = 0.01375 \text{ mol}$$

This is the number of moles of pure Fe. The mass of pure Fe is obtained as follows.

$$n = \frac{m}{M}$$
$$\therefore m = n \times M$$
$$n = 0.01375 \text{ mol}$$
$$M = 56 \text{ g mol}^{-1}$$
$$m = ?$$

Substituting, we have

$$m = 0.01375 \text{ mol} \times \frac{56 \text{ g}}{1 \text{ mol}} = 0.77 \text{ g}$$

Finally, we can now calculate the percent mass of Fe in the sample.

$$\% \text{ mass} = \frac{\text{mass of pure iron}}{\text{mass of impure iron}} \times 100\%$$

Mass of pure iron = 0.77 g

Mass of impure iron = 2.5 g

% mass = ?

Substituting, we have

$$\% \text{ mass} = \frac{0.77 \text{ g}}{2.5 \text{ g}} \times 100\% = 31\%$$

Example 3

It requires 27.50 cm³ of a 0.10 mol dm⁻³ solution of hydrochloric acid, HCl, to reach the endpoint when 25.00 cm³ of a solution containing 1.5 g of impure anhydrous sodium carbonate, Na_2CO_3, per 250 cm³ of solution was titrated against it. Calculate the percent purity of the base.

$$(C = 12, O = 16, Na = 23)$$

Solution

The balanced equation of the reaction is as follows:

$$2HCl(aq) + Na_2CO_3(aq) \rightarrow 2NaCl(aq) + H_2O(l) + CO_2(g)$$

We will begin by calculating the molar concentration of the pure base.

$$\frac{C_a V_a}{C_b V_b} = \frac{n_a}{n_b}$$

$$\therefore C_b = \frac{C_a V_a n_b}{V_b n_a}$$

$C_a = 0.10$ mol dm⁻³

$V_a = 27.50$ cm³

$V_b = 25.00$ cm³

$n_a = 2$

$n_b = 1$

$C_b = ?$

Substituting, we have

$$C_b = \frac{0.10 \text{ mol} \times 27.50 \text{ cm}^3 \times 1}{25.00 \text{ cm}^3 \times 2 \times 1 \text{ dm}^3} = 0.055 \text{ mol dm}^{-3}$$

This is the molar concentration of the pure base. The mass concentration of the pure base is determined as follows.

$$C = \frac{\rho}{M}$$

$$\therefore \rho = C \times M$$

$C = 0.055$ mol dm⁻³

$M = (\{23 \times 2\} + 12 + \{16 \times 3\})$ g mol⁻¹ = 106 g mol⁻¹

$\rho = ?$

Substituting, we have

$$\rho = \frac{0.055 \text{ mol}}{1 \text{ dm}^3} \times \frac{106 \text{ g}}{1 \text{ mol}} = 5.83 \text{ g dm}^{-3}$$

Volumetric Analysis

The mass concentration of the impure base is calculated as follows.

$$\rho = \frac{m}{V}$$
$$m = 1.5\,\text{g}$$
$$V = 250\,\text{cm}^3 = 0.25\,\text{dm}^3$$
$$\rho = ?$$

Substituting, we have

$$\rho = \frac{1.5\,\text{g}}{0.25\,\text{dm}^3} = 6.0\,\text{g dm}^{-3}$$

Finally, the percent purity of the base is calculated using the following relation:

$$\%\text{purity} = \frac{\text{mass concentration of pure base}}{\text{mass concentration of impure base}} \times 100\%$$

Mass concentration of pure base = 5.83 g dm^{-3}

Mass concentration of impure base = 6.0 g dm^{-3}

%purity = ?

Substituting the values, we obtain

$$\%\text{purity} = \frac{5.83\,\text{g dm}^{-3}}{6.0\,\text{g dm}^{-3}} \times 100\% = 97\%$$

PRACTICE PROBLEMS

1. An average titre value of 17.50 cm^3 was obtained when a 0.10 mol dm^{-3} solution of sulfuric acid, H_2SO_4, was titrated with 25.00 cm^3 of a solution containing 8.0 g of impure anhydrous sodium carbonate, Na_2CO_3, per dm^3 of solution. Calculate the percent purity of the base.

$$(C = 12,\ O = 16,\ Na = 23)$$

2. It requires 15.50 cm^3 of a 0.15 mol dm^{-3} solution of potassium permanganate, $KMnO_4$, to reach the endpoint when an acidified solution of sodium oxalate, $Na_2C_2O_4$, containing 1.5 g of the salt was titrated against it. Calculate the percent purity of the sodium oxalate. The equation for the reaction is $5C_2O_4^{2-}(aq) + 2MnO_4^-(aq) + 8H^+(aq) \rightarrow 2Mn^{2+}(aq) + 10CO_2(g) + 8H_2O(l)$.

$$(C = 12,\ O = 16,\ Na = 23)$$

13.1.5 Number of Molecules of Water of Crystallisation

The number of molecules of water of crystallisation in a hydrated substance is determined through titration using any of the following relations.

1.
$$C = \frac{\rho}{(M_{ra} + 18x)\,\text{g mol}^{-1}}$$

where

C = molar concentration of salt
ρ = mass concentration of hydrated salt
M_{ra} = relative molecular mass of anhydrous salt
x = number of molecules of water of crystallisation of the hydrated salt

2.
$$n = \frac{m}{(M_{ra} + 18x)\,\text{g mol}^{-1}}$$

where

n = number of moles of the hydrated salt in a given volume of solution
m = mass of the hydrated salt in a given volume of solution
M_{ra} = relative molecular mass of anhydrous salt
x = number of molecules of water of crystallisation of the hydrated salt

3.
$$\frac{\text{mass concentration of hydrated salt}}{\text{molar mass of hydrated salt}} = \frac{\text{mass concentration of anhydrous salt}}{\text{molar mass of anhydrous salt}}$$

Note: $(M_{ra} + 18x)$ g mol^{-1} = molar mass of the hydrated substance. It is also worth noting that water of crystallisation does not partake in chemical reactions.

Example 1

25.00 cm³ of a hydrated sodium carbonate, $Na_2CO_3 \cdot xH_2O$, solution containing 15.8 g of the hydrated salt per dm³ of solution was titrated against a 0.10 mol dm^{-3} solution of sulfuric acid, H_2SO_4. Determine the number of molecules of water of crystallisation in the base if the average titre value is 17.02 cm³.

$$(C = 12,\ O = 16,\ Na = 23)$$

Solution

The balanced equation of reaction is as follows:

$$H_2SO_4\,(aq) + Na_2CO_3\,(aq) \rightarrow Na_2SO_4\,(aq) + H_2O\,(l) + CO_2\,(g)$$

We will begin by calculating the molar concentration of the base.

$$\frac{C_a V_a}{C_b V_b} = \frac{n_a}{n_b}$$

$$\therefore C_b = \frac{C_a V_a n_b}{V_b n_a}$$

$C_a = 0.10$ mol dm^{-3}

$V_a = 17.02$ cm³

$V_b = 25.00$ cm³

$n_a = 1$

$n_b = 1$

$C_b = ?$

Volumetric Analysis

Substituting, we have

$$C_b = \frac{0.10 \text{ mol} \times 17.02 \text{ cm}^3 \times 1}{25.00 \text{ cm}^3 \times 1 \times 1 \text{ dm}^3} = 0.06808 \text{ mol dm}^{-3}$$

Since water is not involved in the reaction this is also the molar concentration of the hydrated base. The number of molecules of water of crystallisation is obtained as follows.

$$C = \frac{\rho}{(M_{ra} + 18x) \text{ g mol}^{-1}}$$

$$\therefore (M_{ra} + 18x) \text{ g mol}^{-1} = \frac{\rho}{C}$$

$$M_{ra} = (23 \times 2) + 12 + (16 \times 3) = 106$$

$$\rho = 15.8 \text{ g dm}^{-3}$$

$$C = 0.06808 \text{ mol dm}^{-3}$$

$$x = ?$$

Substituting, we have

$$(106 + 18x) \text{ g mol}^{-1} = \frac{15.8 \text{ g dm}^{-3}}{0.06808 \text{ mol dm}^{-3}}$$

So

$$0.06808 \text{ mol dm}^{-3} \times \frac{(106 + 18x) \text{ g}}{1 \text{ mol}} = 15.8 \text{ g dm}^{-3}$$

$$1.22544x \text{ g dm}^{-3} = 15.8 \text{ g dm}^{-3} - 7.21648 \text{ g dm}^{-3}$$

$$1.22544x \text{ g dm}^{-3} = 8.5835 \text{ g dm}^{-3}$$

$$\therefore x = \frac{8.5835 \text{ g dm}^{-3}}{1.22544 \text{ g dm}^{-3}} = 7$$

Alternatively, we can use the following relation:

$$n = \frac{m}{(M_{ra} + 18x) \text{ g mol}^{-1}}$$

$$\therefore (M_{ra} + 18x) \text{ g mol}^{-1} = \frac{m}{n}$$

The number of moles of the hydrated salt is obtained as follows.

$$n = C \times V$$

$$C = 0.06808 \text{ mol dm}^{-3}$$

$$V = 1.0 \text{ dm}^3$$

$$n = ?$$

Substituting, we obtain

$$n = \frac{0.06808 \text{ mol}}{1 \text{ dm}^3} \times 1.0 \text{ dm}^3 = 0.06808 \text{ mol}$$

$$m = 15.8 \text{ g}$$

$$M_{ra} = 106$$

$$x = ?$$

Substituting, we have

$$(106 + 18x) \text{ g mol}^{-1} = \frac{15.8 \text{ g}}{0.06808 \text{ mol}}$$

So

$$0.06808 \text{ mol} \times \frac{(106 + 18x) \text{ g}}{1 \text{ mol}} = 15.8 \text{ g}$$

$$1.22544x \text{ g} = 15.8 \text{ g} - 7.21648 \text{ g}$$

$$1.22544x \text{ g} = 8.58352 \text{ g}$$

$$\therefore x = \frac{8.58352 \text{ g}}{1.22544 \text{ g}} = 7$$

Lastly, we will apply the third relation:

$$\frac{\text{mass concentration of hydrated salt}}{\text{molar mass of hydrated salt}} = \frac{\text{mass concentration of anhydrous salt}}{\text{molar mass of anhydrous salt}}$$

The mass concentration, ρ, of the anhydrous salt is obtained as follows.

$$C = \frac{\rho}{M}$$

$$\therefore \rho = C \times M$$

$$C = 0.06808 \text{ mol dm}^{-3}$$

$$M = 106 \text{ g mol}^{-1}$$

$$\rho = ?$$

Substituting, we have

$$\rho = \frac{0.06808 \text{ mol}}{1 \text{ dm}^3} \times \frac{106 \text{ g}}{1 \text{ mol}} = 7.216 \text{ g mol}^{-1}$$

Mass concentration of hydrated salt = 15.8 g dm^{-3}

Molar mass of hydrated salt = $(106 + 18x) \text{ g mol}^{-1}$

Molar mass of anhydrous salt = 106 g mol^{-1}

Volumetric Analysis

By substituting, we obtain

$$\frac{15.8\,g \times 1\,mol}{(106+18x)\,g \times 1\,dm^3} = \frac{7.216\,g \times 1\,mol}{106\,g \times 1\,dm^3}$$

So

$$(106+18x)\,dm^3 \times 7.216\,mol = 15.8\,mol \times 106\,dm^3$$

$$764.896 + 129.888x = 1674.8$$

$$129.888x = 909.904$$

$$\therefore x = \frac{909.904}{129.888} = 7$$

Thus, the formula of the salt is $Na_2CO_3 \cdot 7H_2O$.

Example 2

20.00 cm³ of a solution of hydrated sodium carbonate, $Na_2CO_3 \cdot xH_2O$, containing 1.5 g of the salt in 150 cm³ of solution was titrated against a 0.15 mol dm⁻³ solution of hydrochloric acid. If the average titre value is 21.50 cm³, determine:

(a) the number of molecules of water of crystallisation of the salt and write its formula;
(b) the percent composition of water of crystallisation in the salt.

$$(C = 12,\ O = 16,\ Na = 23)$$

Solution

(a) The balanced equation for the reaction is as follows:

$$2HCl(aq) + Na_2CO_3(aq) \rightarrow 2NaCl(aq) + H_2O(l) + CO_2(g)$$

We will begin by calculating the molar concentration of the hydrated base.

$$\frac{C_a V_a}{C_b V_b} = \frac{n_a}{n_b}$$

$$\therefore C_b = \frac{C_a V_a n_b}{V_b n_a}$$

$$C_a = 0.15\ mol\ dm^{-3}$$

$$V_a = 21.50\ cm^3$$

$$V_b = 20.00\ cm^3$$

$$n_a = 2$$

$$n_b = 1$$

$$C_b = ?$$

Substituting, we have

$$C_b = \frac{0.15 \text{ mol} \times 21.50 \text{ cm}^3 \times 1}{20.00 \text{ cm}^3 \times 2 \times 1 \text{ dm}^3} = 0.080625 \text{ mol dm}^{-3}$$

We will determine the number of molecules of water of crystallisation as follows.

$$C = \frac{\rho}{(M_{ra} + 18x) \text{ g mol}^{-1}}$$

$$\therefore (M_{ra} + 18x) \text{ g mol}^{-1} = \frac{\rho}{C}$$

$$\rho = \frac{m}{V}$$

$$m = 1.5 \text{ g}$$

$$V = 150 \text{ cm}^3 = 0.15 \text{ dm}^3$$

$$\rho = ?$$

Substituting, we have

$$\rho = \frac{1.5 \text{ g}}{0.15 \text{ dm}^3} = 10 \text{ g dm}^{-3}$$

$$M_{ra} = (23 \times 2) + 12 + (16 \times 3) = 106$$

$$\rho = 10 \text{ g dm}^{-3}$$

$$C = 0.080625 \text{ mol dm}^{-3}$$

$$x = ?$$

Substituting, we have

$$(106 + 18x) \text{ g mol}^{-1} = \frac{10 \text{ g dm}^{-3}}{0.080625 \text{ mol dm}^{-3}}$$

So

$$0.080625 \text{ mol dm}^{-3} \times \frac{(106 + 18x) \text{ g}}{1 \text{ mol}} = 10 \text{ g dm}^{-3}$$

$$8.54625 \text{ g dm}^{-3} + 1.45125x \text{ g dm}^{-3} = 10 \text{ g dm}^{-3}$$

$$1.45125x \text{ g dm}^{-3} = 1.45375 \text{ g dm}^{-3}$$

$$\therefore x = \frac{1.45375 \text{ g dm}^{-3}}{1.45125 \text{ g dm}^{-3}} = 1$$

Thus, the formula of the salt is $Na_2CO_3.H_2O$.
Note that it would have made no difference if we had applied either of the other two relations. We leave it to you to verify this for practice.

Volumetric Analysis

(b) We will apply the relation:

$$\% \text{ composition of water} = \frac{18x}{(M_{ra} + 18x)} \times 100\%$$

$$x = 1$$
$$M_r = 106$$
$$\% \text{ composition of water} = ?$$

Substituting, we have

$$\% \text{ composition of water} = \frac{18}{(106 + 18)} \times 100\% = 15\%$$

PRACTICE PROBLEMS

1. 11.65 cm³ of a 0.15 mol dm⁻³ nitric acid solution neutralises 20.00 cm³ of a solution of hydrated sodium carbonate, $Na_2CO_3.xH_2O$, containing 2.5 g of the salt in 200.00 cm³ of solution.

   Determine:
   (a) the number of molecules of water of crystallisation in the salt;
   (b) the percentage composition of water in the salt.

   $$(C = 12, \ O = 16, \ Na = 23)$$

2. It requires 24.00 cm³ of a 0.25 mol dm⁻³ solution of potassium permanganate, $KMnO_4$, to reach the endpoint when 20.00 cm³ of a solution of hydrated oxalic acid, $H_2C_2O_4.xH_2O$, containing 5.37g of the acid in 250.0 cm³ of solution was titrated against it. The equation of the reaction is:

   $$2MnO_4^-(aq) + 5C_2O_4^{2-}(aq) + 16H^+(aq) \rightarrow 2Mn^{2+}(aq) + 8H_2O(l) + 10CO_2(g).$$

   Determine:
   (a) the number of molecules of water of crystallisation in the salt;
   (b) the percentage composition of water in the salt.

   $$(C = 12, \ O = 16, \ Na = 23)$$

13.1.6 MOLAR MASS

The molar mass of a substance is determined from titration using the following familiar relation:

$$C = \frac{\rho}{M}$$

Example 1

The average titre value when 20.00 cm³ of a 0.150 mol dm⁻³ solution of sodium hydroxide, NaOH, was titrated against an unknown acid, HX, containing 3.65 g of the acid per dm³ of solution is 30.00 cm³. Determine the molar mass of the acid.

Solution

We will apply the relation:

$$C = \frac{\rho}{M}$$

$$\therefore M = \frac{\rho}{C}$$

The equation of reaction is as follows:

$$HX(aq) + NaOH(aq) \rightarrow NaX(aq) + H_2O(l)$$

We will begin by calculating the molar concentration of the acid.

$$\frac{C_a V_a}{C_b V_b} = \frac{n_a}{n_b}$$

$$\therefore C_a = \frac{C_b V_b n_a}{V_a n_b}$$

$$C_b = 0.150 \text{ mol dm}^{-3}$$

$$V_b = 20.00 \text{ cm}^3$$

$$V_a = 30.00 \text{ cm}^3$$

$$n_a = 1$$

$$n_b = 1$$

$$C_a = ?$$

Substituting, we have

$$C_a = \frac{0.150 \text{ mol} \times 20.00 \text{ cm}^3 \times 1}{30.00 \text{ cm}^3 \times 1} = 0.10 \text{ mol dm}^{-3}$$

$$\rho = 3.65 \text{ g}$$

$$M = ?$$

Substituting, we have

$$M = \frac{3.65 \text{ g dm}^{-3}}{0.10 \text{ mol dm}^{-3}} = 36.5 \text{ g mol}^{-1}$$

Volumetric Analysis

Example 2

It requires 15.63 cm³ of a 0.15 mol dm⁻³ solution of sulfuric acid, H_2SO_4, to reach the endpoint when 25.00 cm³ of a solution of an unknown base, XOH, containing 10.5 g of the base per dm³ of solution was titrated against it. Determine the relative atomic mass of X and identify the base.

$$(H = 1,\ O = 16)$$

Solution

As usual, we have to use the relation:

$$C = \frac{\rho}{M}$$

$$\therefore M = \frac{\rho}{C}$$

The equation of reaction is as follows:

$$H_2SO_4(aq) + 2XOH(aq) \rightarrow X_2SO_4(aq) + 2H_2O(l)$$

The molar concentration of the base is determined as follows.

$$\frac{C_a V_a}{C_b V_b} = \frac{n_a}{n_b}$$

$$\therefore C_b = \frac{C_a V_a n_b}{V_b n_a}$$

$$C_a = 0.15\ \text{mol dm}^{-3}$$

$$V_a = 15.63\ \text{cm}^3$$

$$V_b = 25.00\ \text{cm}^3$$

$$n_a = 1$$

$$n_b = 2$$

$$C_b = ?$$

Substituting, we have

$$C_b = \frac{0.15\,\text{mol} \times 15.63\,\text{cm}^3 \times 2}{25.00\,\text{cm}^3 \times 1 \times 1\,\text{dm}^3} = 0.18756\,\text{mol dm}^{-3}$$

$$\rho = 10.5\,\text{g}$$

Let the relative atomic mass of X be x. It then follows that

$$M = (x + 16 + 1)\ \text{g mol}^{-1} = (x + 17)\ \text{g mol}^{-1}$$

Substituting, we have

$$\frac{(x+17)\,\text{g}}{1\,\text{mol}} = \frac{10.5\,\text{g dm}^{-3}}{0.18756\,\text{mol dm}^{-3}}$$

$$0.18756x\ \text{g mol dm}^{-3} + 3.18852\ \text{g mol dm}^{-3} = 10.5\ \text{g mol dm}^{-3}$$

So

$$0.18756x \text{ g mol dm}^{-3} = 7.31148 \text{ g mol dm}^{-3}$$

$$\therefore x = \frac{7.31148 \text{ g mol dm}^{-3}}{0.18756 \text{ g mol dm}^{-3}} = 39$$

This is the relative atomic mass of potassium; hence, the base in question is potassium hydroxide, KOH.

PRACTICE PROBLEMS
1. 20.00 cm³ of a 0.10 mol dm⁻³ solution of anhydrous sodium carbonate, Na_2CO_3, is titrated against an unknown acid, HX, containing 2.5 g of the acid in 250 cm³ of solution. Determine the molar mass of the acid if the average titre value is 25.20 cm³.
2. 20.73 cm³ of a 0.14 mol dm⁻³ solution of nitric acid, HNO_3, is required to reach the endpoint when 25.00 cm³ of an unknown base, XOH, containing 6.5 g of the base per dm³ of solution was titrated against it. Calculate the relative atomic mass of X and write its formula in full.

13.1.7 Mass of Salt, Number of Ions and pH

In titration, just the right amount of reactants is added. Stated differently, the amount of reactants consumed during titration is in proportion. Thus, as we explained in Chapter 5, the amount of a product in the reaction can be obtained from the amount of any of the reactants.

The number of ions in solution is determined by first calculating the number of moles of the ion and then applying the familiar relation:

$$n = \frac{N}{N_A}$$

As usual, the pH of the solution is determined using the relation:

$$pH = -\log\left[H^+\right]$$

Example 1

The average titre value when 25.00 cm³ of anhydrous sodium carbonate, Na_2CO_3, solution was titrated against a 0.20-mol dm⁻³ solution of sulfuric acid, H_2SO_4, is 18.50 cm³. Determine:

(a) the number of sodium ions per dm³ of the base solution;
(b) the mass of salt that would be obtained if the resulting solution is evaporated to dryness.

$$\left(O = 16, Na = 23, S = 32, N_A = 6.02 \times 10^{23} \text{ mol}^{-1}\right)$$

Solution

(a) The very first step is to obtain the molar concentration of the base solution. The balanced equation for the reaction is as follows:

$$H_2SO_4\,(aq) + Na_2CO_3\,(aq) \rightarrow Na_2SO_2\,(aq) + CO_2\,(g) + H_2O\,(l)$$

$$\frac{C_a V_a}{C_b V_b} = \frac{n_a}{n_b}$$

Volumetric Analysis

$$\therefore C_b = \frac{C_a V_a n_b}{V_b n_a}$$

$$C_a = 0.20 \text{ mol dm}^{-3}$$

$$V_a = 18.50 \text{ cm}^3$$

$$V_b = 25.00 \text{ cm}^3$$

$$n_a = 1$$

$$n_b = 1$$

$$C_b = ?$$

Substituting, we have

$$C_b = \frac{0.20 \text{ mol} \times 18.50 \text{ cm}^3 \times 1}{25.00 \text{ cm}^3 \times 1 \times 1 \text{ dm}^3} = 0.148 \text{ mol dm}^{-3}$$

Now, we will apply the relation:

$$n = \frac{N}{N_A}$$

$$\therefore N = n \times N_A$$

Since the molar concentration of the base is $0.148 \text{ mol dm}^{-3}$, it then implies that 1 dm^3 of the base solution contains 0.148 mol of the base. The number of moles of sodium ions, Na^+, in the solution is calculated as follows.

$$Na_2CO_3(s) \rightleftharpoons 2Na^-(aq) + CO_3^{2-}(aq)$$

According to the equation, 1 mol of Na_2CO_3 produces 2 mol of sodium ions, Na^+; hence, 0.148 mol of Na_2CO_3 produces 0.296 mol of Na^+, i.e., $n = 0.296$ mol.

$$N_A = 6.02 \times 10^{23} \text{ mol}^{-1}$$

$$N = ?$$

Substituting, we have

$$N = 0.296 \text{ mol} \times \frac{6.02 \times 10^{23}}{1 \text{ mol}} = 1.8 \times 10^{23}$$

(b) This question is another way of asking us to determine the mass of sodium sulfate, Na_2SO_2, in the reaction mixture. Since reactants are added in stoichiometric proportions, we can determine the amount of the salt from the amount of any of the reactants. The number of moles of a reactant is given by the formula

$$n = C \times V$$

For the acid, we have

$$C = 0.20 \text{ mol dm}^{-3}$$

$$V = 18.50 \text{ cm}^3 = 0.0185 \text{ dm}^3$$

$$n = ?$$

Substituting, we have

$$n = \frac{0.20 \text{ mol}}{1 \text{ dm}^3} \times 0.0185 \text{ dm}^3 = 0.0037 \text{ mol}$$

According to the equation of the reaction, 1 mol of the acid produces 1 mol of Na_2SO_2; hence, 0.0037 mol of the acid would also produce 0.0037 mol of Na_2SO_2. In other words, the number of moles of Na_2SO_4 produced is 0.0037 mol. Finally, we have to convert this amount of Na_2SO_2 to mass as follows.

$$n = \frac{m}{M}$$

$$\therefore m = n \times M$$

$$n = 0.0037 \text{ mol}$$

$$M = (\{23 \times 2\} + 32 + \{16 \times 4\}) \text{ g mol}^{-1} = 142 \text{ g mol}^{-1}$$

$$m = ?$$

Substituting, we have

$$m = 0.0037 \text{ mol} \times \frac{142 \text{ g}}{1 \text{ mol}} = 0.53 \text{ g}$$

It will make no difference if we use the number of moles of the base that was consumed, as shown below.

$$C = 0.148 \text{ mol dm}^{-3}$$

$$V = 25.00 \text{ cm}^3 = 0.025 \text{ dm}^3$$

$$n = ?$$

$$\therefore n = \frac{0.148 \text{ mol}}{1 \text{ dm}^3} \times 0.025 \text{ dm}^3 = 0.0037 \text{ mol}$$

According to the equation, 1 mol of the base produces 1 mol of Na_2SO_4; hence, 0.0037 mol of the base would also produce 0.0037 mol of Na_2SO_4, which is the same as the amount of Na_2SO_4 obtained from the amount of the acid consumed.

Example 2

It requires 22.50 cm³ of a solution of hydrochloric acid, HCl, to reach the endpoint when 20.00 cm³ of a 0.25 mol dm⁻³ solution of sodium hydroxide, NaOH, was titrated against it. Determine:

(a) the pH of the acid solution;
(b) the mass of sodium chloride that would be obtained from the complete reaction of 250.00 cm³ of the base solution with the acid solution;
(c) the number of sodium ions, Na⁺, that would be produced from the complete reaction of 250.00 cm³ of the base solution with the acid solution.

$$(Na = 23, Cl = 35.5, N_A = 6.02 \times 10^{23} \text{ mol}^{-1})$$

Volumetric Analysis

Solution

(a) We have to apply the relation:

$$pH = -\log[H^+]$$

The balanced equation of the reaction is as follows:

$$HCl(aq) + NaOH(aq) \rightarrow NaCl(aq) + H_2O(l)$$

To obtain the concentration of hydrogen ions, H^+, in the acid solution we must first calculate its molar concentration.

$$\frac{C_a V_a}{C_b V_b} = \frac{n_a}{n_b}$$

$$\therefore C_a = \frac{C_b V_b n_a}{V_a n_b}$$

$$C_b = 0.25 \text{ mol dm}^{-3}$$

$$V_b = 20.00 \text{ cm}^3$$

$$V_a = 22.50 \text{ cm}^3$$

$$n_a = 1$$

$$n_b = 1$$

$$C_a = ?$$

Substituting, we have

$$C_a = \frac{0.25 \text{ mol} \times 20.00 \text{ cm}^3 \times 1}{22.50 \text{ cm}^3 \times 1} = 0.22 \text{ mol dm}^{-3}$$

The concentration of hydrogen ions, H^+, in the solution is determined as follows.

$$HCl(aq) \rightleftharpoons H^+(aq) + Cl^-(aq)$$

According to the equation, 1 mol dm^{-3} of hydrochloric acid solution produces 1 mol dm^{-3} of H^+; hence, 0.22 mol dm^{-3} of the acid solution would equally produce 0.22 mol dm^{-3} of H^+.

$$[H^+] = 0.22 \text{ mol dm}^{-3}$$

$$pH = ?$$

Substituting, we have

$$pH = -\log 0.22 = 0.66$$

(b) The first step is to obtain the number of moles of the base in 250.00 cm³ of solution.

$$n = C \times V$$

$$C = 0.25 \text{ mol dm}^{-3}$$

$$V = 250.00 \text{ cm}^3 = 0.25 \text{ dm}^3$$

$$n = ?$$

Substituting, we have

$$n = \frac{0.25\,\text{mol}}{1\,\text{dm}^3} \times 0.25\,\text{dm}^3 = 0.0625\,\text{mol}$$

According to the equation for the reaction, 1 mol of the base produces 1 mol of NaCl; hence, 0.0625 mol of the base would also produce 0.0625 mol of NaCl. In other words, the maximum amount of NaCl that can be obtained from 250.00 cm³ of the base solution is 0.0625 mol. Finally, we have to convert this number of moles of NaCl to mass as follows.

$$n = \frac{m}{M}$$

$$\therefore m = n \times M$$

$$n = 0.0625\,\text{mol}$$

$$M = (23 + 35.5)\,\text{g mol}^{-1} = 58.5\,\text{g mol}^{-1}$$

$$m = ?$$

Substituting, we have

$$m = 0.0625\,\text{mol} \times \frac{58.5\,\text{g}}{1\,\text{mol}} = 3.7\,\text{g}$$

(c) As usual, the number of ions is determined using the relation:

$$n = \frac{N}{N_A}$$

$$\therefore N = n \times N_A$$

Sodium chloride dissociates as follows:

$$\text{NaCl(s)} \rightleftharpoons \text{Na}^+(\text{aq}) + \text{Cl}^-(\text{aq})$$

The number of moles of sodium chloride produced is 0.0625 mol. According to the equation, 1 mol of sodium chloride produces 1 mol of Na⁺; hence, 0.0625 mol of the base would also produce 0.0625 mol of Na⁺.

$$n = 0.0625\,\text{mol}$$

$$N_A = 6.02 \times 10^{23}\,\text{mol}^{-1}$$

$$N = ?$$

Substituting, we have

$$N = 0.0625\,\text{mol} \times \frac{6.02 \times 10^{23}}{1\,\text{mol}} = 3.8 \times 10^{22}$$

PRACTICE PROBLEMS
1. It requires 17.50 cm³ of a sulfuric acid, H_2SO_4, solution for the complete reaction of 25.00 cm³ of a 0.10 mol dm⁻³ solution of sodium hydroxide, NaOH, solution. Calculate:
    (a) the pH of the acid solution;
    (b) the number of moles of sodium ions, Na⁺, in the reaction mixture.

Volumetric Analysis

2. The average titre value when 20.00 cm³ of a solution of potassium hydroxide, KOH, was titrated against a 0.12 mol dm⁻³ solution of nitric acid, HNO₃, is 15.50 cm³. Determine:
   (a) the number of potassium ions in 20.00 cm³ of the base solution;
   (b) the mass of potassium nitrate, KNO₃, that would be produced from 500.00 cm³ of the base.

$$\left(N = 14,\ O = 16,\ K = 39,\ N_A = 6.02 \times 10^{23}\ \text{mol}^{-1}\right)$$

13.1.8 Volume of Gas Evolved

Carbon dioxide is released in titrations involving reagents such as sodium carbonate and oxalic acid. For example,

$$H_2SO_4(aq) + Na_2CO_3(aq) \rightarrow Na_2SO_2(aq) + H_2O(l) + CO_2(g)$$

In this section, we will use our knowledge of stoichiometry to determine the amount of carbon dioxide released during such a titration.

Example 1

It requires 23.10 cm³ of a 0.22 mol dm⁻³ potassium permanganate, KMnO₄, solution to reach the endpoint when 20.00 cm³ of a solution of oxalic acid was titrated against it. Determine the volume of carbon dioxide liberated at STP. The equation of the reaction is 2MnO₄⁻(aq) + 5C₂O₄²⁻(aq) + 16H⁺(aq) → 2Mn²⁺(aq) + 8H₂O(l) + 10CO₂(g).

$$\left(\text{Molar volume of gases at STP} = 22.4\ \text{dm}^3\ \text{mol}^{-1}\right)$$

Solution

You will recall that at STP,

$$n = \frac{V}{22.4\ \text{dm}^3\ \text{mol}^{-1}}$$

$$\therefore V = n \times 22.4\ \text{dm}^3\ \text{mol}^{-1}$$

We will begin by calculating the number of moles of CO₂ released at STP, using the amount of KMnO₄ consumed in the reaction.

$$n = C \times V$$

$$C = 0.22\ \text{mol dm}^{-3}$$

$$V = 23.10\ \text{cm}^3 = 0.0231\ \text{dm}^3$$

$$n = ?$$

Substituting, we have

$$n = \frac{0.22\ \text{mol}}{1\ \text{dm}^3} \times 0.0231\ \text{dm}^3 = 0.005082\ \text{mol}$$

As given by the equation of reaction, 2 mol of $KMnO_4$ produces 10 mol of CO_2; it follows that 0.005082 mol of $KMnO_4$ would produce 0.02541 mol of CO_2. So the number of moles of CO_2 liberated at STP is 0.02541 mol.

$$V = ?$$

Finally, we can substitute to obtain:

$$V = 0.02541\,\text{mol} \times \frac{22.4\,\text{dm}^3}{1\,\text{mol}} = 0.57\,\text{dm}^3$$

We would obtain the same result if the amount of CO_2 were determined from the amount of $H_2C_2O_4$. Of course, we would first have to go through the rigour of calculating its molar concentration. You should be able to verify this for practice.

Example 2

The average titre value of the titration a solution of nitric acid, HNO_3, against 20.00 cm³ of a 0.15 mol dm⁻³ of an anhydrous sodium carbonate solution is 28.50 cm³. Determine the volume of carbon dioxide liberated at RTP.

$$\left(\text{Molar volume of gases at RTP} = 24\,\text{dm}^3\,\text{mol}^{-1}\right)$$

Solution

You will recall that at RTP,

$$n = \frac{V}{24\,\text{dm}^3\,\text{mol}^{-1}}$$

$$\therefore V = n \times 24\,\text{dm}^3\,\text{mol}^{-1}$$

We will calculate the amount of CO_2 released at RTP based on the amount of Na_2CO_3.

$$n = C \times V$$

$$C = 0.15\,\text{mol}\,\text{dm}^{-3}$$

$$V = 20.00\,\text{cm}^3 = 0.020\,\text{dm}^3$$

$$n = ?$$

Substituting, we have

$$n = \frac{0.15\,\text{mol}}{1\,\text{dm}^3} \times 0.020\,\text{dm}^3 = 0.0030\,\text{mol}$$

The balanced equation of reaction is as follows:

$$2HNO_3(aq) + Na_2CO_3(aq) \rightarrow 2NaNO_3(aq) + CO_2(g) + H_2O(l)$$

Volumetric Analysis

According to the equation, 1 mol of Na_2CO_3 produces 1 mol of CO_2; it follows that 0.0030 mol of Na_2CO_3 would also produce 0.0030 mol of CO_2. In other words, the amount of CO_2 liberated at RTP is 0.0030 mol.

$$V = ?$$

Substituting, we have

$$V = 0.0030 \, mol \times \frac{24 \, dm^3}{1 \, mol} = 0.072 \, dm^3$$

PRACTICE PROBLEMS

1. It requires 18.20 cm³ of a 0.25 mol dm⁻³ solution of hydrochloric acid, HCl, to reach the endpoint when 25.00 cm³ of anhydrous sodium carbonate, Na_2CO_3, solution was titrated against it. Determine the volume of carbon dioxide liberated at RTP.

   $$\left(\text{Molar volume of gases at RTP} = 24 \, dm^3 \, mol^{-1}\right)$$

2. The average titre value when 25.00 cm³ of a 0.25 mol dm⁻³ solution of anhydrous sodium carbonate, Na_2CO_3, is titrated against a sulfuric acid solution is 24.50 cm³. Determine the volume of carbon dioxide liberated at STP.

   $$\left(\text{Molar volume of gases at STP} = 22.4 \, dm^3 \, mol^{-1}\right)$$

FURTHER PRACTICE PROBLEMS

1. It requires 14.80 cm³ of a solution of ethanoic acid, CH_3COOH, for the complete reaction of 25.00 cm³ of a 0.18 mol dm⁻³ solution of sodium hydroxide, NaOH. Determine the molar concentration of the acid solution.

2. 20.00 cm³ of a solution of anhydrous sodium carbonate, Na_2CO_3, is titrated against a 0.27 mol dm⁻³ solution of sulfuric acid. Determine the mass concentration of the base solution if the average titre value is 17.70 cm³.

   $$(C = 12, O = 16, Na = 23)$$

3. Calculate the volume of a 0.18 mol dm⁻³ solution of sodium thiosulfate, $Na_2S_2O_3$, required to reach the endpoint when 20.00 cm³ of a 0.10 mol dm⁻³ solution of iodine is titrated against it. The equation for the reaction is $I_2(aq) + 2S_2O_3^{2-}(aq) \rightarrow 2I^-(aq) + S_4O_6^{2-}(aq)$.

4. A student titrated 25.00 cm³ of a solution of iron(II) compound against a 0.21 mol dm⁻³ solution of acidified potassium permanganate, $KMnO_4$. Given that it requires 12.50 cm³ of the potassium permanganate solution to reach the endpoint, determine:
   (a) the molar concentration of the potassium permanganate solution;
   (b) the number of moles of potassium permanganate in 150.00 cm³ of solution.
   The equation reaction is $MnO_4^-(aq) + 5Fe^{2+}(aq) + 8H^+(aq) \rightarrow Mn^{2+}(aq) + 5Fe^{3+}(aq) + 4H_2O(l)$.

5. It requires 28.00 cm³ of a 0.085 mol dm⁻³ solution of sulfuric acid, H_2SO_4, to reach the endpoint when 25.00 cm³ of a saturated solution of potassium hydroxide, KOH, was titrated against it at a certain temperature. Determine:
   (a) the molar concentration of the base solution;
   (b) the solubility of the base per 100 g of water at the prevailing temperature.

6. An anhydrous sodium carbonate, Na_2CO_3, solution was prepared by diluting 50.00 cm³ of a saturated solution of the base to 1.0 dm³ at 15°C. Given that 20.00 cm³ of the solution requires 15.58 cm³ of a 0.28 mol dm⁻³ solution of nitric acid, HNO_3, for complete reaction, determine:
   (a) the number of moles of the salt in the saturated solution;
   (b) the molar solubility of the salt at 15°C;
   (c) the mass solubility of the salt at 15°C.

$$(C = 12, O = 16, Na = 23)$$

7. 29.09 cm³ of a 0.22 mol dm⁻³ solution of nitric acid, HNO_3, neutralises 20.00 cm³ of a 0.16-mol dm⁻³ of a solution of a base. Determine:
   (a) the number of moles of the acid in 29.09 cm³ of the acid solution;
   (b) the number of moles of the base in 20.00 cm³ of the base solution;
   (c) the mole ratio of acid to base.

8. It requires 25.00 cm³ of a 0.15 mol dm⁻³ solution of an acid to reach the endpoint when 20.00 cm³ of a solution containing 1.88 g of sodium hydroxide, NaOH, per 250 cm³ of solution was titrated against it. Determine:
   (a) the number of moles of the acid in 25.00 cm³ of the acid solution;
   (b) the number of moles of the base in 20.00 cm³ of the base solution;
   (c) the mole ratio of acid to base.

$$(H = 1, O = 16, Na = 23)$$

9. It requires 22.50 cm³ of a 0.230 mol dm⁻³ of a solution of acidified potassium permanganate, $KMnO_4$, to reach the endpoint when 25.00 cm³ of a solution containing 10.0 g of an iron(II) salt per dm³ of solution was titrated against it. Determine the percent by mass of iron in the salt. The equation of reaction is $MnO_4^-(aq) + 5Fe^{2+}(aq) + 8H^+(aq) \rightarrow Mn^{2+}(aq) + 5Fe^{3+}(aq) + 4H_2O(l)$.

$$(Fe = 56)$$

10. The average titre value of the titration of 25.00 cm³ of a 0.17 mol dm⁻³ solution of potassium hydroxide, KOH, against a solution containing 2.5 g of impure sulfuric acid per 150.00 cm³ of solution is 15.70 cm³. Determine:
    (a) the mass of the pure acid per dm³ of solution;
    (b) the percentage purity of the acid.

$$(H = 1, O = 16, S = 32)$$

11. 25.00 cm³ of a solution containing 4.50 g of impure oxalic acid per 250.00 cm³ of solution was titrated against a 0.100 mol dm⁻³ solution of potassium permanganate, $KMnO_4$. Determine the percent purity of acid if the average titre value is 10.50 cm³. The equation of reaction is $2MnO_4^-(aq) + 5C_2O_4^{2-}(aq) + 16H^+(aq) \rightarrow 2Mn^{2+}(aq) + 8H_2O(l) + 10CO_2(g)$.

$$(H = 1, C = 12, O = 16)$$

Volumetric Analysis

12. It requires 18.75 cm³ of a 0.25 mol dm⁻³ hydrochloric acid, HCl, solution to reach the endpoint when 25.00 cm³ of a solution containing 5.44 g of hydrated sodium carbonate, $Na_2CO_3.xH_2O$, per 250.00 cm³ of solution was titrated against it. Determine:
    (a) the number of molecules of water of crystallisation in the salt;
    (b) the percentage composition of water in the salt.

$$(C = 12, O = 16, Na = 23)$$

13. A solution of hydrated sodium carbonate, $Na_2CO_3.xH_2O$, contains 0.54 g of the salt in 20.00 cm³ of solution. Given that 20.00 cm³ of the base solution requires 24.10 cm³ of a 0.18 mol dm⁻³ sulfuric acid solution to reach the endpoint. Determine:
    (a) the number of molecules of water of crystallisation in the salt;
    (b) the percentage composition of water in the salt.

$$(C = 12, O = 16, Na = 23)$$

14. The average titre value of the titration of 25.00 cm³ of a 0.25 mol dm⁻³ solution of potassium hydroxide, KOH, against an acid, H_2X, containing 2.55 g of the acid per 250.00 cm³ of solution, is 30.00 cm³. Determine:
    (a) the molar concentration of the acid solution;
    (b) he molar mass of the acid.

15. It requires 16.80 cm³ of a 0.10 mol dm⁻³ of a solution of nitric acid, HNO_3, to reach the endpoint when 20.00 cm³ of a base, XOH, containing 0.71 g of the base per 150.00 cm³ of solution was titrated against it. Determine:
    (a) the molar concentration of the base solution;
    (b) the molar mass of the base;
    (c) the relative atomic mass of the element X. Hence, state the IUPAC name of the base.

$$(H = 1, O = 16)$$

16. 32.20 cm³ of a 0.10 mol dm⁻³ solution of nitric acid, HNO_3, is required for the complete reaction of 25.00 cm³ of a sodium hydroxide, NaOH, solution. Determine:
    (a) the molar concentration of the base solution;
    (b) the number of sodium ions per 25.00 cm³ of the base solution;
    (c) the mass of sodium nitrate produced in the reaction.

$$(N = 14, O = 16, Na = 23, N_A = 6.02 \times 10^{23} \text{ mol}^{-1})$$

17. 20.00 cm³ of a 0.25 mol dm⁻³ potassium hydroxide, KOH, solution is titrated against a sulfuric acid, H_2SO_4, solution. Given that the average titre value is 17.20 cm³. Determine
    (a) the molar concentration of the acid solution;
    (b) the number of hydrogen ions per 17.20 cm³ of the acid solution;
    (c) the pH of the acid solution;
    (d) the mass of potassium sulfate produced in the reaction.

$$(O = 16, S = 32, K = 39, N_A = 6.02 \times 10^{23} \text{ mol}^{-1})$$

18. It requires 18.50 cm³ of a 0.30 mol dm⁻³ sulfuric acid, H_2SO_4, solution to reach the endpoint when 20.00 cm³ of a solution of anhydrous sodium carbonate, Na_2CO_3, was titrated against it. Determine:
    (a) the molar concentration of the base solution;
    (b) the volume of carbon dioxide produced at RTP.

    $$\left(\text{Molar volume of gases at RTP} = 24\,\text{dm}^3\,\text{mol}^{-1}\right)$$

19. 26.70 cm³ of a solution of hydrochloric acid, HCl, is required to reach the endpoint when 20.00 cm³ of a 0.17 mol dm⁻³ anhydrous sodium carbonate, Na_2CO_3, solution was titrated against it. Determine:
    (a) the molar concentration of the acid solution;
    (b) the pH of the acid solution;
    (c) the volume of carbon dioxide produced at STP.

    $$\left(\text{Molar volume of gases at STP} = 22.4\,\text{dm}^3\,\text{mol}^{-1}\right)$$

14 Rates of Chemical Reactions

14.1 INTRODUCTION

Chemical reactions occur at different speeds. For example, rusting is a very slow process that takes several days to occur, while a neutralisation reaction is a fast reaction that happens almost immediately. Combustion and the precipitation of insoluble salts from aqueous solutions are other examples of fast reactions. Some reactions, such as digestion and the hydrolysis of esters, occur at an intermediate speed.

14.2 MEASURING REACTION RATES

The rate of a chemical reaction is the speed at which the reaction occurs. Reaction rates are measured in terms of the change in a measurable property such as mass, concentration or volume of reactants or products over a certain period of time. The reaction rate varies over the course of a reaction, decreasing until it finally becomes zero when all the reactants have been used up. The variation of reaction rate with time is shown graphically as a rate curve (Figure 14.1).

The average rate of reaction over the course of a reaction is given by

$$\text{Rate} = \frac{\text{change in a measurable property of a reactant or product}}{\text{time taken}}$$

$$= \frac{\text{final value of the property} - \text{initial value of the property}}{\text{time taken}}$$

Recall that the Greek letter delta, Δ, is used to denote change in a quantity. Thus, the equation can also be written as follows:

$$\text{Rate} = \frac{\Delta \text{ measurable property of a reactant or product}}{\Delta \text{ time}}$$

Reaction rate is always positive. If the average rate is defined in terms of a property of a reactant, a negative sign must be included in the rate equation to ensure the value of the average reaction rate turns out positive. This is because a measurable property of a reactant decreases with time.

$$\text{Rate} = -\frac{\Delta \text{ measurable property of a reactant}}{\Delta \text{ time}}$$

The unit of reaction rate depends on the specific property used as well as the unit of time. If, for example, concentration is used, the unit reaction rate can be mol dm^{-3} s^{-1}, mol dm^{-3} min^{-1} or mol dm^{-3} hr^{-1}.

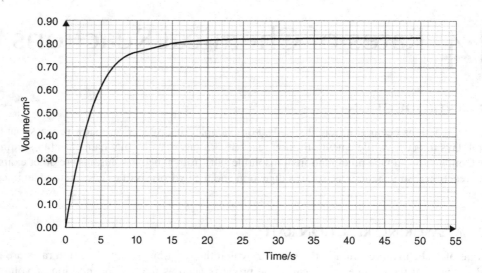

FIGURE 14.1 A rate curve.

Example 1

In the reaction between zinc granules and hydrochloric acid, 5.00 g of zinc granules was consumed in 25 min. Calculate the average rate of reaction in g s^{-1}.

Solution

We will apply the relation:

$$\text{Rate} = -\frac{\Delta m}{\Delta t} = -\frac{m_2 - m_1}{t_2 - t_2}$$

$$m_1 = 5.0 \text{ g}$$

$$m_2 = 0$$

$$t_1 = 25 \text{ min} = 1500 \text{ s}$$

$$t_2 = 0$$

$$\text{Rate} = ?$$

Substituting, we have

$$\text{Rate} = -\frac{(0-5) \text{ g}}{(1500-0) \text{ s}} = 3.33 \times 10^{-3} \text{ g s}^{-1}$$

Example 2

A dilute solution of sulfuric acid, H_2SO_4, was added to a dilute solution of sodium carbonate, Na_2CO_3. Calculate the average rate of formation of sodium sulfate, Na_2SO_4, between 15 s and 120 s, given that the concentration of the salt changes from 0.100 mol dm^{-3} to 0.115 mol dm^{-3} within this period.

Rates of Chemical Reactions

Solution

We have to apply the relation:

$$\text{Rate} = \frac{\Delta[Na_2SO_4]}{\Delta t} = \frac{[Na_2SO_4]_{t_2} - [Na_2SO_4]_{t_1}}{t_2 - t_1}$$

$$[Na_2SO_4]_{t_1} = 0.100 \text{ mol dm}^{-3}$$

$$[Na_2SO_4]_{t_2} = 0.115 \text{ mol dm}^{-3}$$

$$t_1 = 15 \text{ s}$$

$$t_2 = 120 \text{ s}$$

$$\text{Rate} = ?$$

Substituting, we have

$$\text{Rate} = \frac{(0.115 - 0.110) \text{ mol dm}^{-3}}{(120 - 15) \text{ s}} = 4.8 \times 10^{-5} \text{ mol dm}^{-3} \text{ s}^{-1}$$

PRACTICE PROBLEMS

1. It took just 5.0 min for the reaction between 1.50 g of sodium carbonate, Na_2CO_3, and excess dilute hydrochloric acid, HCl, solution to reach completion. Calculate the average rate of reaction in g min^{-1}.
2. 5.00 cm^3 of hydrogen, H_2, was evolved in 10.0 min when hydrochloric acid, HCl, was added to a calcium carbonate solution. Determine the rate of formation of hydrogen in cm^3 s^{-1}.

14.3 REACTION RATES AND CHEMICAL REACTIONS

The rate of formation of a product or consumption of a reactant in a chemical reaction is related to its stoichiometric coefficient in the balanced equation of the reaction. Consider the balanced general equation given below:

$$aA + bB \rightarrow cC + dD$$

The different rates are related as follows:

$$-\frac{1}{a}\frac{\Delta[A]}{\Delta t} = -\frac{1}{b}\frac{\Delta[B]}{\Delta t} = \frac{1}{c}\frac{\Delta[C]}{\Delta t} = \frac{1}{d}\frac{\Delta[D]}{\Delta t}$$

For gas-phase reactions, the rates can also be written in terms of partial pressures.

$$-\frac{1}{a}\frac{\Delta P_A}{\Delta t} = -\frac{1}{b}\frac{\Delta P_B}{\Delta t} = \frac{1}{c}\frac{\Delta P_C}{\Delta t} = \frac{1}{d}\frac{\Delta P_D}{\Delta t}$$

Example 1

Consider the following reaction:

$$2PH_3(g) \rightarrow P_2(g) + 3H_2(g)$$

Write an equation that shows the relationship between the rates of formation of the products and disappearance of the reactant in terms of partial pressures.

Solution

The rates relationship is given as:

$$-\frac{1}{2}\frac{\Delta P_{PH_3}}{\Delta t} = \frac{\Delta P_{P_2}}{\Delta t} = \frac{1}{3}\frac{\Delta P_{H_2}}{\Delta t}$$

Example 2

The following data were obtained in an experiment on the rate of decomposition of dinitrogen pentoxide, N_2O_5.

t (min)	0.00	2.00	4.00	6.00	8.00
$[N_2O_5]$ (mol dm^{-3})	0.250	0.249	0.248	0.247	0.246

The equation of reaction is $2N_2O_5(g) \rightarrow 4NO_2(g) + O_2(g)$. Calculate:

(a) the average rate of decomposition of N_2O_5 over the course of the experiment.
(b) the average rate of formation of NO_2 and O_2 between 4.00 s and 8.00 s.

Solution

We will apply the relation:

$$\text{Rate} = -\frac{\Delta[N_2O_5]}{\Delta t} = -\frac{[N_2O_5]_{t_2} - [N_2O_5]_{t_1}}{t_2 - t_1}$$

$$[N_2O_5]_{t_1} = 0.248 \text{ mol dm}^{-3}$$

$$[N_2O_5]_{t_2} = 0.246 \text{ mol dm}^{-3}$$

$$t_1 = 0.00 \text{ s}$$

$$t_2 = 8.00 \text{ s}$$

$$\text{Rate} = ?$$

Substituting, we have

$$\text{Rate} = -\frac{(0.246 - 0.248) \text{ mol dm}^{-3}}{(8.00 - 0.00) \text{ s}} = 5.00 \times 10^{-4} \text{ mol dm}^{-3} \text{ s}^{-1}$$

(b) We will determine the rates from the average rate of decomposition of N_2O_5 over the stated period of time.

$$\text{Rate} = -\frac{\Delta[N_2O_5]}{\Delta t} = -\frac{[N_2O_5]_{t_2} - [N_2O_5]_{t_1}}{t_2 - t_1}$$

Rates of Chemical Reactions

$$[N_2O_5]_{t_1} = 0.248 \text{ mol dm}^{-3}$$

$$[N_2O_5]_{t_2} = 0.246 \text{ mol dm}^{-3}$$

$$t_1 = 4.00 \text{ s}$$

$$t_2 = 8.00 \text{ s}$$

$$\text{Rate} = ?$$

Substituting, we have

$$\text{Rate} = -\frac{(0.246 - 0.248) \text{ mol dm}^{-3}}{(8.00 - 0.00) \text{ s}} = 7.50 \times 10^{-4} \text{ mol dm}^{-3} \text{ s}^{-1}$$

According to the equation of reaction, the rates relationship is given by

$$-\frac{1}{2}\frac{\Delta[N_2O_5]}{\Delta t} = \frac{1}{4}\frac{\Delta[NO_2]}{\Delta t} = \frac{\Delta[O_2]}{\Delta t}$$

Thus, for NO_2, we have

$$\text{Rate} = \frac{\Delta[NO_2]}{\Delta t} = -2\frac{\Delta[N_2O_5]}{\Delta t}$$
$$= 2 \times 7.50 \times 10^{-4} \text{ mol dm}^{-3} \text{ s}^{-1} = 1.50 \times 10^{-3} \text{ mol dm}^{-3} \text{ s}^{-1}$$

For O_2, we have

$$\text{Rate} = \frac{\Delta[O_2]}{\Delta t} = -\frac{1}{2}\frac{\Delta[N_2O_5]}{\Delta t}$$

$$= \frac{1}{2} \times 7.50 \times 10^{-4} \text{ mol dm}^{-3} \text{ s}^{-1} = 3.75 \times 10^{-4} \text{ mol dm}^{-3} \text{ s}^{-1}$$

PRACTICE PROBLEMS
1. Write the equation for the rates relationship of the reaction $2SO_2(g) + O_2(g) \rightarrow 2SO_3(g)$.
2. Consider the gas-phase reaction:

$$N_2(g) + 3H_2(g) \rightarrow 2NH_3(g)$$

Determine the average rates of disappearance of the reactants and formation of the product, given that the concentration of hydrogen decreases from 0.450 mol dm^{-3} to 0.215 mol dm^{-3} within 15.5 s.

14.4 COLLISION THEORY

The *collision theory* assumes that there must be collisions between the particles of the reactants for a chemical reaction to occur. However, not all collisions result in chemical reaction. The collisions that lead to a chemical reaction are called *effective collisions*. For effective collisions to occur, the particles of reactants must have the right orientation in space and collide with enough energy to overcome the bond energy or *energy barrier* of the reactants.

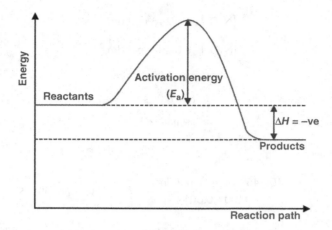

FIGURE 14.2 Energy profile of exothermic reaction.

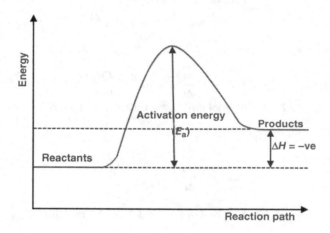

FIGURE 14.3 Energy profile of endothermic reaction.

The minimum energy with which particles of reactants must collide in order to overcome the energy barrier of reactants is called *activation energy*, E_a. For example, the activation energy for the decomposition of dinitrogen pentoxide, N_2O_5, is 88 kJ mol^{-1} at 30°C. The diagrams showing changes in the energy contents of reactants during a chemical reaction are called energy profiles (see Figures 14.2 and 14.3).

14.5 FACTORS AFFECTING THE RATE OF CHEMICAL REACTION

Reaction rates are influenced by many factors, the main ones being temperature, the nature of reactants, concentration, pressure, surface area of the reactants, light and catalysts.

14.5.1 Temperature

The particles of reactants gain more kinetic energy when the temperature is increased. This increases the speed and frequency of collisions, leading to an increase in the reaction rate. On the other hand, reducing the temperature of a reaction slows down the reaction rate.

Rates of Chemical Reactions

14.5.2 Nature of Reactants

Different substances have varying energy contents; hence, the reaction rate depends on the nature of the substances involved. Some reactions are inherently slow, while others occur very quickly. For example, when a piece of calcium metal is dropped into water, hydrogen gas is evolved slowly. In contrast, the reaction between potassium and water is so fast that it occurs explosively.

14.5.3 Concentration

An increase in the concentration of a reactant increases the reaction rate by raising the frequency of particle collisions, leading to more effective collisions. Conversely, if the concentration of a reactant decreases, the particles collide less frequently, reducing the reaction rate.

14.5.4 Pressure

Changes in pressure influence the rates of gas-phase reactions. Increasing the pressure raises the concentration of gaseous reactants, leading a higher reaction rate. Conversely, reducing the pressure lowers the reaction rate.

14.5.5 Surface Area of Reactants

Increasing the surface area of solid reactants enhances the reaction rate by allowing more frequent collisions of reactant particles. The surface area of a solid reactant can be increased by cutting it into smaller pieces or grinding it into a fine powder. The reaction between iron and hydrochloric acid, for instance, proceeds much faster when iron filings are used instead of iron lumps. For reactants that are not in solid form, greater contact between reactant particles can be achieved through thorough mixing.

14.5.6 Light

Certain reactions cannot occur without light. These reactions are called *photochemical reactions*. Examples include photosynthesis, the substitution reaction between chlorine and methane, and the decomposition of hydrogen peroxide. Light influences the reaction rate by providing the energy required to activate the reacting particles.

14.5.7 Catalyst

A catalyst is a substance that alters the rate of a chemical reaction without being being consumed or undergoing any permanent change in the process. It accelerates the reaction rate by providing an alternative pathway with a lower activation energy. The effect of a catalyst on a reaction is shown in Figure 14.4.

For example, the production of oxygen by the thermal decomposition of potassium chlorate, $KClO_3$, is catalysed by manganese dioxide, MnO_2. Other examples of catalysts include platinum, used in the decomposition of hydrogen peroxide, H_2O_2; finely divided nickel, used in the hydrogenation of oil; finely divided iron, used in the production of ammonia, NH_3, by the Haber process; vanadium(V) oxide, V_2O_5, used in the conversion of sulfur dioxide, SO_2, to sulfur trioxide, SO_3; and aluminium oxide, Al_2O_3, used in the production of sulfur from hydrogen sulfide, H_2S, and so on.

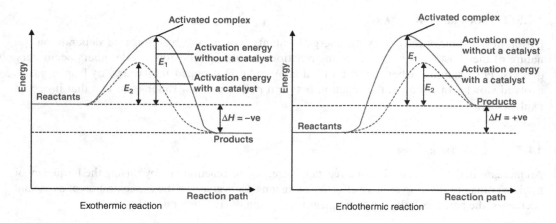

FIGURE 14.4 Effect of a catalyst on reaction rate.

14.6 RATE LAW AND CONCENTRATION

A *rate law* is an equation expressing reaction rate in terms of a constant and the concentration of the reactants. Consider the general reaction:

$$aA + bB \rightarrow \text{products}$$

The rate law or reaction rate can be expressed as follows:

$$r = k[A]^x[B]^y$$

Here, the constant k is called the *rate constant*. The rate constant of a reaction depends on the temperature and the catalyst used. The power to which the concentration of a reactant is raised in a rate law is called the *order of reaction* with respect to that reactant. The sum of the orders for all the reactants in the rate law is referred to as the *overall order* of the reaction. In the above reaction, for example, the overall order of the reaction is $x + y$.

If a reaction is zero-order with respect to a reactant, the reaction rate is independent of the concentration of that reactant. Consequently, the concentration of that reactant does not appear in the rate law. An example is the decomposition of ammonia:

$$2NH_3(g) \rightarrow N_2(g) + 3H_2(g) \qquad r = k[NH_3]^0 = k$$

A reaction is said to be first-order with respect to a reactant if the concentration of the reactant is raised to the first power in the rate law. A typical example is the decomposition of dinitrogen pentoxide:

$$2N_2O_5(g) \rightarrow 4NO_2(g) + O_2(g) \qquad r = k[N_2O_5]$$

This reaction is first-order in N_2O_5 and first-order overall. Another example is the formation reaction between hydrogen and iodine:

$$H_2(g) + I_2(g) \rightarrow 2HI(g) \qquad r = k[H_2][I_2]$$

This reaction is first-order in both H_2 and I_2 and second-order overall.

Rates of Chemical Reactions

A reaction is said to be second-order with respect to a reactant if the concentration of the reactant is raised to an exponent of two in the rate law. An example is the reaction:

$$NO_2(g) + CO(g) \rightarrow NO(g) + CO_2(g) \qquad r = k[NO_2]^2$$

This reaction is second-order in NO_2 and second-order overall. The reduction of nitrogen dioxide to nitrogen is another example:

$$2NO(g) + 2H_2(g) \rightarrow 2N_2(g) + 2H_2O(g) \qquad r = k[NO]^2[H_2]$$

This reaction is second-order in NO, first-order in H_2, and third-order overall.

There is no correlation between the order of a reaction and its stoichiometry. The order of reaction is determined experimentally by varying the concentration of one reactant at a time and observing the change in the reaction rate. A reaction is zero-order with respect to a reactant if the reaction rate does not change when the concentration of the reactant is varied. If a reaction is first-order with respect to a reactant, the reaction rate increases by the same factor as the concentration of that reactant. In a reaction that is second-order with respect to a reactant, the reaction rate increases by the square of the factor by which the concentration of that reactant is raised.

In general, if a reaction is of order a with respect to a particular reactant, then multiplying the concentration of that reactant by a factor x will multiply the reaction rate by a factor x^a. In other words, the reaction rate increases by a factor f, which is given by

$$x^a = f$$

Example 1

Consider the reaction:

$$CH_3OH(aq) + CH_3CH_2OCOCH_3(aq) \rightarrow CH_3OCOCH_3(aq) + CH_3CH_2OH(aq) \qquad r = k[CH_3OH]$$

State:

(a) the order of reaction with respect to each reactant;
(b) the overall order of reaction.

Solution

(a) The concentration of CH_3OH is raised to the first power in the rate law, while the concentration of $CH_3CH_2OCOCH_3$ does not appear at all. Hence, the reaction is of the first- and zero-order with respect to CH_3OH and $CH_3CH_2OCOCH_3$, respectively.
(b) The overall order of reaction = 1 + 0 = 1.

Example 2

The rate law for the oxidation of nitrogen monoxide is given by $r = k[NO]^2[O_2]$. Determine:

(a) the order of reaction in each reactant and the overall order of reaction;
(b) the unit of the rate constant if the unit of the reaction rate is mol dm^{-3} s^{-1}.

Solution

(a) The order with respect to NO = 2, while the order with respect to O_2 = 1. Hence, the overall order = 2 + 1 = 3

(b) We know that the rate law is given by

$$r = k[NO]^2[O_2]$$

$$k = \frac{r}{[NO]^2[O_2]}$$

The units of reaction rate and concentration are mol dm^{-3} s^{-1} and mol dm^{-3}, respectively. Substituting into the equation, we have

$$\therefore k = \frac{\text{mol dm}^{-3}\text{ s}^{-1}}{(\text{mol dm}^{-3})^2 \times \text{mol dm}^{-3}} = \text{mol}^{-2}\text{ dm}^6\text{ s}^{-1}$$

Example 3

Consider the reaction represented by the equation 2A + B → 2C. The following data were obtained when the initial rate of reaction was measured at different concentrations of A and B.

EXP.	[A] (mol dm^{-3})	[B] (mol dm^{-3})	Rate (mol dm^{-3} s^{-1})
I	0.15	0.25	0.10
II	0.20	0.25	0.13
III	0.20	0.30	0.16

(a) Determine the order of reaction with respect to each reactant.
(b) Write the rate law for the reaction.
(c) State the overall order of the reaction.
(d) Calculate the rate constant at the temperature at which the experiment was performed.

Solution

(a) Let the rate law be $r = k[A]^a[B]^b$
For A, we have

$$x^a = f$$

From the first two rows of the table, we can see that the [A] increases by the factor

$$x = \frac{0.20}{0.15} = 1.3$$

This raises the reaction rate by the factor

$$f = \frac{0.13}{0.10} = 1.3$$

So

$$1.3^a = 1.3$$

$$\therefore a = 1$$

Thus, the reaction is first-order with respect to A because the reaction rate increases by the same factor as the [A] is raised.
Similarly, for B, we have

$$x^b = f$$

The [B] increases by the factor

$$x = \frac{0.30}{0.25} = 1.2$$

This raises the reaction rate by the factor

$$f = \frac{0.16}{0.13} = 1.2$$

So

$$1.20^b = 1.20$$

$$\therefore b = 1$$

Thus, the reaction is also first-order with respect to B because the reaction rate increases by the same factor as the [B] is raised.

(b) Substituting the order of reaction with respect to each reactant into the rate law, we have

$$r = k[A][B]$$

(c) The overall order of reaction is the sum of the powers of [A] and [B] in the rate law.

$$\text{Overall order} = 1+1 = 2$$

(d) The rate constant, k, is obtained as follows.

$$r = k[A][B]$$

$$\therefore k = \frac{r}{[A][B]}$$

We can use data from any row of the table. For example, using data from the first row, we have

$$r = 0.980 \text{ mol dm}^{-3} \text{ s}^{-1}$$

$$[A] = 0.150 \text{ mol dm}^{-3}$$

$$[B] = 0.250 \text{ mol dm}^{-3}$$

$$k = ?$$

Substituting, we have

$$k = \frac{0.98 \text{ mol dm}^{-3} \text{ s}^{-1}}{0.15 \text{ mol dm}^{-3} \times 0.25 \text{ mol dm}^{-3}} = 26 \text{ dm}^3 \text{ mol}^{-1} \text{s}^{-1}$$

Example 4

Consider the following reaction:

$$2NO(g) + O_2(g) \rightarrow 2NO_2(g)$$

The following data were obtained in the measurement of the initial rate of the reaction at different concentrations of NO and O_2.

EXP.	[NO] (mol dm^{-3})	[O_2] (mol dm^{-3})	Rate (mol dm^{-3} s^{-1})
I	0.10	0.20	3.00
II	0.30	0.20	27.00
III	0.30	0.25	33.75

(a) Obtain the rate law.
(b) Determine the overall order of the reaction.
(c) Evaluate the rate constant.

Solution

(a) Let the rate law be $r = k[NO]^a[O_2]^b$
For NO, we have

$$x = \frac{0.30}{0.10} = 3$$

$$f = \frac{27.00}{3.00} = 9$$

So

$$3^a = 9$$

$$3^a = 3^2$$

$$\therefore a = 2$$

For O_2, we have

$$x^b = f$$

$$x = \frac{0.25}{0.20} = 1.25$$

$$f = \frac{33.75}{27.00} = 1.25$$

So

$$1.25^b = 1.25$$

$$b = 1$$

$$\therefore r = k[NO]^2[O_2]$$

Rates of Chemical Reactions

(b) Overall order = 2 + 1 = 3
(c) The rate constant, k, is obtained as follows.

$$r = k[NO]^2[O_2]$$

$$\therefore k = \frac{r}{[NO]^2[O_2]}$$

Using data from the first row, we have

$$r = 3.00 \text{ mol dm}^{-3} \text{ s}^{-1}$$

$$[NO] = 0.10 \text{ mol dm}^{-3}$$

$$[O_2] = 0.20 \text{ mol dm}^{-3}$$

$$k = ?$$

Substituting, we have

$$k = \frac{3.00 \text{ mol dm}^{-3} \text{ s}^{-1}}{(0.10 \text{ mol dm}^{-3})^2 \times 0.20 \text{ mol dm}^{-3}} = 1.50 \times 10^3 \text{ mol}^2 \text{ dm}^6 \text{ s}^{-1}$$

PRACTICE PROBLEM

The following data were obtained during the measurement of the initial rate of the reaction shown below.

$$2H_2O_2(g) \rightarrow 2H_2O(l) + O_2(g)$$

EXP.	$[H_2O_2]$ (mol dm^{-3})	Rate (mol dm^{-3} s^{-1})
I	0.25	0.50
II	0.30	0.60

(a) Obtain the rate law.
(b) Determine the order of the reaction.
(c) Evaluate the rate constant at the prevailing temperature.

14.7 INSTANTANEOUS RATE OF REACTION

The *instantaneous rate of reaction* is the rate at a specific moment in time. It can be calculated using either a rate curve or an integrated rate law.

14.7.1 GRAPHICAL DETERMINATION OF INSTANTANEOUS RATE

To determine the instantaneous rate from a rate curve, we have to calculate the slope or gradient of the tangent to the point of interest. The following example illustrate this method.

Example

The rate curve for the consumption of hydrogen in the reaction $H_2(g) + I_2(g) \rightarrow 2HI(g)$ at 600 K is given in Figure 14.5. Determine the instantaneous rate of reaction at 12 s.

Solution

The instantaneous rate of reaction at $t = 12$ s is obtained by determining the slope or gradient of the tangent to the curve at this point as shown in Figure 14.6.

$$\text{Rate} = -\frac{\Delta[H_2]}{\Delta t} = -\frac{[H_2]_{t_2} - [H_2]_{t_1}}{t_2 - t_1}$$

$$= -\frac{(0.96 - 0.36) \text{ mol dm}^{-3}}{(5 - 22)\text{s}} = 0.035 \text{ mol dm}^{-3}\text{ s}^{-1}$$

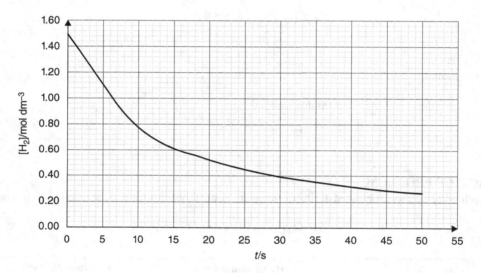

FIGURE 14.5 Rate curve for the disappearance of hydrogen

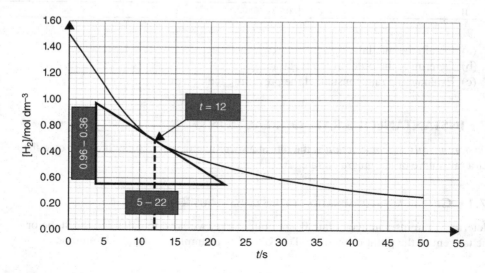

FIGURE 14.6 Determination of the instantaneous rate of reaction of hydrogen.

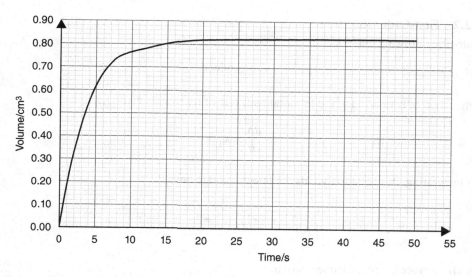

FIGURE 14.7 Rate curve for the formation of oxygen.

PRACTICE PROBLEM

The graph in Figure 14.7 shows the volume of oxygen released over time during the decomposition of hydrogen peroxide, H_2O_2. Determine the instantaneous rate of reaction at 22 s.

14.7.2 Integrated Rate Laws

An integrated rate law relates the initial concentration of a substance to its concentration at any point in time. The form of an integrated rate law depends on the order of reaction. Only the integrated rate laws involving the decomposition of a substance into one or more products shall be considered.

14.7.2.1 Zero-Order Reaction

Consider the following reaction:

$$A \rightarrow \text{products}$$

If the reaction is zero-order, then the instantaneous rate of reaction is given by

$$-\frac{dA}{dt} = k[A]^0 = k$$

On integrating, this equation yields the integrated rate law:

$$[A]_t = [A]_0 - kt$$

where

$[A]_t$ = concentration of A at time t
$[A]_0$ = initial concentration of A
k = rate constant
t = time

A plot of $[A]_t$ against t yields a straight-line graph, with a slope of $-k$ and a y-intercept of $[A]_0$.

14.7.2.2 First-Order Reaction

Consider the following reaction:

$$A \rightarrow products$$

If the reaction is first-order, the instantaneous rate of reaction is given by

$$-\frac{dA}{dt} = k[A]$$

On integrating, this equation yields the integrated rate law:

$$\ln \frac{[A]_0}{[A]_t} = kt \text{ or } \ln[A]_t = \ln[A]_0 - kt$$

where all parameters are as defined earlier.

A plot of $\ln[A]_t$ against t yields a straight-line graph with a slope of $-k$ and a y-intercept of $\ln[A]_0$.

14.7.2.3 Second-Order Reaction

Consider the following reaction:

$$A \rightarrow products$$

If the reaction is second-order, the instantaneous rate of reaction is given by

$$-\frac{dA}{dt} = k[A]^2$$

On integrating, this equation yields the integrated rate law:

$$\frac{1}{[A]_t} = \frac{1}{[A]_0} + kt$$

where all parameters are as defined earlier.

A plot of $\frac{1}{[A]_t}$ against t yields a straight-line graph with a slope of k and a y-intercept of $\frac{1}{[A]_0}$.

14.7.2.4 Half-Life

The time required for the concentration of a reacting substance to decrease to half of its initial value is called *half-life*, $t_{1/2}$. The mathematical expression for the half-life of a reaction is obtained by substituting $[A]_0 = \frac{1}{2}[A]_0$ at $t = t_{1/2}$ into its integrated rate law.

For a zero-order reaction, the half-life is given by

$$t_{1/2} = \frac{[A]_0}{2k}$$

Rates of Chemical Reactions

For a first-order reaction, the half-life is given by

$$t_{1/2} = \frac{0.693}{k}$$

For a second-order reaction, the half-life is given by

$$t_{1/2} = \frac{1}{k[A]_0}$$

Example 1

Ammonia, NH_3, undergoes a zero-order decomposition. At a certain temperature, it was found that the concentration of ammonia changed from 1.10 mol dm^{-3} to 0.850 mol dm^{-3} in 15.5 s. Determine:

(a) the rate constant;
(b) the concentration of ammonia after 20.5 s;
(c) the time required for the concentration of ammonia to decrease to 10% of its initial value.

Solution

We will apply the zero-order integrated rate law:

$$[A]_t = [A]_0 - kt$$

(a)
$[A]_0 = 1.10$ mol dm^{-3}
$[A]_t = 0.850$ mol dm^{-3}
$t = 15.5$ s
$k = ?$

Substituting, we have

$$0.850 \text{ mol dm}^{-3} = 1.10 \text{ mol dm}^{-3} - k \times 15.5 \text{ s}$$

So

$$k \times 15.5 \text{ s} = 0.25 \text{ mol dm}^{-3}$$

$$\therefore k = \frac{0.25 \text{ mol dm}^{-3}}{15.5 \text{ s}} = 0.0161 \text{ mol dm}^{-3} \text{ s}^{-1}$$

(b)
$[A]_0 = 1.10$ mol dm^{-3}
$t = 20.5$ s
$k = 0.0161$ mol dm^{-3} s^{-1}
$[A]_t = ?$

Substituting, we have

$$[A]_t = 1.10 \text{ mol dm}^{-3} - \left(\frac{0.0161 \text{ mol dm}^{-3}}{1 \text{s}} \times 20.5 \text{ s}\right) = 0.770 \text{ mol dm}^{-3}$$

(c) Since the initial concentration of NH_3 is 1.10 mol dm⁻³, then 10% of this value is 0.110 mol dm⁻³.

$$[A]_0 = 1.10 \text{ mol dm}^{-3}$$

$$[A]_t = 0.110 \text{ mol dm}^{-3}$$

$$k = 0.0161 \text{ mol dm}^{-3} \text{ s}^{-1}$$

$$t = ?$$

Substituting, we have

$$0.110 \text{ mol dm}^{-3} = 1.10 \text{ mol dm}^{-3} - 0.0161 \text{ mol dm}^{-3} \text{s}^{-1} \times t$$

So

$$0.0161 \text{ mol dm}^{-3} \text{s}^{-1} \times t = 0.990 \text{ mol dm}^{-3}$$

$$\therefore t = \frac{0.990 \text{ mol dm}^{-3} \times 1\text{s}}{0.0161 \text{ mol dm}^{-3}} = 61.5 \text{s}$$

Example 2

The following data were obtained during an experiment on the rate of decomposition of dinitrogen pentoxide, N_2O_5, at 337 K.

t (s)	0	50	100	150	200	250	300
$[N_2O_5]$ (mol dm⁻³)	0.500	0.393	0.309	0.243	0.191	0.150	0.118

Determine:

(a) whether the reaction is zero, first or second-order;
(b) the rate constant;
(c) the concentration of N_2O_5 left after 280 s.

Solution

(a) We have to plot three graphs, one for each order of kinetics, and determine which one results in a straight line. To test for zero-order kinetics, a plot of $[N_2O_5]_t$ against t is required. For first-order, a plot of $\ln[N_2O_5]_t$ against t is required. For second-order kinetics kinetics, a plot of $\frac{1}{[N_2O_5]_t}$ against t is required.

The very first step is to prepare the table of data.

t (s)	0	50	100	150	200	250	300
$[N_2O_5]$ (mol dm⁻³)	0.500	0.393	0.309	0.243	0.191	0.150	0.118
$\ln[N_2O_5]$	-0.693	-0.934	-1.17	-1.41	-1.66	-1.90	-2.14
$\frac{1}{[N_2O_5]_t}$ (dm³ mol⁻¹)	2.00	2.54	3.24	4.11	5.24	6.67	8.47

The three graphs are shown in Figure 14.8.
The graph of $\ln[N_2O_5]_t$ against t (Figure 14.8b) is a straight line; hence, the reaction is first-order.

(b) The slope of $\ln[N_2O_5]_t$ against t is the negative value of the rate constant.
From the graph in Figure 14.9,

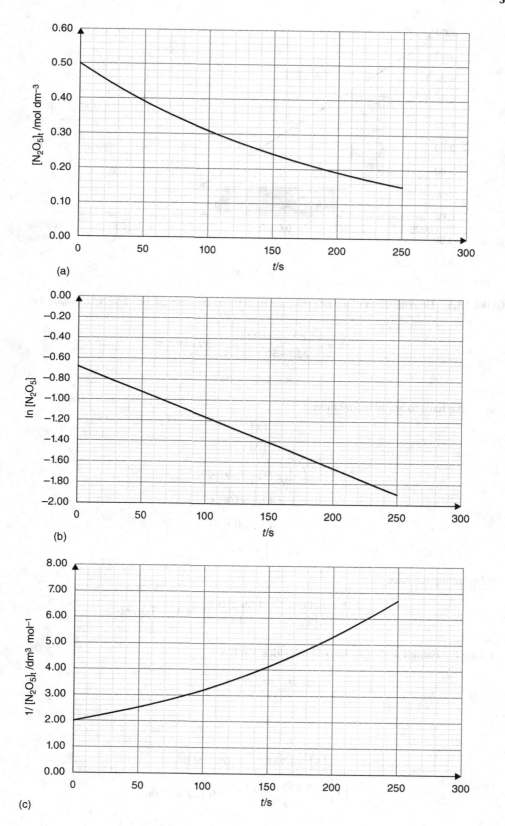

FIGURE 14.8 (a) Graph of $[N_2O_5]_t$ against t. (b) Graph of $\ln[N_2O_5]_t$ against t. (c) Graph of $1/[N_2O_5]_t$ against t.

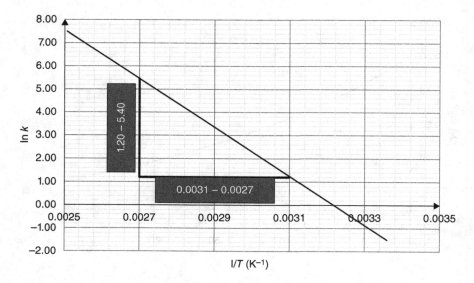

FIGURE 14.9 Determination of the rate constant from the slope of the graph of $\ln[N_2O_5]_t$ against t.

$$-k = \frac{-1.80-(-0.92)}{230-50 \text{ s}} = -4.89 \times 10^{-3} \text{ s}^{-1}$$

$$\therefore k = 4.89 \times 10^{-3} \text{ s}^{-1}$$

(c) We have to apply the relation:

$$\ln\frac{[A]_0}{[A]_t} = kt$$

$$[A]_0 = 0.500 \text{ mol dm}^{-3}$$

$$k = 4.89 \times 10^{-3} \text{ s}^{-1}$$

$$t = 280 \text{ s}$$

$$[A]_t = ?$$

Substituting, we have

$$\ln\frac{0.500 \text{ mol dm}^{-3}}{[A]_t} = \frac{4.89 \times 10^{-3}}{1 \text{ s}} \times 280 \text{ s} = 1.3692$$

Taking the natural antilogarithm of both sides, we have

$$\frac{0.500 \text{ mol dm}^{-3}}{[A]_t} = e^{1.3692}$$

So

$$[A]_t \times 3.9322 = 0.500 \text{ mol dm}^{-3}$$

$$\therefore [A]_t = \frac{0.500 \text{ mol dm}^{-3}}{3.9322} = 0.127 \text{ mol dm}^{-3}$$

Rates of Chemical Reactions

Example 3

The decomposition of nitrogen(IV) oxide, NO_2, is second-order. At 592 K, the rate constant is 0.500 dm³ mol⁻¹ s⁻¹. Calculate the time required for 0.500 mol dm⁻³ of NO_2 to decompose to 15% of the initial value.

Solution

We apply the relation:

$$\frac{1}{[A]_t} = \frac{1}{[A]_0} + kt$$

$$[A]_0 = 0.500 \text{ mol dm}^{-3}$$

$$k = 0.500 \text{ dm}^3 \text{ mol}^{-1} \text{ s}^{-1}$$

$$[A]_t = \frac{15}{100} \times 0.500 \text{ mol dm}^{-3} = 0.0750 \text{ mol dm}^{-3}$$

$$t = ?$$

Substituting the values, we have

$$\frac{1}{0.0750 \text{ mol dm}^{-3}} = \frac{1}{0.500 \text{ mol dm}^{-3}} + 0.500 \text{ dm}^3 \text{ mol}^{-1} \text{ s}^{-1} \times t$$

So

$$0.500 \text{ dm}^3 \text{ mol}^{-1} \text{ s}^{-1} \times t = 11.33 \text{ dm}^3 \text{ mol}^{-1}$$

$$\therefore t = \frac{11.33 \text{ dm}^3 \text{ mol}^{-1} \times 1\text{s}}{0.500 \text{ dm}^3 \text{ mol}^{-1}} = 22.7 \text{ s}$$

Example 4

The decomposition of nitrogen dioxide, NO_2, to nitrogen, N_2, and oxygen, O_2, is a second-order reaction. Calculate the rate constant for the reaction, given that the half-life of 0.120 mol dm⁻³ of NO_2 is 1400 s.

Solution

The equation for the half-life of a second-order reaction is given by

$$t_{1/2} = \frac{1}{k[A]_0}$$

$$t_{1/2} = 1400 \text{ s}$$

$$[A]_0 = 0.120 \text{ mol dm}^{-3}$$

$$k = ?$$

Substituting, we have

$$1400 \text{ s} = \frac{1}{k \times 0.120 \text{ mol dm}^{-3}}$$

So

$$168 \times k \text{ mol dm}^{-3} \text{ s} = 1$$

$$\therefore k = \frac{1}{168 \text{ mol dm}^{-3} \text{ s}} = 5.95 \times 10^{-3} \text{ dm}^3 \text{ mol}^{-1} \text{ s}^{-1}$$

PRACTICE PROBLEM

1. The fermentation of sucrose, $C_{12}H_{22}O_{11}$, to ethanol, C_2H_5OH, and carbon dioxide, CO_2, follows a first-order kinetics. If the half-life of the reaction is 2250 s at a certain temperature, calculate:
   (a) the rate constant;
   (b) the concentration of 0.175 mol dm^{-3} of $C_{12}H_{22}O_{11}$ after 1500 s.
2. Consider the following rate data for the hypothetical reaction A → B + C.

t (s)	0	10	20	30	40	50	60
[A] (mol dm^{-3})	0.750	0.261	0.158	0.113	0.0882	0.0723	0.0612

   (a) Determine whether the reaction is first- or second-order.
   (b) Calculate the rate constant for the reaction.
   (c) The half-life of the reaction.

14.8 REACTION RATES AND TEMPERATURE

As mentioned earlier, the rate constant is temperature-dependent. The variation of rate constant with temperature is given by the Arrhenius equation:

$$\ln k = \ln A - \frac{E_a}{RT} \quad \text{or} \quad k = Ae^{-E_a/RT}$$

where

k = rate constant
A = frequency or pre-exponential factor
E_a = activation energy
T = absolute temperature
R = universal gas constant = 8.314 J mol^{-1} K^{-1}

A plot of $\ln k$ against $1/T$ produces a straight-line graph with a slope of $-E_a/R$ and a y-intercept of $\ln A$.

The Arrhenius equation allows us to relate the rate constants of a reaction at two different temperatures as follows:

$$\ln \frac{k_2}{k_1} = -\frac{E_a}{R}\left(\frac{1}{T_2} - \frac{1}{T_1}\right)$$

where k_1 and k_2 are the rate constants at temperatures T_1 and T_2, respectively.

Rates of Chemical Reactions

Example 1

The rate constant for the second-order decomposition of nitrogen dioxide, NO_2, into nitrogen monoxide, NO, and oxygen, O_2, is 235 mol dm^{-3} s^{-1} at 527°C. Calculate the frequency factor if the activation energy is 115 kJ mol^{-1}.

$$\left(R = 8.314 \text{ J mol}^{-1}\text{K}^{-1}\right)$$

Solution

We will apply the Arrhenius equation:

$$k = Ae^{-E_a/RT}$$

$$k = 235 \text{ mol dm}^{-3}\text{s}^{-1}$$

$$E_a = 115 \text{ kJ mol}^{-1} = 115000 \text{ J mol}^{-1}$$

$$T = (273 + 527) \text{ K} = 800 \text{ K}$$

$$R = 8.314 \text{ J mol}^{-1} \text{ K}^{-1}$$

$$A = ?$$

Substituting, we have

$$235 \text{ mol dm}^{-3}\text{s}^{-1} = Ae^{-\frac{115000 \text{ J mol}^{-1} \times 1\text{K}}{8.314 \text{ J mol}^{-1} \times 800 \text{ K}}}$$

So

$$235 \text{ mol dm}^{-3}\text{s}^{-1} = A \times 3.097 \times 10^{-8}$$

$$\therefore A = \frac{235 \text{ mol dm}^{-3} \text{ s}^{-1}}{3.097 \times 10^{-8}} = 7.59 \times 10^{9} \text{ mol dm}^{-3} \text{ s}^{-1}$$

Note that the frequency factor of a reaction has the same unit as the rate constant.

Example 2

The following data were collected during an experiment on the decomposition of N_2O_5 at 30°C.

T (K)	298	318	338	358	378	398
k (dm^3 mol^{-1})	0.236	2.21	15.8	91.0	434	1780

$$\left(R = 8.314 \text{ J mol}^{-1} \text{ K}^{-1}\right)$$

Use these data to obtain the activation energy of the reaction.

Solution

The first step is to plot the graph of ln k against 1/T. The table of values is follows.

T (K)	298	318	338	358	378	398
k (dm^3 mol^{-1} s^{-1})	0.236	2.21	15.8	91.0	434	1780
1/T (K^{-1})	0.00336	0.00314	0.00296	0.00279	0.00264	0.00251
ln k	−1.44	0.793	2.76	4.51	6.07	7.48

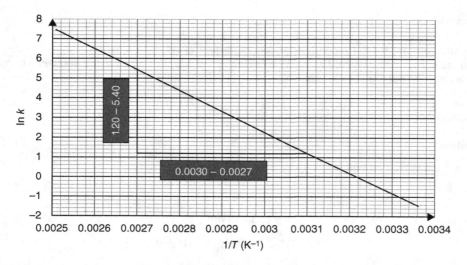

FIGURE 14.10 Determination of the activation energy from the graph of ln k against $1/T$.

The graph of ln k against $1/T$ is shown in Figure 14.10. From the graph, the slope can be used to calculate activation energy as follows.

$$\text{Slope} = \frac{(1.20 - 5.40) \times 1\,K}{0.00310 - 0.00270} = -\frac{E_a}{R}$$

$$R = 8.314\,J\,mol^{-1}\,K^{-1}$$

$$E_a = ?$$

So

$$-\frac{E_a}{8.314\,J\,mol^{-1}\,K^{-1}} = -10500\,K$$

$$\therefore E_a = 10500\,K \times \frac{8.314\,J\,mol^{-1}}{1\,K} = 87.3\,kJ\,mol^{-1}$$

Example 3

The rate constants for the reaction $H_2(g) + I_2(g) \rightarrow 2HI(g)$ are 4.40×10^{-4} dm^3 mol^{-1} s^{-1} and 4.69×10^{-3} dm^3 mol^{-1} s^{-1} at 600 K and 750 K, respectively. Calculate the activation energy of the reaction.

$$\left(R = 8.314\,J\,mol^{-1}\,K^{-1}\right)$$

Solution

We apply the relation:

$$\ln\frac{k_2}{k_1} = -\frac{E_a}{R}\left(\frac{1}{T_2} - \frac{1}{T_1}\right)$$

$$k_1 = 4.40 \times 10^{-4}\,dm^3\,mol^{-1}\,s^{-1}$$

Rates of Chemical Reactions

$$k_2 = 4.69 \times 10^{-3} \, dm^3 \, mol^{-1} \, s^{-1}$$
$$T_1 = 600 \, K$$
$$T_2 = 750 \, K$$
$$R = 8.314 \, J \, K^{-1} \, mol^{-1}$$
$$E_a = ?$$

Substituting, we have

$$\ln \frac{4.69 \times 10^{-3} \, dm^3 \, mol^{-1} \, s^{-1}}{4.40 \times 10^{-4} \, dm^3 \, mol^{-1} \, s^{-1}} = -\frac{E_a \times 1 \, K}{8.314 \, J \, mol^{-1}} \left(\frac{1}{750 \, K} - \frac{1}{600 \, K} \right)$$

So

$$2.3664 = \frac{E_a \times 3.33 \times 10^{-4}}{8.314 \, J \, mol^{-1}}$$

$$\therefore E_a = \frac{19.674 \, J \, mol^{-1}}{3.33 \times 10^{-4}} = 59.1 \, kJ \, mol^{-1}$$

Example 4

The rate constant for the decomposition of dinitrogen monoxide, N_2O, is $1.20 \times 10^{-3} \, s^{-1}$ at 823 K. Calculate the rate constant at 900 K, given that the activation energy is 250 kJ mol^{-1}.

$$\left(R = 8.314 \, J \, mol^{-1} \, K^{-1} \right)$$

Solution

We will apply the relation:

$$\ln \frac{k_2}{k_1} = -\frac{E_a}{R} \left(\frac{1}{T_2} - \frac{1}{T_1} \right)$$

$$k_1 = 1.20 \times 10^{-3} \, s^{-1}$$
$$R = 8.314 \, J \, K^{-1} \, mol^{-1}$$
$$E_a = 250 \, kJ \, mol^{-1} = 250000 \, J \, mol^{-1}$$
$$T_1 = 823 \, K$$
$$T_2 = 900 \, K$$
$$k_2 = ?$$

Substituting, we have

$$\ln \frac{k_2}{1.20 \times 10^{-3} \, s^{-1}} = -\frac{250000 \, J \, mol^{-1} \times 1 \, K}{8.314 \, J \, mol^{-1}} \left(\frac{1}{900 \, K} - \frac{1}{823 \, K} \right)$$

So

$$\ln \frac{k_2}{1.20 \times 10^{-3} \, \text{s}^{-1}} = 3.126$$

Taking the natural antilogarithm of both sides, we have

$$\frac{k_2}{1.20 \times 10^{-3} \, \text{s}^{-1}} = e^{3.126}$$

$$\therefore k_2 = 22.78 \times 1.20 \times 10^{-3} \, \text{s}^{-1} = 0.0273 \, \text{s}^{-1}$$

PRACTICE PROBLEMS

1. The following data were obtained in a study of the decomposition $2N_2O_5(g) \rightarrow 2N_2O_4(g) + O_2(g)$ at various temperatures.

T (K)	338	358	378	398	418	438
k (s^{-1})	0.00240	0.0186	0.116	0.603	2.67	10.3

   Determine the activation energy of the reaction.

   $$\left(R = 8.314 \, \text{J mol}^{-1} \, \text{K}^{-1} \right)$$

2. The rate constants for the hydrolysis of starch are 36.0 dm³ mol⁻¹ s⁻¹ and 1700 dm³ mol⁻¹ s⁻¹ at 30°C and 60°C, respectively. Determine the activation energy of the reaction.

   $$\left(R = 8.314 \, \text{J mol}^{-1} \, \text{K}^{-1} \right)$$

FURTHER PRACTICE PROBLEMS

1. When zinc granules were added to a dilute solution of hydrochloric acid, HCl, a total of 25 cm³ of hydrogen was collected in 10.5 min. Calculate the average rate of formation of hydrogen in cm³ s⁻¹.

2. In the laboratory preparation of ethyl ethanoate, $CH_3COOC_2H_5$, the concentration of an ethanoic acid, CH_3COOH, solution decreased by 0.050 mol dm⁻³ in 2.5 min when it was mixed with ethanol solution. Calculate the average rate of reaction in mol dm⁻³ s⁻¹.

3. A total of 10.5 g of calcium carbonate was added to excess dilute nitric acid, HNO_3, resulting in the evolution of carbon dioxide, CO_2. Calculate the average rate of reaction in mol s⁻¹, given that the reaction took 10 min to complete.

   $$\left(C = 12, \, O = 16, \, Ca = 40 \right)$$

4. (a) State the collision theory.
   (b) State the effect of each of the following on a reaction at equilibrium:
      (i) Temperature, (ii) Pressure, (iii) Catalyst, (iv) Concentration

5. Consider the following reaction.

   $$2BrF_3(g) \rightarrow Br_2(g) + 3F_2(g)$$

At a particular temperature, the rate of formation of F_2 is 1.50×10^{-3} mol dm^{-3} s^{-1}.
(a) Calculate the rate of formation of Br_2.
(b) Calculate the rate of decomposition of BrF_3.

6. The following data were obtained for the decomposition of hydrogen peroxide, H_2O_2, at a particular temperature.

t (s)	0	50	100	150	200	250	300	350
$[H_2O_2]$ (mol dm^{-3})	1.000	0.366	0.132	0.0490	0.0180	0.00700	0.00300	0.00300

(a) Plot the rate curve for the reaction.
(b) Determine the instantaneous rate at 110 s.

7. The following data were obtained during the measurement of the initial rate of the following reaction at a certain temperature.

$$2ICl(g) + H_2(g) \rightarrow I_2(g) + 2HCl(g)$$

EXP.	[ICl] (mol dm^{-3})	[H$_2$] (mol dm^{-3})	Rate (mol dm^{-3} s^{-1})
I	0.20	0.15	0.0120
II	0.30	0.15	0.0180
III	0.30	0.25	0.0300

(a) Determine the rate law.
(b) Find the overall order of the reaction.
(c) Calculate the rate constant at the prevailing temperature.

8. The following table shows the data obtained during an experiment to determine the initial rate of the reaction between nitrogen dioxide, NO_2, and carbon monoxide, CO, at different concentrations.

EXP.	[NO$_2$] (mol dm^{-3})	[CO] (mol dm^{-3})	Rate (mol dm^{-3} s^{-1})
I	0.10	0.20	0.065
II	0.30	0.20	0.585
III	0.30	0.40	0.585

(a) Determine the order of reaction with respect to each reactant.
(b) Find the overall order of reaction.
(c) Calculate the rate constant at the prevailing temperature.

9. The hydrolysis of starch follows a second-order kinetics.
(a) Calculate the rate constant of the hydrolysis of a 0.110 mol dm^{-3} starch solution, given that the half-life is 535 s.
(b) How long will it take for the concentration of the starch solution to fall to 15% of its initial value?

10. The decomposition of ammonia follows a zero-order kinetics. Calculate the half-life of 0.750 mol dm^{-3} of ammonia solution at a temperature where the rate constant is 2.50×10^{-4} mol dm^{-3} s^{-1}.

11. The following data were obtained in an experiment on the decomposition of dimethyl ether, C_2H_6O, to methane, CH_4, hydrogen, H_2, and carbon monoxide, CO, at 750 K.

t (s)	0	500	1000	1500	2000	2500
$[C_2H_6O]$ (mol dm^{-3})	1.50	1.42	1.34	1.27	1.20	1.13

(a) Determine whether the reaction is first- or second-order.
(b) Calculate the rate constant of the reaction.
(c) Find the time required for a 0.760 mol dm^{-3} dimethyl ether solution to dissociate to 10% of the initial value.

12. The following data were obtained in a study of the decomposition of dinitrogen monoxide, N_2O, into nitrogen, N_2, and oxygen, O_2, at a particular temperature.

t (s)	0	20	40	60	80	100
[N_2O] mol dm^{-3}	1.000	0.670	0.448	0.299	0.199	0.132

(a) Determine whether the reaction is zero-order or first-order.
(b) Calculate the rate constant.
(b) Calculate the half-life of the reaction.

13. The decomposition of cyclopropane follows a first-order kinetics. What percentage of a 2.50 mol dm^{-3} cyclopropane solution would remain after one hour at a temperature where the rate constant is 5.50×10^{-4} s^{-1}?

14. The rate constant for the hydrolysis of sucrose, $C_{12}H_{22}O_{11}$, into glucose, $C_6H_{12}O_6$, and fructose, $C_6H_{12}O_6$, is 1.00×10^{-4} s^{-1} at 25°C. Determine the activation energy of the reaction if the frequency factor is 8.54×10^{14} s^{-1}.

$$\left(R = 8.314 \text{J mol}^{-1} \text{K}^{-1}\right)$$

15. The activation energy for the reaction $H_2(g) + I_2(g) \rightarrow 2HI(g)$ is 120 kJ mol^{-1}. Calculate the frequency factor of the reaction if the rate constant is 7.00×10^{-3} dm^3 mol^{-1} s^{-1} at 25°C.

$$\left(R = 8.314 \text{J mol}^{-1} \text{K}^{-1}\right)$$

16. The activation energy for the decomposition of ethanal, CH_3CHO, into methane, CH_4, and carbon monoxide, CO, is 170 kJ mol^{-1}. Calculate the rate constant at 500°C if the frequency factor is 4.31×10^8 s^{-1}.

$$\left(R = 8.314 \text{J mol}^{-1} \text{K}^{-1}\right)$$

17. The following data were obtained in an experiment on the effect of temperature on the decomposition of ammonia.

T (K)	420	440	460	480	500	520
k (s^{-1})	0.139	10.91	585	2.25×10^4	6.46×10^5	1.43×10^7

Determine the activation energy of the reaction.

$$\left(R = 8.314 \text{J mol}^{-1} \text{K}^{-1}\right)$$

18. The rate constants for the decomposition of hydrogen peroxide, H_2O_2, are 1.30×10^{-4} s^{-1} and 0.0572 s^{-1} at 25°C and 100°C, respectively. Determine the activation energy of the reaction.

$$\left(R = 8.314 \text{J mol}^{-1} \text{K}^{-1}\right)$$

19. The activation energy of a reaction is 88 kJ mol^{-1}. Determine the rate constant at 75°C if the rate constant is 9.14×10^{-3} at 27°C.

$$(R = 8.314 \, \text{J K}^{-1} \, \text{mol}^{-1})$$

20. The rate constants for the catalysed decomposition of ethanol, C_2H_5OH, into ethene, C_2H_4, and steam, H_2O, are 4.00×10^{-5} dm^3 mol^{-1} s^{-1} and 0.0306 dm^3 mol^{-1} s^{-1} at 324°C and 550°C, respectively.
    (a) Calculate the activation energy of the reaction.
    (b) At what temperature is the rate constant of the reaction be 1.25×10^{-3} dm^3 mol^{-1} s^{-1}?

$$(R = 8.314 \, \text{J K}^{-1} \, \text{mol}^{-1})$$

15 Nuclear Chemistry

15.1 RADIOACTIVITY

Radioactivity is the random, spontaneous emission of radiations from the nuclei of atoms. It occurs due to the instability of a nucleus, which lacks sufficient binding energy to hold its nucleons together.

Every element has at least one radioactive isotope. An isotope that undergoes radioactive decay is called a *radioactive isotope* or *radioisotope*. Examples of radioisotopes include carbon-14, phosphorus-32, cobalt-57, hydrogen-3, potassium-40 and lead-202.

All isotopes of elements with atomic numbers greater than 83—i.e., beyond bismuth—are radioactive. This elements are called radioactive elements. Examples include uranium, plutonium, polonium, radium and thorium.

An unstable nucleus will emit radiation to release excess energy until a stable nuclide or element is formed. During this process, several intermediate nuclides or elements may be produced. This sequence of transformations is known as a *radioactive decay series*. For example, as shown in Figure 15.1,

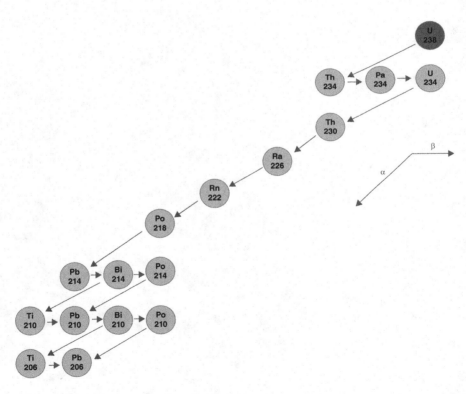

FIGURE 15.1 The decay series of uranium-238.

Nuclear Chemistry

the uranium-238 decay series begins with uranium-238 and ends with lead-206, a stable isotope of lead. The new nuclide formed when a nuclide emits radiation is called a *daughter nuclide*, while the original nuclide is called the *parent nuclide*.

15.1.1 Types of Radioactive Decay

There are three main types of radiation: *alpha (α) particles or rays, beta (β) particles or rays and gamma (γ) rays*. These are known as *ionising radiations* because they have the ability to alter physiological processes and molecular structure.

As shown in Table 15.1, alpha particles are helium nuclei (positively charged), while beta particles are electrons (negatively charged). Consequently, as illustrated in Figure 15.2, alpha particles

TABLE 15.1
Properties of Nuclear Radiations

Property	α Particle	β Particle	γ Rays
Nature	Helium nuclei ($^4_2$He)	Electron ($^0_{-1}$e)	Electromagnetic radiation
Mass	4 amu	Mass of electron	None
Charge	+2	−1	None
Ionising power	Very high	Moderate	Very low
Penetrating power	Low	Moderate	Very high

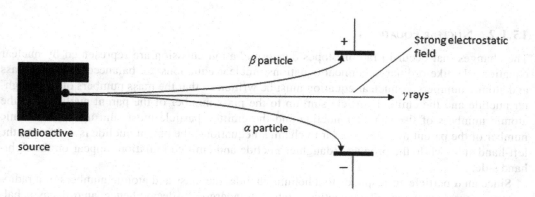

FIGURE 15.2 Behaviour of the three types of radiations in electrostatic field showing alpha particles as positively charged, beta particles as negatively charged and gamma rays as uncharged.

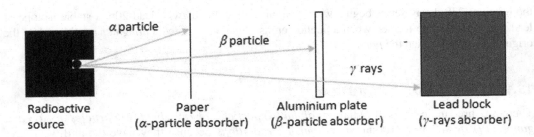

FIGURE 15.3 The penetrating powers of nuclear radiations.

are deflected towards a negatively charged plate in an electrostatic field, while beta particles are attracted by a positively charged plate. Gamma rays, on the other hand, are not particles but rather high-energy electromagnetic radiation, and they remain unaffected by an electrostatic field.

The penetrating and ionising powers of the different types of radiation discussed vary significantly (see Figure 15.3). Alpha particles are the most ionising but the least penetrating. In fact, they can be stopped by a sheet of paper. Beta particles have moderate ionising and penetrating power and can be absorbed by an aluminium plate. Gamma rays are highly penetrating and require shielding with a thick block of lead or concrete.

As mentioned earlier, ionising radiations pose serious risks to living systems, especially when absorbed above safe limits. Therefore, individuals working with or near radioactive sources must take appropriate precautions to minimise exposure. These precautions include wearing protective clothing, wearing lapel radiation badges to monitor exposure to radiation, shielding radiation sources with dense materials—such as lead and concrete—and undergoing routine medical checks.

15.1.2 Nuclear Equations

The changes that occur in radioisotopes during radiation emission are represented by nuclear equations. Unlike ordinary chemical equations, nuclear equations are balanced based on mass and atomic numbers. A nuclear equation must be written so that the mass numbers of the daughter nuclide and the emitted particle sum up to the mass number of the parent nuclide, and the atomic numbers of the daughter nuclide and the emitted particle must sum up to the atomic number of the parent nuclide. Similar to chemical equations, the parent nuclide is written on the left-hand side, while the products (daughter nuclide and emitted radiation) appear on the right-hand side.

Since an α particle corresponds to a helium nuclide, the mass and atomic numbers of a radioisotope decreases by 4 and 2, respectively, when it undergoes α-decay; hence, an α-decay is balanced with a helium nuclide. The atomic number of a radioisotope increases by 1 when it undergoes β-decay, but its mass number remains unchanged; hence, a β-decay is balanced with an electron. The identity and properties of a radioisotope remain unaltered when it emits γ radiation.

Nuclear Chemistry

Example 1

Uranium-238 undergoes α-decay to form thorium. Write the balanced nuclear equation for the decay. Uranium has an atomic number of 92.

Solution

Since an α particle is emitted, the atomic and mass numbers of thorium, the daughter nuclide, are as follows:

$$A = 238 - 4 = 234$$
$$Z = 92 - 2 = 90$$

Thus, the balanced nuclear equation is given by

$$^{238}_{92}U \rightarrow \,^{234}_{90}Th + \,^{4}_{2}He$$

Example 2

Lead-214 decays into bismuth by emitting a β particle. Write the balanced nuclear equation for the decay. The atomic number of lead is 82.

Solution

Since a β particle is emitted, the atomic and mass numbers of thorium are as follows:

$$A = 214 + 0 = 214$$
$$Z = 82 + 1 = 83$$

Thus, the balanced nuclear equation is given by

$$^{214}_{82}Pb \rightarrow \,^{214}_{83}Bi + \,^{0}_{-1}e$$

PRACTICE PROBLEMS

Write the balanced nuclear equations for the following decays.
1. U-234 → Th-230
2. Bi-214 → Tl-210
3. Hg-206 → Tl-206
4. Rn-222 → Po-218
5. Pa-234 → U-234

15.1.3 THE LAW OF RADIOACTIVE DECAY AND HALF-LIFE

The rate of decay of a radioactive substance is proportional to the number or mass of atoms or nuclides in the substance. This relationship is expressed by the radioactive decay rate equation:

$$\frac{dN}{dt} = -\lambda N \text{ or } \frac{dm}{dt} = -\lambda m$$

where

$\dfrac{dN}{dt}$ = rate of decay
N = the number of atoms or nuclides in a radioisotope at time t
m = the mass of atoms or nuclides in a radioisotope at time t
λ = decay constant

The negative sign indicates that the mass or number of atoms or nuclides in a radioactive substance decreases over time. The decay constant, λ, represents the probability that a radioisotope will decay within a specified time. For example, the decay constants of carbon-15 and strontium-90 are 0.288 s^{-1} and 0.0247 yr^{-1}, respectively. When integrated, the radioactive decay rate equation becomes:

$$N = N_o e^{-\lambda t} \text{ or } m = m_o e^{-\lambda t}$$

where

N_o = the original number of atoms or nuclides present in a radioisotope
m_o = the original mass of atoms or nuclides originally present in a radioisotope

The above equation shows that a radioisotope undergoes exponential decay at a rate dependent only on mass or number of atoms or nuclides.

Another important parameter that describes how fast a radioisotope decays is its *half-life*. The half-life, $t_{1/2}$, of a radioisotope is the time required for a sample to decay to half of its original mass or number of atoms or nuclides (see Figure 15.4). Different radioisotopes have varying half-lives, ranging from a few seconds to billions of years. For instance, the half-lives of polonium-215 and uranium-238 are 0.0018 s and 4.5 billion years, respectively. The relationship between half-life and decay constant is given by

$$t_{1/2} = \dfrac{0.693}{\lambda}$$

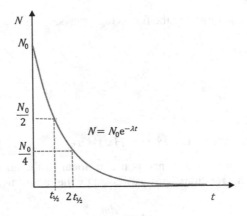

FIGURE 15.4 Rate of radioactive decay showing half-life.

Nuclear Chemistry

The half-life equation is obtained by substituting $N = \frac{1}{2} N_o$ at $t = t_{1/2}$ into the integrated decay rate equation.

The mass or number of atoms or nuclides of a radioisotope after any given time can be determined from its half-life using the equation:

$$N = \left(\frac{1}{2}\right)^n \times N_o \text{ or } m = \left(\frac{1}{2}\right)^n \times m_o$$

where n is the number of half-lives that have elapsed, given by $n = \dfrac{\text{time}}{\text{half-life}}$

Example 1

A radioisotope containing 200000 nuclides is kept for four half-lives. Determine the amount of nuclides left.

Solution

One method is to use the equation:

$$N = \left(\frac{1}{2}\right)^n \times N_o$$

$N_o = 200000$

$n = 4$

$N = ?$

Substituting, we have

$$N = \left(\frac{1}{2}\right)^4 \times 200000 = 12500$$

An alternative method is to successively divide the original number of nuclides by 2 for each half-life as, follows:

Nuclides after first half-life = 100000
Nuclides after second half-life = 50000
Nuclides after third half-life = 25000
Nuclides after fourth half-life = 12500

Example 2

The decay constant of carbon-15 is 0.288 s^{-1}. Calculate the mass of 3000.0g of the radioisotope that would be left after 20.0 s.

Solution

$$m = m_o e^{-\lambda t}$$

$m_o = 3000.0 \text{ g}$

$\lambda = 0.288 \text{ s}^{-1}$

$t = 20.0 \text{ s}$

$m = ?$

Substituting, we have

$$m = 3000.0 \text{ g} \times e^{-0.288 \times 20.0} = 9.45 \text{ g}$$

Alternatively, we can determine the required mass by calculating the number of half-lives that have elapsed after 20.0 s.

$$m = \left(\frac{1}{2}\right)^n \times m_o$$

The number of half-lives, n, that have elapsed after 20.0 s is obtained as follows.

$$t_{1/2} = \frac{0.693}{\lambda}$$

$$\lambda = 0.288 \text{ s}^{-1}$$

$$t_{1/2} = ?$$

Substituting, we have

$$t_{1/2} = \frac{0.693 \times 1 \text{ s}}{0.288} = 2.406 \text{ s}$$

$$\therefore n = \frac{20.0 \text{ s}}{2.406 \text{ s}} = 8.31$$

$$m = ?$$

Now, we can calculate the required mass:

$$m = \left(\frac{1}{2}\right)^{8.31} \times 3000.0 \text{ g} = 9.45 \text{ g}$$

Example 3

0.550 g of a sample of radioisotope was left after 25.0 ms. Determine the original mass of the sample if it has a half-life of 3.30 ms.

Solution

One method is to use the relation:

$$m = m_o e^{-\lambda t}$$

$$\therefore m_o = \frac{m}{e^{-\lambda t}}$$

The decay constant is calculated as follows.

$$t_{1/2} = \frac{0.693}{\lambda}$$

$$\therefore \lambda = \frac{0.693}{t_{1/2}}$$

Nuclear Chemistry

$$t_{1/2} = 3.30 \text{ ms} = 0.00330 \text{ s}$$

$$\lambda = ?$$

So

$$\lambda = \frac{0.693}{0.00330 \text{ s}} = 210 \text{ s}^{-1}$$

$$m = 0.550 \text{ g}$$

$$t = 25.0 \text{ ms} = 0.0250 \text{ s}$$

$$m_o = ?$$

Substituting, we have

$$m_o = \frac{0.550 \text{ g}}{e^{-\frac{210}{1\text{s}} \times 0.0250 \text{ s}}} = 105 \text{ g}$$

Alternatively, we can determine the required mass from the number of half-lives that have elapsed after 25 ms.

$$m = \left(\frac{1}{2}\right)^n \times m_o$$

$$\therefore m_o = \frac{m}{\left(\frac{1}{2}\right)^n}$$

$$n = \frac{25 \text{ ms}}{3.3 \text{ ms}} = 7.58$$

$$m = 0.550 \text{ g}$$

$$m_o = ?$$

Substituting, we have

$$m_o = \frac{0.550 \text{ g}}{\left(\frac{1}{2}\right)^{7.58}} = 105 \text{ g}$$

Example 4

The half-life of sodium-24 is 15.0 h. How long would it take 5.00 g of the radioisotope to decay to 0.625 g?

Solution

Our first approach is to apply the relation:

$$m = m_o e^{-\lambda t}$$

Taking the natural logarithm of both sides and solving for t, we have

$$t = -\frac{1}{\lambda} \ln \frac{m}{m_o}$$

The decay constant is calculated as follows.

$$t_{1/2} = \frac{0.693}{\lambda}$$

$$\therefore \lambda = \frac{0.693}{t_{1/2}}$$

$$t_{1/2} = 15\,h$$

$$\lambda = ?$$

So

$$\lambda = \frac{0.693}{15\,h} = 0.0462\,h^{-1}$$

$$m = 5.00\,g$$

$$m_o = 0.625\,g$$

$$t = ?$$

Substituting, we have

$$t = -\frac{1\,h}{0.0462} \ln \frac{0.625\,g}{5.00\,g} = 45\,h$$

Alternatively, we can obtain the same result by determining how many half-lives it would take for 5.00 g of the radioisotope to decay to 0.625 g.

$$m = \left(\frac{1}{2}\right)^n \times m_o$$

Dividing both sides by m_o, we have

$$\left(\frac{1}{2}\right)^n = \frac{m}{m_o}$$

Taking the common logarithm of both sides, we have

$$n \log \frac{1}{2} = \log \frac{m}{m_o}$$

$$\therefore n = -3.32 \log \frac{m}{m_o}$$

$$m = 5.00\,g$$

$$m_o = 0.625\,g$$

$$n = ?$$

Substituting, we have

$$n = -3.32 \log \frac{0.625 \text{ g}}{5.00 \text{ g}} = 3$$

$$\therefore t = 3 \times 15 \text{ h} = 45 \text{ h}$$

Example 5

A radioactive source initially emits 6603 counts per minute. Given that emission rate decreases to 1500 counts per minute after 1.00 min, determine:

(a) its decay constant in s^{-1};
(b) its half-life.

Solution

(c) We use the exponential decay equation:

$$N = N_o e^{-\lambda t}$$

Taking the natural logarithm of both sides and solving for λ, we have

$$\lambda = -\frac{1}{t} \ln \frac{N}{N_o}$$

$$t = 1.00 \text{ min} = 60 \text{ s}$$
$$N = 1500$$
$$N_o = 6603$$
$$\lambda = ?$$

Substituting, we have

$$\lambda = -\frac{1}{60.0 \text{ s}} \ln \frac{1500}{6603} = 0.0247 \text{ s}^{-1}$$

(d) The half-life is given by

$$t_{\frac{1}{2}} = \frac{0.693}{\lambda}$$

$$\lambda = 0.0247 \text{ s}^{-1}$$
$$t_{\frac{1}{2}} = ?$$

Substituting, we have

$$t_{\frac{1}{2}} = \frac{0.693 \times 1 \text{ s}}{0.0247} = 28 \text{ s}$$

PRACTICE PROBLEMS

1. Determine the half-life of a radioactive element with a decay constant of 0.30 s^{-1}.
2. Determine the decay constant of a radioisotope with a half-life of 194 years.
3. The half-life of iodine-131 is 8.07 days. How long will it take for 2.5 g of the radioisotope to decay to 0.3125 g?
4. The half-life of cobalt-60 is 5.26 years. How many nuclides will remain in a sample that originally contained 150000 nuclides if it is stored for 52.6 years?
5. The emission rate of a radioactive source decreases from 500 counts per minute to 168 counts per minute in 6 days. Determine the decay constant of the source.

15.2 Nuclear Fission and Fusion

Nuclear fission is the process in which a heavy nucleus splits into two nuclei of roughly equal mass, with a release of energy and radiation. Nuclear fission can occur spontaneously due to the oscillation of a heavy nucleus. However, a nucleus can also be induced to undergo nuclear fission by bombarding it with neutrons. A nuclide that can be induced to undergo nuclear fission is called a *fissionable nucleus*.

An example of a fissionable nucleus is uranium-235, which produces fragments of krypton and barium—along with energy and radiation—when bombarded with neutrons:

$$^{235}_{92}\text{U} + ^{1}_{0}\text{n} \rightarrow ^{141}_{56}\text{Ba} + ^{92}_{36}\text{Kr} + 3^{1}_{0}\text{n} + \text{energy}$$

As we can see in this equation, neutrons are released during nuclear fission. These released neutrons can induce fission in other nuclei, making the process self-sustaining. This type of nuclear fission is called a *chain reaction* (see Figure 15.5).

The reactions in *nuclear reactors* or *atomic* piles are controlled chain reactions. The minimum amount of fissionable material required to sustain a chain reaction is called the *critical mass*. In a nuclear reactor, electricity is generated from the enormous energy produced in carefully controlled chain reactions. The principle of a nuclear or atomic bomb, however, is based on an uncontrolled chain reaction.

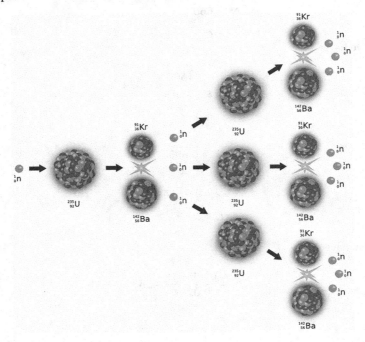

FIGURE 15.5 Nuclear chain reaction.

Nuclear Chemistry

Nuclear fusion is the process in which two light nuclei fuse or bind together to form a heavier nucleus, releasing energy and radiation. An example of nuclear fusion is the formation of helium nucleus from deuterium and tritium:

$$^{2}_{1}H + ^{3}_{1}H \rightarrow ^{4}_{2}He + ^{1}_{0}n + \text{energy}$$

Much more energy is generated in nuclear fusion than in nuclear fission. Nuclear fusion is an example of a thermonuclear reaction, i.e., reactions that take place under extremely high temperatures. The vast amount of energy generated by the Sun and other stars is due to nuclear fusion in their cores.

While nuclear fusion promises to be a source of cheap and unlimited energy, scientists have so far been unable to devise a practical method for achieving it due to the extremely high temperatures required to bind the two nuclei together.

Unlike ordinary chemical reactions, mass is not conserved in nuclear reactions, as some mass is converted into energy. The difference between the total mass of the products and the total mass of the reactants is called *mass defect*. The energy generated in a nuclear reaction is calculated with the Einstein's mass-energy equation:

$$E = \Delta mc^2$$

where

E = energy released (in kJ)
Δm = mass defect (in kg)
c = velocity of light = 3.00×10^8 m s^{-1}

As stated in Chapter 1, the masses of nuclei and sub-atomic particles are given in atomic mass unit, amu. To convert the mass defect to the SI unit of kg, we have to use the following conversion factor:

$$1 \text{ amu} = 1.6605 \times 10^{-27} \text{ kg}$$

Example

The fission of uranium-235 is given by the following nuclear equation:

$$^{235}_{92}U + ^{1}_{0}n \rightarrow ^{141}_{56}Ba + ^{92}_{36}Kr + 3 ^{1}_{0}n + \text{energy}$$

Calculate the amount of energy released when 1.0 g of uranium-235 undergoes nuclear fission.
(U-235 = 235.040 amu, Ba-141 = 140.910 amu, Kr-92 = 91.910 amu, ^{1}n = 1.009 amu, N_A = 6.02×10^{23} mol^{-1}, c = 3.00×10^8 m s^{-1})

Solution

We will apply Einstein's mass-energy relation:

$$E = \Delta mc^2$$

Δm = mass of reactants − mass of products

Mass of reactants = mass of U-235 + mass of ^{1}n

$$U\text{-}235 = 235.040 \text{ amu}$$

$$^1n = 1.009 \text{ amu}$$

$$\therefore \text{Mass of reactants} = 235.040 \text{ amu} + 1.009 \text{ amu} = 236.049 \text{ amu}$$

$$\text{Mass of products} = \text{mass of Ba-141} + \text{mass of Kr-92} + 3(\text{mass of } ^1n)$$

$$\text{Ba-141} = 140.910 \text{ amu}$$

$$\text{Kr-92} = 91.910 \text{ amu}$$

$$^1n = 1.009 \text{ amu}$$

$$\therefore \text{Mass of products} = 140.910 \text{ amu} + 91.910 \text{ amu} + 3(1.009 \text{ amu}) = 235.847 \text{ amu}$$

$$m = \Delta m$$

Substituting, we have

$$\Delta m = 236.049 - 235.847 \text{ amu} = 0.202 \text{ amu}$$

In the SI unit, we have

$$\Delta m = 0.202 \times 1.6607 \times 10^{-27} \text{ kg} = 3.354614 \times 10^{-28} \text{ kg}$$

$$c = 3.00 \times 10^8 \text{ ms}^{-1}$$

$$E = ?$$

We now substitute the values into Einstein's equation to obtain:

$$E = 3.354614 \times 10^{-28} \text{ kg} \times \left(3.00 \times 10^8 \text{ ms}^{-1}\right)^2$$

So

$$E = 3.354614 \times 10^{-28} \text{ kg} \times 9.00 \times 10^{16} \text{ m}^2 \text{ s}^{-2} = 3.0192 \times 10^{-11} \text{ J}$$

Note: $1 \text{ kg m}^2 \text{ s}^{-2} = 1 \text{ J}$.

This is the energy released in the nuclear fission of a single U-235 atom. To calculate the energy released in the nuclear fission of 1.0 g of U-235, we must multiply this value by the number of U-235 atoms in 1.0 g of the nuclide. This is done as follows.

$$n = \frac{m}{M}$$

$$m = 1.0 \text{ g}$$

$$M = 235 \text{ g mol}^{-1}$$

$$n = ?$$

Substituting, we have

$$n = \frac{1.0 \text{ g} \times 1 \text{ mol}}{235 \text{ g}} = 0.0042553 \text{ mol}$$

Nuclear Chemistry

The number of atoms is calculated as follows.

$$n = \frac{N}{N_A}$$

$$\therefore N = n \times N_A$$

$$n = 0.0042553 \text{ mol}$$

$$N_A = 6.02 \times 10^{23} \text{ mol}^{-1}$$

$$n = ?$$

Substituting, we have

$$n = 0.0042553 \text{ mol} \times \frac{6.02 \times 10^{23}}{1 \text{ mol}} = 2.56169 \times 10^{21}$$

$$\therefore E = 2.56169 \times 10^{21} \times 3.0192 \times 10^{-11} \text{ J}$$
$$= 7.73 \times 10^{7} \text{ kJ}$$

The generation of this huge amount of energy from the nuclear fission of just 1.0 g of U-235 highlights the tremendous energy produced in nuclear fission.

PRACTICE PROBLEM

The nuclear fission of uranium-235 generates 3.02×10^{-11} J of energy per atom. Calculate the total energy released in the nuclear fission of 2.5 g of the radioisotope.

$$\left(N_A = 6.02 \times 10^{23} \text{ mol}^{-1}\right)$$

15.3 BINDING ENERGY

Binding energy is the energy required to hold a nucleus together. It is equivalent to the energy released when nucleons fuse to form a nucleus. Another way to describe binding energy is the energy needed to completely break a nucleus apart or separate its nucleons. The mass of a nucleus is usually slightly lower than the sum of the masses of its individual nucleons due to the energy released when they bind together to form the nucleus.

The binding energy of a nucleus is obtained from its mass defect using Einstein's equation. The binding energy per nucleon of a nucleus is its binding energy divided by its atomic mass or number of nucleons.

Example

Calculate the nuclear binding energy of Helium-4 atom. Hence, determine:

(a) the binding energy in kJ mol^{-1};
(b) the binding in MeV.
(c) the binding per nucleon in MeV.

(He-4 = 4.003 amu, ^{1}n = 1.009 amu, ^{1}p = 1.007 amu; $c = 3.00 \times 10^8$ m s^{-1}, 1 eV = 1.602×10^{-19} J)

Solution

The binding energy is calculated with the relation:

$$E = \Delta m c^2$$

$$\Delta m = \text{mass of nucleons} - \text{mass of nucleus (He-4)}$$

The atomic and mass numbers of He-4 are 2 and 4, respectively; hence, it contains two protons and two neutrons.

$$\text{Mass of nucleons} = 2(\text{mass of }^1p) + 2(\text{mass of }^1n)$$

$$^1p = 1.007 \text{ amu}$$

$$^1n = 1.009$$

$$\therefore \text{Mass of nucleons} = 2(1.007 \text{ amu}) + 2(1.009 \text{ amu}) = 4.032 \text{ amu}$$

$$\text{He-4} = 4.003 \text{ amu}$$

$$\Delta m = ?$$

So

$$\Delta m = 4.032 \text{ amu} - 4.003 \text{ amu} = 0.029 \text{ amu}$$

$$\therefore \Delta m = 0.029 \times 1.6607 \times 10^{-27} \text{ kg} = 4.81603 \times 10^{-29} \text{ kg}$$

$$c = 3.00 \times 10^8 \text{ m s}^{-1}$$

$$E = ?$$

Substituting, we have

$$E = 4.81603 \times 10^{-29} \text{ kg} \times (3.00 \times 10^8 \text{ ms}^{-1})^2 = 4.33 \times 10^{-15} \text{ kJ}$$

(a) The binding energy in kJ mol^{-1} is calculated by multiplying the binding energy by the Avogadro's number.

$$E = 4.33 \times 10^{-15} \text{ kJ} \times N_A = 6.02 \times 10^{23} \text{ mol}^{-1}$$
$$= 2.61 \times 10^9 \text{ kJ mol}^{-1}$$

(b) To calculate the binding energy in MeV—i.e., mega electron volts—we need to use the following conversion factor:

$$1 \text{ eV} = 1.602 \times 10^{-19} \text{ J}$$

$$\therefore E = \frac{4.33 \times 10^{-12} \text{ J} \times 1 \text{ eV}}{1.602 \times 10^{-19} \text{ J}} = 2.70 \times 10^7 \text{ eV} = 27.0 \text{ MeV}$$

(c) Since there are 4 nucleons in He-4 atom, then

$$E = \frac{27.0 \text{ MeV}}{4} = 6.75 \text{ MeV}$$

Nuclear Chemistry

PRACTICE PROBLEMS

1. Calculate the binding energy of Helium-3 in kJ mol^{-1}.
   (He-3 = 3.016 amu, ^{1}n = 1.009 amu, ^{1}p = 1.007 amu; c = 3.00 × 10^8 m s^{-1})

2. Determine the binding energy per nucleon of Lithium-7 in MeV.
   (Li-7 = 6.941 amu, ^{1}n = 1.009 amu, ^{1}p = 1.007 amu; c = 3.00 × 10^8 m s^{-1}, 1 eV = 1.602 × 10^{-19} J)

FURTHER PRACTICE PROBLEMS

1. Write the balanced nuclear equation for each of the following radioactive decays:
   (a) Bi-210 → Po-210
   (b) Po-214 → Pb-210
   (c) Ra-226 → Rn-222

2. The half-life of Pb-210 is 22 years. Determine the mass of 2.0 g of the radioisotope that would remain after 110 years.

3. How long would it take for 4.0 g of Po-214 decay to 0.25 g if the half-life of the radioisotope is 22 min?

4. Consider the following nuclear equation:

   $$^{239}_{94}\text{Pu} + ^{1}_{0}\text{n} \rightarrow ^{100}_{42}\text{Mo} + ^{134}_{52}\text{Te} + 5^{1}_{0}\text{n}$$

   Determine the energy release in the nuclear fission of 1.0 g of Plutonium-239.
   (Pu-239 = 239.052 amu, Mo-100 = 99.907 amu, Te-134 = 133.911 amu, ^{1}n = 1.009 amu; N_A = 6.02 × 10^{23} mol^{-1}, c = 3.00 × 10^8 m s^{-1})

5. Calculate the binding energy of Carbon-12 in kJ mol^{-1}.
   (C-12 = 12.000 amu, ^{1}n = 1.009 amu, ^{1}p = 1.007 amu; c = 3.00 × 10^8 m s^{-1}, N_A = 6.02 × 10^{23} mol^{-1})

6. Determine the binding energy per nucleon of deuterium (D) in MeV.
   (D = 2.014 amu, ^{1}n = 1.009 amu, ^{1}p = 1.007 amu; c = 3.00 × 10^8 m s^{-1}, 1 eV = 1.602 × 10^{-19} J)

Appendix

A. CONVERSION FACTORS

Quantity	Common Units	SI Unit
Length	1 μ (micron)	10^{-6} m
	1 Å (Angstrom)	10^{-8} cm
	A light year = 1.52×10^{13} mi	9.46×10^{12} km
Mass	1 ton = 2000 lb	1000 kg
	1 amu = 1.66×10^{-24} lb	1.66×10^{-27} kg
	10 oz	28.4 g
	2.205 lb	1 kg
Area	1 ft² = 144 in²	0.0929 m²
	1 in² = 0.00694 ft²	6.452 cm²
Volume	1.05679 t (quartz) = 0.353 ft³	1 dm³ = 1 L
Time	1 min = 0.0167 h	60 s
	1 h = 60 min	3600 s
	1 day = 24 hr = 1440 min	86400 s
	1 wk = 7 days = 168 hr = 10080 min	6.048×10^5 s
	1 year = 52 wks = 365 days = 8760 hr = 525600 min	3.2×10^7 s
Velocity	1 mi hr⁻¹ = 1.62 km hr⁻¹ = 1.47 ft s⁻¹	0.447 m s⁻¹
	1 mi min⁻¹ = 88 ft s⁻¹ = 60 mi hr⁻¹	26.82 m s⁻¹
Acceleration	1 ft s⁻² = 5280 ft s⁻²	0.3048 m s⁻²
Energy	1 ft lb = 1.36×10^{-7} erg	1.3557 J
	1 erg	1.0×10^{-7} J
	1 ev = 1.6×10^{-12} erg = 3.28×10^{-20} Ca	1.60×10^{-19} J
	1 Ca = 2.6×10^{-19} eV	4.184 J
	1 amu = 931.5 MeV	1.5×10^{-16} J
Power	1 ft lb s⁻¹ = 0.00182 hp	764 W (J s⁻¹)
	1 erg s⁻¹ = 7.3×10^{-8} ft lb s⁻¹	1.0×10^{-7} W
Pressure	1 atm = 760 mmHg = 760 Torr	101.325 kPa (N m⁻²)
	1 bar = 1.013 atm = 770.1 mmHg = 14.89 lb in⁻²	100 kPa (N m⁻²)
	1 atm = 14.70 lb in⁻² = 1.0×10^{16} dynes cm⁻³	101.325 kPa
Temperature	(5/9°F − 32)°C = (9/5°C + 32) °F	(°C + 273.17) K

B. SI BASE UNITS

Quantity	Unit	Symbol of Unit
Length	Metre	m
Mass	Kilogram	Kg
Time	Second	s
Electric current	Ampere	A
Temperature	Kelvin	K
Amount of substance	Mole	mol
Luminous intensity	Candela	Cd

Appendix

C. MULTIPLES OF SI BASE UNITS

Prefix	Symbol	Value
Deca-	Da	10^1
Hecto-	H	10^2
Kilo-	K	10^3
Mega-	M	10^6
Giga-	G	10^9
Tera-	T	10^{12}
Peta-	P	10^{15}
Exa-	E	10^{18}
Zetta-	Z	10^{21}
Yotta-	Y	10^{24}

D. SUB-MULTIPLES OF SI BASE UNITS

Prefix	Symbol	Value
Deci-	d	10^{-1}
Centi-	c	10^{-2}
Milli-	m	10^{-3}
Micro-	μ	10^{-6}
Nano-	n	10^{-9}
Pico-	p	10^{-12}
Femto-	f	10^{-15}
Atto-	a	10^{-18}
Zepto-	z	10^{-21}
Yocto-	y	10^{-24}

E. SOME SCIENTIFIC CONSTANTS

Constant	Symbol	Value
Gravitational constant	G	6.670×10^{-11} N m^2 kg^{-2}
Velocity of light	c	2.9979×10^8 m s^{-1}
Atomic mass unit	amu	1.6605×10^{-27} kg
Avogadro's number	N_A	6.0221×10^{23} mol^{-1}
Boltzmann's constant	k	1.3807×10^{-23} J K^{-1}
Mass of electron	m_e	9.109×10^{-31} kg
Electron charge	$-e$	1.602218×10^{-19} C
Electron radius	r_e	2.817939×10^{-15} m
Mass of neutron	m_n	1.6726×10^{-27} kg
Neutron radius	r_n	1.11492×10^{-13} m
Mass of proton	m_p	1.6726×10^{-27} kg
Proton charge	$+e$	1.602218×10^{-19} C
Proton radius	r_p	1.113386×10^{-15} m
Faraday constant	F	9.64870×10^4 C mol^{-1}

(*Continued*)

Constant	Symbol	Value
Electron charge	E	9.1096×10^{-31} kg
Gas constant	R	8.314 J K^{-1} mol^{-1}
		0.08206 dm^3 atm K^{-1} mol^{-1}
		62.37 dm^3 Torr K^{-1} mol^{-1}
Molar volume of gases at STP	V_m	22.4136 dm^{-3} mol^{-1}
Molar volume of gases at RTP	V_m	24.0 dm^3
Electron radius		2.8177×10^{-15} m
Rydberg's constant	R_H	1.0974 m^{-1}
Acceleration due to gravity	g	9.8066 m s^{-2}
Boltzmann constant	k_B	1.38×10^{-23} J K^{-1}
Planck constant	H	6.62×10^{-34} J s
Coulomb constant	k	8.99×10^9 N m^2 C^{-2}
Faraday's constant	F	96485.3399 C mol^{-1}

F. STANDARD REDUCTION POTENTIALS

Reduction Half-equation	E^0 (V)
$H_4XeO_6 + 2H^+ + 2e^- \rightarrow XeO_3 + 3H_2O$	+3.00
$F_2 + 2e^- \rightarrow 2F^-$	+2.87
$O_3 + 2H^+ + 2e^- \rightarrow O_2 + H_2O$	+2.07
$S_2O_8^{2-} + 2e^- \rightarrow 2SO_4^{2-}$	+2.05
$Ag^{2+} + e^- \rightarrow Ag^+$	+1.98
$Co^{2+} + e^- \rightarrow Co^+$	+1.81
$H_2O_2 + 2H^+ + 2e^- \rightarrow 2H_2O$	+1.78
$Au^+ + e^- \rightarrow Au$	+1.69
$Pb^{4+} + 2e^- \rightarrow Pb^{2+}$	+1.67
$2HClO + 2H^+ + 2e^- \rightarrow Cl_2 + 2H_2O$	+1.63
$Ce^{4+} + e^- \rightarrow Ce^{3+}$	+1.61
$2HBrO + 2H^+ + 2e^- \rightarrow Br_2 + 2H_2O$	+1.60
$MnO_4^- + 8H^+ + 5e^- \rightarrow Mn^{2+} + H_2O$	+1.51
$Mn^{3+} + e^- \rightarrow Mn^{2+}$	+1.40
$Cl_2 + 2e^- \rightarrow 2Cl^-$	+1.36
$Cr_2O_7^- + 14H^+ + 6e^- \rightarrow 2Cr^{3+} + 7H_2O$	+1.33
$O_3 + H_2O + 2e^- \rightarrow O_2 + 2OH^-$	+1.24
$O_2 + 4H^+ + 4e^- \rightarrow 2H_2O$	+1.23
$ClO_4^- + 2H^+ + 2e^- \rightarrow + H_2O$	+1.23
$Pt^{2+} + 2e^- \rightarrow Pt$	+1.20
$Br_2 + 2e^- \rightarrow 2Br^-$	+1.09
$Pu^{4+} + e^- \rightarrow Pu^{3+}$	+0.97
$NO_3 + 4H^- + 3e^- \rightarrow NO + 2H_2O$	+0.96
$2Hg^{2+} + 2e^- \rightarrow Hg_2^{2+}$	+0.92
$ClO^- + H_2O + 2e^- \rightarrow Cl^- + 2OH^-$	+0.89
$NO_3^- + 2H^+ + e^- \rightarrow NO_2 + H_2O$	+0.80
$Ag + e^- \rightarrow Ag$	+0.80
$Hg_2^{2+} + 2e^- \rightarrow 2Hg$	+0.79
$AgF + e^- \rightarrow Ag + F$	+0.78

(Continued)

Appendix

Reduction Half-equation	E^0 (V)
$Fe^{3+} + e^- \rightarrow Fe^{2+}$	+0.77
$BrO^- + H_2O + 2e^- \rightarrow Br^- + 2OH^-$	+0.76
$MnO_4^{2-} + 2H_2O + 2e^- \rightarrow MnO_2 + 4OH^-$	+0.76
$MnO_4^{2-} + 2H_2O + 2e^- \rightarrow MnO_2 + 4OH^-$	+0.60
$MnO_4^- + e^- \rightarrow MnO_4^{2-}$	+0.56
$I_2 + 2e^- \rightarrow 2I^-$	+0.54
$Cu^+ + e^- \rightarrow Cu$	+0.52
$I_3^- + 2e^- \rightarrow 3I^-$	+0.53
$Ni(OH)_3 + H_2O + e^- \rightarrow Ni(OH)_2 + OH^-$	+0.49
$IO^- + H_2O + 2e^-$	+0.49
$O_2 + 2H_2O + 4e^- \rightarrow 4OH^-$	+0.40
$ClO_4^- + H_2O + 2e^- \rightarrow ClO_3^- + 2OH^-$	+0.36
$Cu^{2+} + 2e^- \rightarrow Cu$	+0.34
$BiO^+ + 2H^+ + 3e^- \rightarrow Bi + H_2O$	+0.32
$Hg_2Cl_2 + 2e^- \rightarrow 2Hg + 2Cl^-$	+0.27
$AgCl + e^- \rightarrow Ag + Cl^-$	+0.22
$Bi^{3+} + 3e^- \rightarrow Bi$	+0.20
$SO_4^{2-} + 4H^+ + 2e^- \rightarrow H_2SO_3 + H_2O$	+0.17
$Cu^{2+} + e^- \rightarrow Cu^+$	+0.15
$Sn^{4+} + 2e^- \rightarrow Sn^{2+}$	+0.15
$AgBr + e^- \rightarrow Ag + Br^-$	+0.07
$NO_3^- + H_2O + 2e^- \rightarrow NO_2^- + 2OH^-$	+0.01
$Ti^{4+} + e^- \rightarrow Ti^{3+}$	0.00
$2H^+ + 2e^- \rightarrow H_2$	0 (Reference electrode)
$Fe^{3+} + 3e^- \rightarrow Fe$	−0.04
$O_2 + H_2O + 2e^- \rightarrow HO_2^- + OH^-$	−0.08
$Pb^{2+} + 2e^- \rightarrow Pb$	−0.13
$In^+ + e^- \rightarrow In$	−0.14
$Sn^{2+} + 2e^- \rightarrow Sn$	−0.14
$AgI + e^- \rightarrow Ag + I^-$	−0.15
$Ni^{2+} + 2e^- \rightarrow Ni$	−0.23
$Ti^{3+} + e^- \rightarrow Ti^{2+}$	−0.37
$Cd^{2+} + 2e^- \rightarrow Cd$	−0.40
$2H_2O + 2e^- \rightarrow H_2 + 2OH^-$	−0.83
$SnO_2^- + H_2O + 2e^-$	−0.91
$Cr^{2+} + 2e^- \rightarrow Cr$	−0.91
$Ce^{3+} + 3e^- \rightarrow Ce$	−2.48
$Mn^{2+} + 2e^- \rightarrow Mn$	−1.18
$V^{2+} + 2e^- \rightarrow V$	−1.19
$Zn(OH)_2 + 2e^- \rightarrow Zn + 2OH^-$	−1.25
$Ti^{2+} + 2e^- \rightarrow Ti$	−1.63
$Al^{3+} + 3e^- \rightarrow Al$	−1.66
$U^{3+} + 3e^- \rightarrow U$	−1.79
$Be^{2+} + 2e^- \rightarrow Be$	−1.85
$Mg^{2+} + 2e^- \rightarrow Mg$	−2.36
$La^{3+} + 3e^- \rightarrow La$	−2.52
$Mg(OH)_2 + 2e^- \rightarrow Mg + 2OH^-$	−2.69

(*Continued*)

Reduction Half-equation	E^θ (V)
$Na^+ + e^- \rightarrow Na$	−2.71
$Ca^{2+} + 2e^- \rightarrow Ca$	−2.87
$Ba^{2+} + 2e^- \rightarrow Ba$	−2.91
$Ra^{2+} + 2e^- \rightarrow Ra$	−2.92
$Cs^+ + e^- \rightarrow Cs$	−2.93
$Ba(OH)_2 + 8H_2O + 2e^-$	−2.97
$Sr^{2+} + 2e^- \rightarrow Sr$	−2.98
$K^+ + e^- \rightarrow K$	−2.98
$Sr(OH)_2 + 8H_2O + 2e^-$	−2.99
$Ca(OH)_2 + 2e^-$	−3.03
$Li^+ + e^- \rightarrow Li$	−3.05

G. STANDARD THERMODYNAMIC DATA OF SELECTED SUBSTANCES

Substance	Enthalpy of Formation ΔH_f^θ (kJ mol^{-1})	Free energy of Formation ΔG_f^θ (kJ mol^{-1})	Entropy S^θ (J mol^{-1} K^{-1})
Aluminium			
Al(s)	0	0	28.3
Al^{3+}(aq)	−524.7	−481.2	−321.7
Al(g)	330.0	289.4	164.6
Al$_2$O$_3$(s)	−1675.7	−1582.3	50.9
Al(OH)$_3$(s)	−1276	–	–
AlCl$_3$(s)	−704.2	−628.8	110.7
Antimony			
SbH$_3$(g)	145.11	147.8	232.8
SbCl$_3$(g)	−313.8	−301.2	337.8
SbCl$_5$(g)	−394.3	−334.3	401.9
Arsenic			
As(s, gray)	0	0	35.1
As$_2$S$_3$(s)	−169.0	−168.6	163.6
AsO$_4^{3-}$(aq)	−888.1	−648.4	−162.8
Barium			
Ba(s)	0	0	62.8
Ba^{2+}(aq)	−537.6	−560.8	9.6
BaO(s)	−553.5	−525.1	70.42
BaCO$_3$(s)	−1216.3	−1137.6	112.1
BaCO$_3$(aq)	−1214.8	−1088.6	−47.3
BaSO$_4$(s)	−1473.2	−1362.2	132.2
Beryllium			
Be(s)	0.0	0.0	62.5
Be(g)	324.0	286.6	136.3
Be(OH)$_2$(s)	−902.5	−815.0	45.5
BeO(s)	−609.4	−580.1	13.8
Bismuth			
Bi(s)	0	0	56.7
Bi(g)	207.1	168.2	187.0

(*Continued*)

Substance	Enthalpy of Formation ΔH_f^θ (kJ mol^{-1})	Free energy of Formation ΔG_f^θ (kJ mol^{-1})	Entropy S^θ (J mol^{-1} K^{-1})
Boron			
B(s)	0	0	5.9
B_2O_3(s)	−1272.8	−1193.7	54.0
BF_3(g)	−1137.0	−1120.3	254.1
Bromine			
Br_2(l)	0	0	152.2
Br_2(g)	30.9	3.11	245.5
Br(g)	111.9	82.40	175.0
Br$^-$(aq)	−121.6	−104.0	82.4
HBr(g)	−36.40	−53.5	198.7

Substance	Enthalpy of Formation ΔH_f^θ (kJ mol^{-1})	Free energy of Formation ΔG_f^θ (kJ mol^{-1})	Entropy S^θ (J mol^{-1} K^{-1})
Cadmium			
Cd(s)	0	0	51.8
Cd(g)	111.8	−	167.7
CdO(s)	−258.0	−228.0	55.0
$CdCl_2$(s)	−391.5	−343.9	115.3
CdS(s)	−161.9	−156.5	64.9
$CdSO_4$(s)	−935.0	−823.0	123.0
Calcium			
Ca(s)	0	0	41.4
Ca(g)	178.2	144.3	154.9
Ca^{2+}(aq)	−542.8	−553.6	−53.1
CaO(s)	−635.1	−604.0	39.8
CaS	−482.4	−477.4	56.5
Ca(OH)$_2$(s)	−986.1	−898.5	83.4
Ca(OH)$_2$(aq)	−1002.8	−868.1	−74.5
$CaCO_3$(s, calcite)	−1206.9	−1128.8	92.9
$CaCO_3$(s, aragonite)	−1207.1	−1127.8	88.7
$CaCO_3$(aq)	−1220.0	−1081.4	−110.0
CaF_2(s)	−1219.6	−1167.3	68.9
CaF_2(aq)	−1208.1	−1111.2	−80.8
$CaCl_2$(s)	−795.8	−748.1	104.6
$CaCl_2$(aq)	−877.1	−816.0	59.8
$CaBr_2$(s)	−682.8	−663.6	130
CaC_2(s)	−59.8	−64.9	70.0
$Ca_3(PO_4)_2$(s)	−4126.0	−3890.0	241.0
$CaSO_4$(s)	−1434.1	−1321.8	106.7
$CaSO_4$(aq)	−1452.10	−298.1	−33.1
Carbon			
C(s, graphite)	0	0	5.7
C(s, diamond)	1.9	2.9	2.4
CO_2(g)	−393.5	−394.4	213.7

(*Continued*)

Substance	Enthalpy of Formation $\Delta H_f^\ominus$ (kJ mol^{-1})	Free energy of Formation $\Delta G_f^\ominus$ (kJ mol^{-1})	Entropy $S^\ominus$ (J mol^{-1} K^{-1})
C(s, fullerene—C_{70})	2555.0	2537.0	464.0
C(g, fullerene—C_{60})	2502.0	2442.0	544.0
C(g, fullerene—C_{70})	2755.0	2692.0	614.0
CBr_4(s)	29.4	47.7	212.5
CBr_4(g)	83.9	67.0	358.1

Substance	Enthalpy of Formation $\Delta H_f^\ominus$ (kJ mol^{-1})	Free energy of Formation $\Delta G_f^\ominus$ (kJ mol^{-1})	Entropy $S^\ominus$ (J mol^{-1} K^{-1})
CCl_2O(g)	−219.1	−204.9	283.5
CCl_4(l)	−128.2	−62.6	216.2
CCl_4(g)	−95.7	−53.6	309.9
CF_4(g)	−933.6	−888.3	261.6
$CHCl_3$(l)	−134.1	−73.7	201.7
$CHCl_3$(g)	−102.7	6.0	295.7
CH_2Cl_2(l)	−124.2	—	177.8
CH_2Cl_2(g)	−95.4	−68.9	270.2
CH_3Cl(g)	−81.9	−58.5	234.6
CH_4(g)	−74.6	−50.5	186.3
CH_3COOH(l)	−484.3	−389.9	159.8
CH_3OH(l)	−239.2	−166.6	126.8
CH_3OH(g)	−201.0	−162.3	239.9
CH_3NH_2(l)	−47.3	35.7	150.2
CH_3NH_2(g)	−22.5	32.7	242.9
CH_3CN(l)	40.6	86.5	149.6
CH_3CN(g)	74.0	91.9	243.4
CO(g)	−110.5	−137.2	197.7
CO_2(g)	−393.5	−394.4	213.8
CS_2(l)	89.0	64.6	151.3
CS_2(g)	116.7	67.1	237.8
C_2H_2(g)	227.4	209.9	200.9
C_2H_4(g)	52.4	68.4	219.3
C_2H_6(g)	−84.0	−32.0	229.2
C_3H_8(g)	−103.8	−23.4	270.3
$C_3H_6O_3$(s)	−694.1	−522.9	142.3
C_6H_6(l)	49.1	124.5	173.4
C_6H_6(g)	82.9	129.7	269.2
C_6H_{12}(g)	-167.2	-0.3	388.82
$C_6H_{12}O_6$(s)(glucose)	−1273.3	−910.4	212.1
C_2H_5OH(l)	−277.6	−174.8	160.7
C_2H_5OH(g)	−234.8	−167.9	281.6
$(CH_3)_2O$(l)	−203.3	—	—
$(CH_3)_2O$(g)	−184.1	−112.6	266.4
$CH_3CO_2^-$	−486.0	−369.3	86.6
n-$C_{12}H_{26}$(l)	−350.9	28.1	490.6

Substance	Enthalpy of Formation ΔH_f^θ (kJ mol^{-1})	Free energy of Formation ΔG_f^θ (kJ mol^{-1})	Entropy S^θ (J mol^{-1} K^{-1})
CO_3^{2-}(aq)	−677.1	−527.8	−56.9
H_2CO_3(aq)	−698.7	−335.0	91.6
CCl_4(l)	−135.4	−65.2	216.4
HCOOH(g)	−363.0	−351.0	249.0
CH_3CHO(g)	−166.0	−129.0	250.0
CS_2(l)	89.7	65.3	153.3
HCN(g)	135.1	124.7	201.8
HCN(l)	108.9	125.0	112.8
C_2H_4O(g)	−52.0	−13.0	242.0
CH_2=CHCN(l)	152.0	190.0	274.0
Caesium			
Cs(s)	0	0	85.2
Cs(g)	76.5	49.6	175.6
CsCl(s)	−443.0	−414.5	101.2
Cerium			
Ce(s)	0	0	72.0
Ce^{3+}(aq)	−696.2	−672.0	−205.0
Ce^{4+}(aq)	−537.2	−503.8	−301.0
Chlorine			
Cl_2(g)	0	0	223.1
Cl(g)	121.7	105.7	165.2
Cl^-(aq)	−167.2	−131.2	56.5
HCl(g)	−92.3	−95.3	186.9
HCl(aq)	−167.2	−131.2	56.5
ClF_3(g)	−163.2	−123.0	281.6
Chromium			
Cr(s)	0	0	23.8
Cr(g)	396.6	351.8	174.5
$CrCl_3$(s)	−556.5	−486.1	123.0
CrO_3(g)	−292.9	—	266.2
Cr_2O_3(s)	−1139.7	−1058.1	81.2
Cobalt			
Co(s)	0	0	30.0
Co(g)	424.7	380.3	179.5
$CoCl_2$(s)	−312.5	−269.8	109.2
Copper			
Cu(s)	0	0	33.2
Cu^+(aq)	71.7	50.0	40.6
Cu^{2+}(aq)	64.8	65.5	−99.6

Substance	Enthalpy of Formation ΔH_f^θ (kJ mol^{-1})	Free energy of Formation ΔG_f^θ (kJ mol^{-1})	Entropy S^θ (J mol^{-1} K^{-1})
Cu_2O(s)	−168.6	−146.0	93.1
CuO(s)	−157.3	−129.7	42.6

(Continued)

Substance	Enthalpy of Formation ΔH_f^θ (kJ mol^{-1})	Free energy of Formation ΔG_f^θ (kJ mol^{-1})	Entropy S^θ (J mol^{-1} K^{-1})
CuSO$_4$(s)	−771.4	−661.8	109.0
CuCl(s)	−137.2	−119.9	86.2
CuCl$_2$(s)	−220.1	−175.7	108.1
CuS(s)	−53.1	−53.6	66.5
Cu$_2$S(s)	−79.5	−86.2	120.9
CuCN(s)	96.2	111.3	84.5
Fluorine			
F(g)	79.4	62.3	158.8
F$^-$(aq)	−332.6	−278.8	−13.8
F$_2$(g)	0	0	202.8
HF(g)	−273.3	−275.4	173.8
HF(aq)	−332.6	−278.8	−13.8
Hydrogen			
H(g)	218.0	203.3	114.7
H$_2$(g)	0	0	130.7
H$^+$(aq)	0	0	0
H$_2$O(l)	−285.8	−237.1	69.9
H$_2$O(g)	−241.8	228.6	188.8
H$_2$O$_2$(l)	−187.8	−120.4	109.6
H$_2$O$_2$(aq)	−191.2	−134.0	143.9
D$_2$(g)	0	0	145.0
D$_2$O(l)	−294.6	−243.4	75.9
D$_2$O(g)	−249.2	−234.5	198.3
Iodine			
I(g)	106.8	70.2	180.8
I$^-$(aq)	−55.2	−51.6	111.3
I$_2$(s)	0	0	116.1
I$_2$(g)	62.4	19.3	260.7
HI(g)	26.5	1.7	206.6
HI(aq)	−55.2	−51.6	111.3
Iron			
Fe(s)	0	0	27.3
Fe(g)	416.3	370.7	180.5
Fe^{2+}(aq)	−89.1	−78.9	−137.7
Fe^{3+}(aq)	−48.5	−4.7	−315.9
FeCl$_2$(s)	−341.8	−302.3	118.0
FeCl$_3$(s)	−399.5	−334.0	142.3
FeO(s)	−272.0	−251.4	60.7

Substance	Enthalpy of Formation ΔH_f^θ (kJ mol^{-1})	Free energy of Formation ΔG_f^θ (kJ mol^{-1})	Entropy S^θ (J mol^{-1} K^{-1})
Fe$_2$O$_3$(s, haematite)	−824.2	−742.2	87.4
Fe$_2$O$_3$(s, magnetite)	−1118.4	−1015.4	164.4
FeS$_2$(s)	−178.2	−166.9	52.9

(*Continued*)

Substance	Enthalpy of Formation ΔH_f^θ (kJ mol^{-1})	Free energy of Formation ΔG_f^θ (kJ mol^{-1})	Entropy S^θ (J mol^{-1} K^{-1})
FeS(s, α)	−100.0	−100.4	60.29
FeS(aq)	−	6.9	−
Lead			
Pb(s)	0	0	64.8
Pb(g)	195.2	162.2	175.4
Pb^{2+}(aq)	−1.7	−24.43	10.5
PbO(s, red or litharge)	−219.0	−188.9	66.5
PbO(s, yellow or massicot)	−217.3	−187.9	68.7
PbO$_2$(s)	−277.4	−217.3	68.6
PbCl$_2$(s)	−359.4	−314.1	136.0
PbS(s)	−100.4	−98.7	91.2
PbSO$_4$(s)	−920.0	−813.0	148.5
PbCO$_3$(s)	−699.1	−625.5	131.0
Pb(NO$_3$)$_2$(s)	−451.9	−	−
Pb(NO$_3$)$_2$(aq)	−416.3	−246.9	303.3
PbBr$_2$(s)	−278.7	−261.9	161.5
PbBr$_2$(aq)	−244.8	−232.3	175.3
Lithium			
Li(s)	0	0	29.1
Li(g)	159.3	126.6	138.8
Li$^+$(aq)	−278.5	−293.3	13.4
LiCl(s)	−408.6	−384.4	59.3
Li$_2$O(s)	−597.9	−561.2	37.6
Magnesium			
Mg(s)	0	0	32.7
Mg(g)	147.1	112.5	148.6
Mg^{2+}(aq)	−466.9	−454.8	−138.1
MgCl$_2$(s)	−641.3	−591.8	89.6
MgO(s)	−601.6	−569.3	27.0
Mg(OH)$_2$(s)	−924.5	−833.5	63.2
MgSO$_4$(s)	−1284.9	−1170.6	91.6
Substance	Enthalpy of Formation ΔH_f^θ (kJ mol^{-1})	Free energy of Formation ΔG_f^θ (kJ mol^{-1})	Entropy S^θ (J mol^{-1} K^{-1})
MgS(s)	−346.0	−341.8	50.3
MgBr$_2$(s)	−524.3	−503.8	117.2
Manganese			
Mn(s)	0	0	32.0
Mn(g)	280.7	238.5	173.7
MnCl$_2$(s)	−481.3	−440.5	118.2
MnO(s)	−385.2	−362.9	59.7
MnO$_2$(s)	−520.0	−465.1	53.1
MnO$_4^-$(aq)	−541.4	−447.2	191.2

(Continued)

Substance	Enthalpy of Formation $\Delta H_f^\ominus$ (kJ mol^{-1})	Free energy of Formation $\Delta G_f^\ominus$ (kJ mol^{-1})	Entropy $S^\ominus$ (J mol^{-1} K^{-1})
KMnO$_4$(s)	−837.2	−737.6	171.7
Mn$_3$O$_4$(s)	−1387.0	−1280.0	149.0
Mn$_2$O$_3$(s)	−971.0	−893.0	110.0
Mercury			
Hg(l)	0	0	75.9
Hg(g)	61.4	31.8	175.0
HgCl$_2$(s)	−224.3	−178.6	146.0
Hg$_2$Cl$_2$(s)	−265.4	−210.7	191.6
HgO(s)	−90.8	−58.5	70.3
HgS(s, red)	−58.2	−50.6	82.4
Hg$_2$(g)	108.8	68.2	288.1
Molybdenum			
Mo(s)	0	0	28.7
Mo(g)	658.1	612.5	182.0
MoO$_2$(s)	−588.9	−533.0	46.3
MoO$_3$(s)	−745.1	−668.0	77.7
Nickel			
Ni(s)	0	0	29.9
Ni(g)	429.7	384.5	182.2
NiCl$_2$(s)	−305.3	−259.0	97.7
Ni(OH)$_2$(s)	−529.7	−447.2	88.0
NiO(s)	−239.7	−211.7	38.0
NiS(s)	−93.0	−90.0	53.0
Nitrogen			
N(g)	472.7	455.5	153.3
N$_2$(g)	0	0	191.6
NH$_3$(g)	−45.9	−16.4	192.8
NH$_4^+$(aq)	−132.5	−79.3	113.4
N$_2$H$_4$(l)	50.6	149.3	121.2
N$_2$H$_4$(g)	95.4	159.4	238.5
NH$_4$Cl(s)	−314.4	−202.9	94.6

Substance	Enthalpy of Formation $\Delta H_f^\ominus$ (kJ mol^{-1})	Free energy of Formation $\Delta G_f^\ominus$ (kJ mol^{-1})	Entropy $S^\ominus$ (J mol^{-1} K^{-1})
NH$_4$OH(l)	−361.2	−254.0	165.6
NH$_4$NO$_3$(s)	−365.6	−183.9	151.1
NH$_4$NO$_3$(aq)	−339.9	−190.6	259.8
(NH$_4$)$_2$SO$_4$(s)	−1180.9	−901.7	220.1
NO(g)	91.3	87.6	210.8
NO$_2$(g)	33.2	51.3	240.1
N$_2$O(g)	81.6	103.7	220.0
N$_2$O$_4$(l)	−19.5	97.5	209.2
N$_2$O$_4$(g)	11.1	99.8	304.4
HNO$_2$(g)	−79.5	−46.0	254.1

(*Continued*)

Appendix

Substance	Enthalpy of Formation ΔH_f^θ (kJ mol^{-1})	Free energy of Formation ΔG_f^θ (kJ mol^{-1})	Entropy S^θ (J mol^{-1} K^{-1})
HNO$_3$(l)	−174.1	−80.7	155.6
HNO$_3$(g)	−133.9	−73.5	266.9
HNO$_3$(aq)	−207.4	−111.3	146.4
NF$_3$(g)	−132.1	−90.6	260.8
HCN(l)	108.9	125.0	112.8
HCN(g)	135.1	124.7	201.8
N$_2$H$_3$CH$_3$(l)	54.0	180.0	166.0
NH$_4$Cl(s)	−314.4	−202.9	94.6
NH$_4$Cl(aq)	−299.7	−210.5	169.9
NH$_4$ClO$_4$(s)	−295.0	−89.0	186.0
NOCl(g)	51.7	66.1	261.8
Osmium			
Os(s)	0	0	32.6
Os(g)	791.0	745.0	192.6
OsO$_4$(s)	−394.1	−304.9	143.9
OsO$_4$(g)	−337.2	−292.8	293.8
Oxygen			
O(g)	249.2	231.7	161.1
O$_2$(g)	0	0	205.2
O$_3$(g)	142.7	163.2	238.9
OH$^-$(aq)	−230.0	−157.2	−10.8
H$_2$O(l)	−285.8	−237.1	70.0
H$_2$O(g)	−241.8	−228.6	188.8
H$_2$O$_2$(l)	−187.8	−120.4	109.6
H$_2$O$_2$(g)	−136.3	−105.6	232.7
Phosphorus			
P(s, white)	0	0	41.1
P(s, red)	−17.6	−12.5	22.8
P(s, black)	−39.3	−	−

Substance	Enthalpy of Formation ΔH_f^θ (kJ mol^{-1})	Free energy of Formation ΔG_f^θ (kJ mol^{-1})	Entropy S^θ (J mol^{-1} K^{-1})
P(g, white)	316.5	280.1	163.2
P$_2$(g)	144.0	103.5	218.1
P$_4$(g)	58.9	24.4	280.0
PCl$_3$(l)	−319.7	272.3	217.1
PCl$_3$(g)	−287.0	−267.8	311.8
POCl$_3$(l)	−597.1	−520.8	222.5
POCl$_3$(g)	−558.5	−512.9	325.5
PCl$_5$(g)	−374.9	−305.0	364.6
PCl$_5$(s)	−443.5	−	−
PH$_3$(g)	5.4	13.5	210.2
H$_3$PO$_4$(s)	−1284.4	−1124.3	110.5
H$_3$PO$_4$(l)	−1271.7	−1123.6	150.8

(Continued)

Substance	Enthalpy of Formation ΔH_f^θ (kJ mol⁻¹)	Free energy of Formation ΔG_f^θ (kJ mol⁻¹)	Entropy S^θ (J mol⁻¹ K⁻¹)
$P_4O_{10}(s)$	−2984.0	−2697.7	228.9
Potassium			
K(s)	0	0	64.7
K(g)	89.0	60.5	160.3
KBr(s)	−393.8	−380.7	95.9
KCl(s)	−436.5	−408.5	82.6
$KClO_3(s)$	−397.7	−296.3	143.1
$K_2O(s)$	−361.5	−322.1	94.1
$K_2O_2(s)$	−494.1	−425.1	102.1
$KNO_2(s)$	−369.8	−306.6	152.1
$KNO_3(s)$	−494.6	−394.9	133.1
KSCN(s)	−200.2	−178.3	124.3
$K_2CO_3(s)$	−1151.0	−1063.5	155.5
$K_2SO_4(s)$	−1437.8	−1321.4	175.6
KOH (s)	−424.8	−379.1	78.9
KOH (aq)	−482.4	−440.5	91.6
Silicon			
Si(s)	0	0	18.8
Si(g)	450.0	405.5	168.0
$SiCl_4(l)$	−687.0	−619.8	239.7
$SiCl_4(g)$	−657.0	−617.0	330.7
$SiH_4(g)$	34.3	56.9	204.6
SiC(s, cubic)	−65.3	−62.8	16.6
SiC(s, hexagonal)	−62.8	−60.2	16.5
SiO_2 (s, quartz)	−910.9	−856.6	41.8
Silver			
Ag(s)	0	0	42.6
Ag(g)	284.9	246.0	173.0

Substance	Enthalpy of Formation ΔH_f^θ (kJ mol⁻¹)	Free energy of Formation ΔG_f^θ (kJ mol⁻¹)	Entropy S^θ (J mol⁻¹ K⁻¹)
$Ag^+(aq)$	105.6	77.1	72.7
AgBr(s)	−100.4	−96.9	107.1
AgCl(s)	−127.0	−109.8	96.3
$AgNO_3(s)$	−124.4	−33.4	140.9
$Ag_2O(s)$	−31.1	−11.2	121.3
$Ag_2S(s)$	−32.6	−40.7	144.0
AgCN(s)	146.0	164.0	84.0
$Ag_2CrO_4(s)$	−712.0	−622.0	217.0
AgI(s)	−62.0	−66.0	115.0
Ag_2S (s)	−32.0	−40.0	146.0
Sodium			
Na(s)	0	0	51.3
Na(g)	107.5	77.0	153.7

(*Continued*)

Substance	Enthalpy of Formation ΔH_f^θ (kJ mol^{-1})	Free energy of Formation ΔG_f^θ (kJ mol^{-1})	Entropy S^θ (J mol^{-1} K^{-1})
Na$^+$(g)	609.4	—	—
Na$^+$(aq)	−240.1	−261.9	59.0
NaF(s)	−576.6	−546.3	51.1
NaF(aq)	−572.8	−540.7	45.2
NaCl(s)	−411.2	−384.1	72.1
NaI(s)	−287.8	−286.1	98.5
NaBr(s)	−361.1	−349.0	86.8
NaBr(g)	−143.1	−177.1	241.2
NaBr(aq)	−361.7	−365.8	141.4
NaO$_2$(s)	−260.2	−218.4	115.9
Na$_2$O(s)	−414.2	−375.5	75.1
Na$_2$O$_2$(s)	−510.9	−447.7	95.0
NaCN(s)	−87.5	−76.4	115.6
NaNO$_3$(aq)	−447.5	−373.2	205.4
NaNO$_3$(s)	−467.9	−367.0	116.5
NaN$_3$(s)	21.7	93.8	96.9
Na$_2$CO$_3$(s)	−1130.7	−1044.4	135.0
NaHCO$_3$(s)	−948.0	−852.0	102.0
Na$_2$SO$_4$(s)	−1387.1	−1270.2	149.6
NaOH(s)	−425.6	−379.5	64.5
NaOH(aq)	−470.1	−419.2	48.1
Sulfur			
S(s, rhombic)	0	0	32.1
S(g, rhombic)	277.2	236.7	167.8
S$_8$(g)	102.0	50.0	431.0
SO$_2$(g)	−296.8	−300.1	248.2

Substance	Enthalpy of Formation ΔH_f^θ (kJ mol^{-1})	Free energy of Formation ΔG_f^θ (kJ mol^{-1})	Entropy S^θ (J mol^{-1} K^{-1})
SO$_3$(g)	−395.7	−371.1	256.8
SO$_4^{2-}$	−909.3	−744.5	20.1
SOCl$_2$(g)	−212.5	−198.3	309.8
H$_2$S(g)	−20.6	−33.4	205.8
H$_2$S(aq)	−39.7	−27.8	121.0
H$_2$SO$_4$(aq)	−909.3	−744.5	20.1
H$_2$SO$_4$(l)	−814.0	−690.0	156.9
S$_2$Cl$_2$(g)	−18.4	−31.8	331.5
SF$_6$(g)	−1209.0	−1105.3	291.8
Tin			
Sn(s, white)	0	0	51.2
Sn(s, gray)	−2.1	0.1	44.1
Sn(g, white)	301.2	266.2	168.5
SnCl$_4$(l)	−511.3	−440.1	258.6
SnCl$_4$(g)	−471.5	−432.2	365.8

(*Continued*)

Substance	Enthalpy of Formation $\Delta H_f^\ominus$ (kJ mol^{-1})	Free energy of Formation $\Delta G_f^\ominus$ (kJ mol^{-1})	Entropy $S^\ominus$ (J mol^{-1} K^{-1})
SnO$_2$(s)	−557.6	−515.8	49.0
SnO(s)	−285.0	−257.0	56.0
Sn(OH)$_2$(s)	−561.0	−492.0	155.0
Titanium			
Ti(s)	0	0	30.7
Ti(g)	473.0	428.4	180.3
TiCl$_2$(s)	−513.8	−464.4	87.4
TiCl$_3$(s)	−720.9	−653.5	139.7
TiCl$_4$(l)	−804.2	−737.2	252.3
TiCl$_4$(g)	−763.2	−726.3	353.2
TiO$_2$(s)	−944.0	−888.8	50.6
Uranium			
U(s)	0	0	50.2
U(g)	533.0	488.4	199.8
UO$_2$(s)	−1085.0	−1031.8	77.0
UO$_2$(g)	−465.7	−471.5	274.6
UF$_4$(s)	−1914.2	−1823.3	151.7
UF$_4$(g)	−1598.7	−1572.7	368.0
UF$_6$(s)	−2197.0	−2068.5	227.6
UF$_6$(g)	−2147.4	−2063.7	377.9
U$_3$O$_8$(s)	−3575.0	−3393.0	282.0
UO$_3$(s)	−1230.0	−1150.0	99.0
Vanadium			
V(s)	0	0	28.9
V(g)	514.2	754.4	182.3
VCl$_3$(s)	−580.7	−511.2	131.0
VCl$_4$(l)	−569.4	−503.7	255.0
VCl$_4$(g)	−525.5	−492.0	362.4
V$_2$O$_5$(s)	−1550.6	−1419.5	131.0
Xenon			
Xe(g)	0	0	170.0
XeF$_2$(g)	−108.0	−48.0	254.0
XeF$_4$(s)	−251.0	−121.0	146.0
XeF$_6$(g)	−294.0	–	–
XeO$_3$(s)	402.0	–	–
Zinc			
Zn(s)	0	0	41.6
Zn(g)	130.4	94.8	161.0
Zn^{2+}(aq)	−153.9	−147.1	−112.1
ZnCl$_2$(s)	−415.1	−369.4	111.5
Zn(NO$_3$)$_2$(s)	−483.7	–	–

(*Continued*)

Substance	Enthalpy of Formation ΔH_f^θ (kJ mol^{-1})	Free energy of Formation ΔG_f^θ (kJ mol^{-1})	Entropy S^θ (J mol^{-1} K^{-1})
ZnS (s, wurtzite)	−206.0	−201.3	57.7
ZnS (s, zinc blende)	−206.0	−201.3	57.7
ZnSO$_4$(s)	−982.8	−871.5	110.5
ZnO (s)	−348.3	−318.3	43.6
Zn(OH)$_2$(s)	−642.0	–	–
Zirconium			
Zr(s)	0.0	0.0	39.0
Zr(g)	608.8	566.5	181.4
ZrCl$_2$(s)	−502.0	−386.0	110.0
ZrCl$_4$(s)	−980.5	−889.9	181.6

H. CHEMISTRY FUN AND FASCINATION

Chemistry can be both fascinating and fun, and there is always something new to learn! While safety is always the top priority in chemistry experiments, there are a few amusing and interesting experiments that you can try at home or in a lab with proper safety precautions.

Here are a few examples:

ELEPHANT TOOTHPASTE

One popular demonstration in chemistry involves creating 'elephant toothpaste.' This involves mixing hydrogen peroxide and potassium iodide to create a foam that looks like toothpaste. The reaction is exothermic and produces a lot of heat and oxygen gas, causing the foam to rapidly expand.

DANCING FLAMES

Another fun demonstration is creating a 'dancing flame' by adding metal salts to a Bunsen burner flame. Different metals produce different colours, creating a colorful display that looks like the flames are dancing.

EXPLODING PRINGLES

Did you know that you can make a potato chip explode using a microwave? If you place a Pringle chip in the microwave and heat it for a few seconds, it will create a plasma that causes the chip to burst into flames.

MAGIC SAND

Magic sand is a hydrophobic material that repels water. It is created by coating sand with a hydrophobic substance, such as silicone or wax. When magic sand is placed in water, it forms a layer of air around each grain of sand, causing it to float.

BLUE BOTTLE EXPERIMENT

The Blue Bottle Experiment is a popular demonstration of a chemical oscillator. It involves mixing a solution of glucose, sodium hydroxide and methylene blue. The solution turns blue when it is exposed to air and then turns clear again when shaken. This cycle repeats several times, creating a mesmerising effect.

Self-Healing Materials

Scientists have developed self-healing materials that can repair themselves when damaged. These materials are made using a combination of polymers and microcapsules that release a healing agent when the material is damaged.

Invisible Ink

You can make your own invisible ink using lemon juice or baking soda. Write a message on paper using the lemon juice or baking soda solution, and then let it dry. When you hold the paper up to a heat source, such as a light bulb or iron, the message will become visible.

Mentos and Soda

This is a classic experiment that creates a large fountain of soda. You will need a 2-dm^3 bottle of soda and some Mentos candy. Drop the Mentos into the soda and then stand back. The reaction between the Mentos and soda creates a large amount of carbon dioxide gas, which causes the soda to shoot out of the bottle.

pH Indicator

You can make your own pH indicator using red cabbage. Boil some chopped red cabbage in water, strain out the solids and then add some vinegar or baking soda. The solution will change colour depending on whether it is acidic or basic.

Rainbow Flame

This experiment creates a colourful flame by burning different metal salts. You will need a heat source, such as a Bunsen burner or propane torch, and different metal salts, such as copper sulfate, strontium chloride and potassium chloride. Dip a Q-tip in each metal salt and then hold it in the flame. Each metal will produce a different colour flame.

Homemade Lava Lamp

Fill a jar with water and add a few drops of food colouring. Then, add vegetable oil until the jar is almost full. Finally, add an Alka-Seltzer tablet, and watch as the oil and water mix and create a lava lamp effect.

Magic Milk

Poursome milk into a dish and add a few drops of food colouring. Then, add a drop of dish soap, and watch as the colours swirl and mix together. This happens because the dish soap breaks the surface tension of the milk, causing the colours to move.

Burning Money

Soak a dollar note in a mixture of alcohol and salt. Then, light the note on fire (in a safe area) and watch as the flame turns green. This is due to the presence of copper in the note's ink.

Appendix

Dancing Raisins

Adding raisins to a glass of carbonated water causes them to bounce around as the bubbles attach to them and rise to the surface, then pop and release them back down.

Alka-Seltzer Rocket

Mixing Alka-Seltzer tablets with water in a film canister produces a build-up of gas pressure that causes the canister to launch like a rocket.

Crystal Garden

This experiment involves growing crystal formations using borax and hot water. When borax dissolves in hot water, it creates a supersaturated solution that allows crystals to grow on a pipe cleaner or string placed in the solution. You can experiment with different shapes and colours to create unique crystal gardens.

Fireworks in a Glass

This experiment shows the interaction between oil and water, which creates a layer of oil on top of the water. When you add food colouring and an effervescent tablet, it creates a chemical reaction that produces bubbles that rise to the surface, creating a 'fireworks' effect.

Gummy Bear Osmosis

Place gummy bears in different solutions, such as water, salt water or vinegar, and see how they change and expand or contract over time.

These are just a few examples of the many funny and interesting experiments in chemistry. Remember to always be cautious and follow proper safety precautions when conducting any chemistry experiment.

Answers

CHAPTER 1

2b. (i) chemical, (ii) physical, (iii) physical, (iv) chemical, (v) chemical, (vi) chemical, (vii) physical, (viii) chemical, (ix) physical, (x) chemical.

3. (i) physical, (ii) chemical, (iii) chemical, (iv) physical, (v) physical, (vi) chemical, (vii) physical

5. (i) chemical, (ii) chemical, (iii) physical, (iv) physical, (v) chemical, (vi) chemical, (vii) physical, (viii) chemical, (ix) physical, (x) physical, (xi) chemical.

9. (a) curium, Cm; mendelevium, Md; Gadolinium, Gd; Einsteinium, Es; Fermium, Fm; nobelium, No; lawrencium, Lr; bohrium, Bh; rutherfordium, Rf; flerovium, Fl; seaborgium, sg; meitnerium, Mt; roentgenium, Rg; copernicium, Cn; oganesson, Og (b) polonium, Po; americium, Am; Francium, Fr; gallium (from Gaul, the ancient name for France) Ga; Indium, In; ruthenium (from Ruthenia, Latin name for Russia), Ru; nihonium (from Nihon, a Japanese name for Japan); thulium (from Thule, referring to either Greenland or Iceland), Tm; germanium, Ge; copper (from cuprum, Latin name for Cyprus (c) Berkelium, Bk; tennessine, Ts; californium, Cf (d) polonium, Po; francium, Fr; ruthenia, Ru; copper, Cu; gallium, Ga (e) beryllium, Be; potassium, K; sodium, Na; silicon, Si; calcium, Ca; nickel, Ni (f) americium, Am; europium, Eu.

10. (a) polonium, (b) zirconium, (c) lead, (d) mercury, (e) germanium, (f) tellurium, (g) lawrencium, (h) sodium, (i) tungsten, (j) barium.

11. (a) He, (b) Sn, (c) Sb, (d) Fe, (e) Cd, (f) U, (g) Rn, (h) Fr, (i) Ti, (j) I.

12. (i) mixture, (ii) mixture, (iii) compound, (iv) element, (v) element, (vi) compound, (vii) element, (viii) compound, (ix) compound, (x) mixture, (xi) element, (xii) mixture, (xiii) mixture, (xiv) compound, (xv) element.

14b. (i) homogeneous, (ii) homogeneous, (iii) heterogeneous, (iv) homogeneous, (v) heterogeneous, (vi) heterogeneous, (vii) heterogeneous, (viii) homogeneous, (ix) homogeneous, (x) heterogeneous.

15. (a) To separate the iron filings, use magnetic separation. Dissolve the rest of the mixture in water and use decantation to recover the sand. The salt is then recovered by evaporating the decanted liquid to dryness. (b) Separate petrol from the mixture using a separating funnel, then recover ethanol from the remaining solution by fractional distillation. (c) Use fractional distillation. (d) Separate ammonium chloride by sublimation, then recover the chalk particles by filtration. Recover sodium chloride from the filtrate by evaporating it to dryness. (e) Dissolve the mixture in water and recover calcium carbonate by filtration. The sodium carbonate is then recovered from the filtrate by crystallisation. (f) Use fractional crystallisation. (g) Use chromatography.

16. (a) Difference in particle sizes, (b) difference in boiling points, (c) solubility difference at different temperatures, (d) difference in movement rates over a porous absorbent medium, (e) difference in settling rates under gravity, (f) density difference, (g) magnetic property, (h) ability of a substance to sublime, (i) difference in solubility in two miscible liquids, (j) difference in density.

Answers

17. (a) Determine the melting point, (b) use chromatography, (c) determine the boiling point, (d) determine the melting point, (e) determine the melting point.

20. (a) Due to the existence of isotopes.

22. (a)

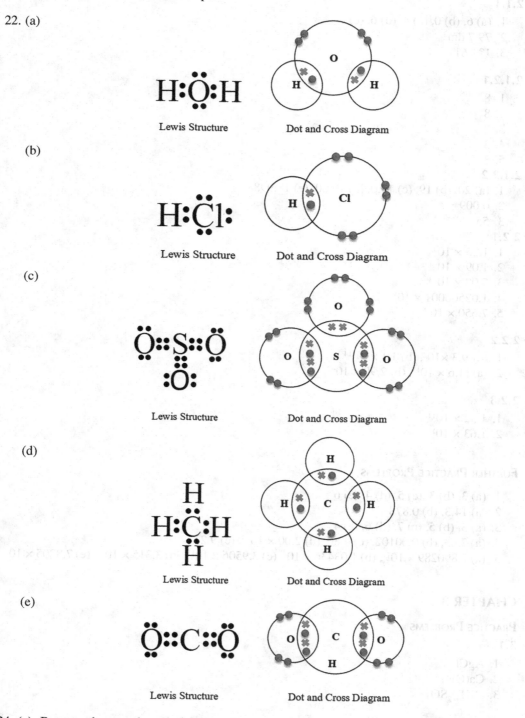

24. (c) Because they conduct electricity in the liquid state or when dissolved in water.

CHAPTER 2

PRACTICE PROBLEMS

2.1.1
1. (a) 6, (b) 0, (c) 4, (d) 6, (e) 2
2. 79.7 dm^3
3. 121.61

2.1.2.1
1. 8
2. 8
3. 8
4. 3
5. 3

2.1.2.2
1. (a) 20, (b) 19, (c) 19.0, (d) 19.00, (e) 18.998
2. 0.009
3. 55

2.2.1
1. 1.25×10^0
2. 1.09×10^{-1}
3. 5.08×10^{-3}
4. 1.02500001×10^8
5. 7.650×10^{-8}

2.2.2
1. (a) 9.3×10^2, (b) 1.1×10^{-14}
2. (a) 11.6×10^{13}, (b) 2.90×10^{-2}

2.2.3
1. 1.32×10^{11}
2. 1.63×10^{11}

FURTHER PRACTICE PROBLEMS

1. (a) 3, (b) 3, (c) 5, (d) 4, (e) 6
2. (a) 14.3, (b) 0.878
3. (a) 3, (b) 5, (c) 7, (d) 6, (e) 4
4. (a) 73.3, (b) 0.00102, (c) 4.80, (d) 2.00×10^{-3}, (e) 9.21
5. (a) 2.890289×10^{11}, (b) 1.03436×10^8, (c) 3.9506×10^{11}, (d) 3.315×10^{13}, (e) 7.8205×10^7

CHAPTER 3

PRACTICE PROBLEMS

3.1
1. AgCl
2. Ca(OH)$_2$
3. (NH$_4$)$_2$SO$_4$

Answers

4. FeCl$_2$
5. Mg(NO$_3$)$_2$

3.2

1. 28 g mol^{-1}
2. 56 g mol^{-1}
3. 17 g mol^{-1}
4. 44 g mol^{-1}
5. 111 g mol^{-1}
6. 100 g mol^{-1}
7. 62 g mol^{-1}
8. 74 g mol^{-1}
9. 80 g mol^{-1}
10. 278 g mol^{-1}

3.3

1. C = 0.64 g; O = 0.86 g
2. C = 4.3 g; H = 0.71 g
3. C = 2.6 g; H = 0.65 g, O = 1.8 g
4. (a) N = 82.4%, H = 17.6% (b) C = 79.9%, O = 20.1% (c) Cu = 88.8%, O = 11.2% (d) Ca = 40%, C = 12%, O = 48% (e) H = 2.0%, S = 32.7%, O = 65.3% (f) Ca = 54.1% O = 43.2%, H = 2.7% (g) N = 21.2%, H = 6.1%, S = 24.2%, O = 48.5%.

Further Practice Problems

1. (a) CaHPO$_4$, (b) K$_2$Cr$_2$O$_7$, (c) Fe$_2$(SO$_4$)$_3$, (d) HClO$_4$, (e) Mn(NO$_3$)$_2$, (f) (NH$_4$)$_2$CO$_3$
2. (a) 364 gmol^{-1}, (b) 342 gmol^{-1}, (c) 249.5 gmol^{-1}, (d) 286 gmol^{-1}, (e) 474 gmol^{-1}
3. (a) C = 40%, H = 6.7%, O = 53.3%; (b) C = 76.6%, H = 7.8%, N = 2.8%, O = 12.8%

CHAPTER 4

Practice Problems

4.1.1

1. (a) 0.016 mol, (b) 0.00071 mol, (c) 0.012 mol.
2. (a) 5.5 g (b) 1.1 g, (c) 133.8 g.
3. (a) 74 g mol^{-1}, (b) 2 g mol^{-1}, (c) 36.5 g mol^{-1}.
4. Na$_2$CO$_3$.H$_2$O.

4.1.2

1. (a) 7.56 mol, (b) 0.000498 mol, (c) 1.51 mol.
2. (a) 1.5 × 10^{24}, (b) 6.6 × 10^{22}, (c) 3.0 × 10^{24}.
3. (a) 3.65 g, (b) 4.93 g, (c) 45. 2 g.
4. (a) 4 g mol^{-1}, (b) 65 g mol^{-1}, (c) 16 g mol^{-1}.

4.1.3

1. (a) 0.00028 mol, (b) 0.038 mol, (c) 0.0020 mol.
2. (a) 11 dm^3, (b) 0.033 dm^3, (c) 0.65 dm^3.
3. (a) 1.0 mol dm^{-3}, (b) 6.0 mol dm^{-3}, (c) 0.66 mol dm^{-3}.
4. (a) 9.0 × 10^{22}, (b) 5.0 × 10^{22}, (c) 1.7 × 10^{22}
5. (a) 106 g mol^{-1}, (b) 40 g mol^{-1}, (c) 98 g mol^{-1}

4.1.4

1. (a) 0.023 mol, (b) 0.010 mol, (c) 0.063 mol.
2. (a) 56 dm^3, (b) 3.4 dm^3, (c) 28 dm^3.
3. (a) 1.4×10^{23}, (b) 3.8×10^{21}, (c) 6.3×10^{21}.
4. (a) 0.9 g, (b) 3.1 g, (c) 0.83 g.
5. (a) 28 g mol^{-1}, (b) 34 g mol^{-1}, (c) 16 g mol^{-1}.

4.2.1

1. (a) C_5H_8, (b) C_2H_5, (c) $CaCO_3$, (d) C_2H_5COOH, (e) $C_4H_5N_2O$
2. (a) CH_3, (b) CO_2, (c) C_2H_6O, (d) $CuSO_4.5H_2O$, (e) $C_9H_8O_2$

4.2.2

1. (a) 0.023 mol (b) 3.01×10^{23}
2. 157 g (to 3 significant figures)
3. 0.023 mol
4. 0.57 dm^3
5. C_3H_7COOH, propanoic acid
6. (a) 276 g mol^{-1}, (b) C_5H_9, (c) $C_{20}H_{36}$
7. (a) 536 g mol^{-1}, (b) C = 89.6%, H = 10.4%, (c) C_5H_7, (d) $C_{40}H_{56}$
8. (a) 72 g mol^{-1}, (b) C = 1.25 g, H = 0.139 g, O = 1.11 g, (c) $C_3H_4O_2$, (d) $C_3H_4O_2$

CHAPTER 5

PRACTICE PROBLEMS
5.1.1

1. $2H_2O_2(g) \rightarrow 2H_2O(l) + O_2(g)$
2. $CH_4(g) + 2O_2(g) \rightarrow CO_2(g) + 2H_2O(g)$
3. Balanced as written
4. Balanced as written
5. $2Na(s) + 2H_2O(l) \rightarrow NaOH(aq) + H_2(g)$
6. $C_5H_{12}(g) + 8O_2(g) \rightarrow 5CO_2(g) + 6H_2O(g)$
7. $3Ca(OH)_2(aq) + H_3PO_4(aq) \rightarrow Ca_3(PO_4)_2(aq) + H_2O(l)$
8. $2AgI(aq) + Na_2S(s) \rightarrow Ag_2S(s) + 2NaI(aq)$
9. Balanced as written
10. Balanced as written

5.2

1. 8.3 mol
2. 6.0 mol
3. 66.9 g
4. 186.3 g
5. 17.8 g
6. 0.833 dm^3 (833 cm^3)
7. 8.1 g
8. 0.99 g
9. 0.25 dm^3 (250 cm^3)
10. 53.5 g

5.2.1

1. Cu
2. H_2

Answers

3. 9.7 g of C
4. 113.9 g

5.2.2

1. 25 cm^3
2. 66.6 cm^3
3. 250 cm^3
4. 225 cm^3

5.2.3

1. 69.6%
2. 68.3%

5.2.4

1. A = 60.3%, B = 60%; the law of definite proportions
2. FeCl$_2$ = 3.93 g, FeCl$_3$ = 2.59 g; the law of multiple proportions

Further Practice Problems

1. (a) $4Na(s) + O_2(g) \rightarrow 2Na_2O(s)$ (b) $4NH_3(g) + 3O_2(g) \rightarrow 6H_2O(g) + 2N_2(g)$ (c) $4Al(NO_3)_3(s) \rightarrow 2Al_2O_3(s) + 3O_2(g) + 12NO_2(g)$ (d) $2Al(OH)_3(s) + 3H_2SO_4(aq) \rightarrow Al_2(SO_4)_3(aq) + 6H_2O(l)$ (e) $2H_2S(g) + 3O_2(g) \rightarrow 2H_2O(l) + 2SO_2(g)$
2. 20 g
3. 82.8 g
4. 0.042 dm^3 (42 cm^3)
5. 1.12 dm^3 (1120 cm^3)
6. 0.54 g
7. 0.71 g
8. Fe$_2$O$_3$, 0.81 g of H$_2$
9. 0.32 g of PbS
10. 30 cm^3
11. 175 cm^3
12. 72.5%
13. 53.5%
14. Each of the two samples contains 80% of zinc
15. SnO = 7.3 g, SnO$_2$ = 3.7 g. The different masses of tin which reacts with a fixed mass of oxygen are in a simple multiple ratio of 2:1.

CHAPTER 6

Practice Problems
6.2.1

1. (a) +1 (b) +2
2. (a) +5 (b) +3 (c) −3 (d) +4 (e) +1
3. (a) +3 (b) +5 (c) +5 (d) +4 (e) +4

6.2.1.1

1. No
2. Yes
3. No
4. Yes
5. Yes

6.3

1. (a) Phosphorus pentachloride/Phosphorus(V) chloride (b) Silicon tetrafluoride/Silicon(IV) fluoride (c) Carbon sulfide/Carbon(IV) sulfide (d) Tetraphosphorus hexoxide/Phosphorus(III) oxide (e) Aluminium oxide
2. (a) Phosphoric acid (b) Potassium chromate (c) Cobalt(II) chloride (d) Calcium dihydrogen phosphate/Calcium dihydogentetraoxophosphate(V) (e) Dinitrogen pentoxide/Nitrogen(V) oxide
3. (a) Iodite (b) Silicate/Trioxosilicate(IV) (c) Zinc nitrate hexahydrate/Zinc trioxonitrate(V) hexahydrate (d) Iodous acid/Dioxoiodate(III) acid (e) Tin(IV) oxide

6.4

1. (a) $Cl_2(g) \rightarrow 2Cl^-(aq) + 2e^-$ (Oxidation half-equation)
   $2I^-(aq) + 2e^- \rightarrow I_2(g)$ (Reduction half-equation)
   (b) $Cu(s) \rightarrow Cu^{2+}(aq) + 2e^-$ (Oxidation half-equation)
   $2Ag^+(aq) + 2e^- \rightarrow 2Ag(s)$ (Reduction half-equation)
   (c) $Fe(s) \rightarrow Fe^{2+}(aq) + 2e^-$ (Oxidation half-equation)
   $2Ag^+(aq) + 2e^- \rightarrow 2Ag(s)$ (Reduction half-equation)
2. $6Fe^{2+}(aq) + Cr_2O_7^{2-}(aq) + 14H^+(aq) \rightarrow 6Fe^{2+}(aq) + 2Cr^{3+}(aq) + 7H_2O(l)$

FURTHER PRACTICE PROBLEMS

1. (a) +4 (b) +2 (c) +5 (d) +4 (e) +2
2. (a) Yes (b) Yes (c) Yes (d) Yes (e) Yes
3. (a) Silicic acid/Trioxosilicate(IV) acid (b) Sodium thiosulfate/Sodium trioxosulfate(II) (c) Tin(II) oxide (d) Mercury(I) sulfate/Mercury(I) tetraoxosulfate(VI) (e) Iodic acid/Trioxoiodate(V) acid (f) Tetraphosphorusdecoxide/Dinitrogen pentoxide (Based on the empirical formula P_2O_5) (g) Dichromate/Heptaoxodichromate(VI) (h) Copper(II) nitrate trihydrate/Copper(II) trioxonitrate(V) trihydrate (i) Magnesium potassium chloride hexahydrate (j) Bromite
4. (a) $2Fe(s) \rightarrow 2Fe^{3+}(aq) + 6e^-$ (Oxidation half-equation)
   $3Cl_2(g) + 6e^- \rightarrow 6Cl^-(aq)$ (Reduction half-equation)
   (b) $2S_2O_3^{2-}(aq) \rightarrow S_4O_6^{2-}(aq) + 2e^-$ (Oxidation half-equation)
   $I_2(aq) + 2e^- \rightarrow 2I^-(aq)$ (Reduction half-equation)
   (c) $5C_2O_4^{2-}(aq) \rightarrow 10CO_2(g) + 10e^-$ (Oxidation half-equation)
   $2MnO_4^-(aq) + 16H^+(aq) + 10e^- \rightarrow 2Mn^{2+}(aq) + 8H_2O(l)$ (Reduction half-equation)
5. (a) $2Cr(OH)_3(s) + 3ClO^-(aq) + 4OH^-(aq) \rightarrow 2CrO_4^{2-}(aq) + 3Cl^-(aq) + 5H_2O(l)$
   (b) $2MnO_4^-(aq) + 6H^+(aq) + 5H_2O_2(aq) \rightarrow 2Mn^{2+}(aq) + 5O_2(g) + 8H_2O(l)$

CHAPTER 7

PRACTICE PROBLEMS

7.1.6

1. $1s^22s^22p^4$
2. $1s^22s^22p^63s^23p^64s^1$

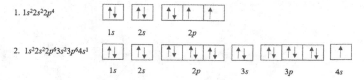

Answers

7.2
1. 16
2. 22
3. 146
4. 90

7.3
1. (a) 79.99 (b) 6.92 (c) 28.11
2. 42%

Further Practice Problems

1. 2
2. 7
3. (a) 14 (b) 16 (c) 7 (d) 0 (e) 28
4.
5.

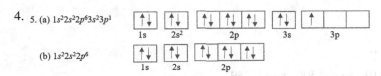

(a) $1s^2 2s^2 2p^6 3s^2 3p^1$

(b) $1s^2 2s^2 2p^6$

5. (a) $1s^2 2s^2 2p^6 3s^2 3p^6 4s^2 3d^2$ (b) $1s^2 2s^2 2p^6 3s^2 3p^6$
6. (a) 12.01 (b) 121.86 (c) 39.14 (d) 16.00 (e) 87.71
7. 81%

CHAPTER 8

Practice Problems

8.1
1. $\chi_{CH_4} = 0.0437$, $\chi_{C_6H_6} = 0.0448$, $\chi_{C_2H_6} = 0.0380$, $\chi_{Ne} = 0.873$
2. 2.42 g

8.2
1. 2250 mmHg
2. 250 mmHg
3. 0.21 atm

8.3
1. 88 cm³
2. 950 mmHg

8.4
1. 1.65 dm³
2. 870 K

8.5
1. 560 cm³
2. 2000 mmHg
3. 249 K

8.6
1. 7.8 dm^3
2. 133 K
3. 6.8 dm^3
4. 36.5 g mol^{-1}

8.7
1. 30
2. 80 g mol^{-1}
3. (a) 8 (b) 32

8.8
1. 3.4 cm^3 s^{-1}
2. 32 g mol^{-1}
3. 1.59 cm^3 s^{-1}
4. 16 s
5. 3.9 s

Further Practice Problems

1. $\chi_{O_2}= 0.304$, $\chi_{CO_2}= 0.101$, $\chi_{Ne}= 0.306$, $\chi_{N_2}= 0.289$
2. 37.2 g
3. 400 mmHg
4. 763 mmHg
5. 175 cm^3
6. 490 cm^3
7. 170 mmHg
8. 59 cm^3
9. 482 K
10. 990 mmHg
11. 470 mmHg
12. 120 K
13. 6.0 atm
14. 0.055 mol
15. 0.48 g
16. 600 K
17. 270 kPa
18. 64 g mol^{-1}
19. 22
20. 640 s
21. 8.6
22. 82 s
23. 690 cm^3
24. 1.5 times
25. 13

CHAPTER 9

Practice Problems
9.2
1. 0.013 mol
2. 0.10 g

Answers

3. (a) 83 g dm^{-3} (b) 0.46 mol dm^{-3}
4. 20 g mol^{-1}

9.3

1. 4.5 cm^3
2. 37 cm^3
3. 0.063 mol dm^{-3}
4. 750 cm^3

9.4

1. 265 g
2. 11 cm^3
3. 0.21 mol dm^{-3}

9.5

1. (a) 37.6 g/100 g H$_2$O, (b) 376 g dm^{-3}
2. (a) 83 g dm^{-3} (b) 0.48 mol dm^{-3}
3. 11 g
4. 1.2 g
5. 4.2 g

9.6

1. 2.6
2. 12.2
3. 11
4. 3.2 × 10^{-6} mol dm^{-3}

9.7

1. (a) 5 (b) CuSO$_4$.5H$_2$O (c) 36%
2. (a) 45% (b) CoCl$_2$.6H$_2$O

FURTHER PRACTICE PROBLEMS

1. (a) 6.0 g dm^{-3} (b) 0.024 mol dm^{-3}
2. 1.2 dm^3
3. 246 g mol^{-1}
4. 0.37 g
5. 10.6 g
6. 25 cm^3
7. 150 cm^3
8. 13 g
9. 15 cm^3
10. 0.36 mol dm^{-3}
11. 0.045 mol
12. (a) 316 g dm^{-3} (b) 3.13 mol dm^{-3}
13. 0.068 mol
14. 0.089 mol
15. (a) 57 g /100 cm^3 (b) 6.5 mol dm^{-3}
16. 2.8
17. 10.5
18. The solution will not turn blue litmus paper red because it is alkaline (with a pH of 9.4)
19. 11

20. 1.0×10^{-8} mol dm^{-3}
21. 5.0×10^{12} mol dm^{-3}
22. 1.2×10^{-10} mol dm^{-3}
23. (a) 5 (b) $MgSO_4.2H_2O$
24. $CaSO_4.3H_2O$
25. (a) 22% (b) $CuCl_2.2H_2O$

CHAPTER 10

Practice Problems

10.3
1. 14.7 kJ
2. -57 kJ mol^{-1}
3. -1300 kJ mol^{-1}

10.4
1. Forward: -46 kJ mol^{-1}; Reverse: 46 kJ mol^{-1}
2. 178.5 kJ mol^{-1}
3. 290 J
4. $\Delta H^\theta_{sub} = 62.4$ kJ mol^{-1}, $\Delta H^\theta_{dep} = -62.4$ kJ mol^{-1}
5. -595 kJ mol^{-1}

10.5
1. -92 kJ mol^{-1}
2. -832 kJ mol^{-1}

10.6
1. -93.2 J K^{-1} mol^{-1}
2. -173 J K^{-1}
3. -105 J K^{-1} mol^{-1}
4. 21.4 kJ

10.7
1. -7.2 kJ mol^{-1}
2. The process requires energy input because the free energy is positive; 130 kJ mol^{-1} to be specific.
3. 0
4. 8.5 kJ

10.8
1. 1.0 kJ of heat is absorbed
2. -650 J
3. 600 J

Further Practice Problems

1. 94.5 kJ
2. 41.2 kJ mol^{-1}
3. 38.6 kJ mol^{-1}
4. -2200 kJ mol^{-1}
5. 17.2 kJ
6. 260 kJ
7. -3037.8 kJ mol^{-1}

Answers

8. −345.4 kJ
9. −2528 kJ mol⁻¹
10. −441 kJ mol⁻¹
11. ΔH^{θ}_{sub}= 178 kJ mol⁻¹, ΔH^{θ}_{dep}= −178 kJ mol⁻¹
12. 1.1 kJ
13. 101 kJ mol⁻¹
14. −1362 kJ mol⁻¹
15. 14.1 kJ
16. 14.3 J K⁻¹
17. 52 kJ
18. −801 kJ
19. 74 kJ
20. 132 kJ
21. 57 K
22. 6.15 kJ
23. 1 kJ
24. −5.5 kJ

CHAPTER 11

PRACTICE PROBLEMS

11.2

1. (a) The forward reaction is favoured (b) The forward reaction is favoured (c) None
2. (a) The forward reaction is favoured (b) The forward reaction is favoured (c) The reverse or backward reaction is favoured (d) None (e) The forward reaction is favoured

11.3

1. $K_c = \dfrac{[Zn^{2+}]}{[Pb^{2+}]}$

2. $K_p = \dfrac{(P_{HI})^2}{P_{H_2} \times P_{I_2}}$

3. $K_c = 0.538$
4. (a) $K_p = 0.192$ (b) $K_c = 2.40 \times 10^{-3}$

11.4

$P_{PCl_3} = 0.277\,atm$, $P_{Cl_2} = 0.327\,atm$, $P_{PCl_5} = 0.0276\,atm$

11.5

1. 1.79×10^{-18}
2. 54 kJ mol⁻¹

11.6

1. (a) 1.72×10^{-4} mol dm⁻³ (b) 2.0×10^{-4}
2. 7.14×10^{-13}g

11.7

1. Yes, more of the solute will dissolve because $Q_{sp} < K_{sp}$.
2. No, a precipitate will not form because $Q_{sp} < K_{sp}$.

11.8

1. 3.11×10^{-5} mol
2. 5.58

FURTHER PRACTICE PROBLEMS

3. (a) $KNO_3(s)$: no effect, i.e., the concentration of a solid plays no role in the position of equilibrium $KNO_3(aq)$: the backward reaction is favoured. (b) Fe: none, i.e., the concentration of a pure solid does not affect the position of equilibrium H_2: the backward reaction is favoured.
4. (a) NO: the forward reaction is favoured.
   O_2: the backward reaction is favoured.
   (b) H_2O: the backward reaction is favoured.
5. (a) The forward reaction is favoured. (b) The backward reaction is favoured.
6. (a) None. (b) The forward reaction is favoured.
7. (a) $K_c = \dfrac{[NO_2]^2}{[N_2O]}$, $K_P = \dfrac{(P_{NO_2})^2}{P_{N_2O}}$

   (b) $K_c = \dfrac{[SO_3]^2}{[SO_2]^2[O_2]}$, $K_P = \dfrac{(P_{SO_3})^2}{(P_{SO_2})^2 P_{O_2}}$

   (c) $K_c = \dfrac{[NO]^2}{[N_2][O_2]}$, $K_P = \dfrac{(P_{NO})^2}{P_{N_2} \times P_{O_2}}$

   (d) $K_c = \dfrac{[CH_3COONa]}{[CH_3COOH][NaOH]}$, K_P is not applicable

   (e) $K_c = \dfrac{[PCl_5]}{[PCl_3][Cl_2]}$, $K_P = \dfrac{P_{PCl_5}}{P_{PCl_3} \times P_{Cl_2}}$

   (f) $K_c = \dfrac{[H_3O^+][BrO^-]}{[HBrO]}$, K_P is not applicable

8. (a) $K_P = 40.2$. (b) $K_P = 0.0248$
9. $K_c = 352.7$
10. 8.54
11. $K_P = 1.02 \times 10^{-4}$
12. $[H_2] = 0.124$ mol dm^{-3}, $[CO_2] = 0.224$ mol dm^{-3}, $[H_2O] = 0.026$ mol dm^{-3} $[CO] = 0.026$ mol dm^{-3}
13. $[H_2] = 0.047$ mol dm^{-3}, $[I_2] = 0.097$ mol dm^{-3}, $[HI] = 0.456$ mol dm^{-3}
14. 0.0239
15. The reaction will shift to the left to reach equilibrium because $Q_c > K_c$.
16. 50%
17. $[N_2O_4] = 0.761$ atm, $[NO_2] = 1.535$ atm
18. (a) $[COCl_2] = 1.5$ atm, $[CO] = 8.5$ atm, $[Cl_2] = 8.5$ atm (b) $K_P = 48.2$
19. 1.08×10^{-14}
20. 220 kJ mol^{-1}
21. 6.41×10^{-6}
22. (a) 2.1×10^{-4} mol dm^{-3} (b) 0.0164 g dm^{-3}
23. 2.90 cm^3
24. 11.34

CHAPTER 12

PRACTICE PROBLEMS
12.1.4

1. 1.56 V
2. −0.11 V

Answers

12.1.5
1. −174 kJ
2. +0.51 V

12.1.6
1. +0.94 V
2. 2.21×10^{24}

12.1.7
1. 0.24 V
2. 0.044 mol dm^{-3}

12.2.2
1. 21600 C
2. 1.3 g
3. 4300 C
4. 2.2 A

12.2.3
1. 0.44 dm^3 (440 cm^3)
2. 110 g

FURTHER PRACTICE PROBLEMS

1. 0.47 V
2. −0.11 V
3. −223 kJ
4. 0.891 V
6. 2.01×10^{-26}
7. +0.94 V
8. 3
9. −0.89 V
10. +0.81
11. 2.4 min
12. 0.83 g
13. 29 min
14. 2.9 A
15. 0.73 dm^3 (730 cm^3)
16. 230 cm^3
17. 2.9 dm^3
18. +2

CHAPTER 13

PRACTICE PROBLEMS
13.1.1
1. 0.18 mol dm^{-3}
2. (a) 0.14 mol dm^{-3} (b) 0.0028 mol (c) 0.11 g

13.1.2
1. 1.2 g/100 g H$_2$O
2. 1.6 mol dm^{-3}

13.1.3
1. 1:2
2. 1:1; NaOH or KOH

13.1.4
1. 93%
2. 52%

13.1.5
1. (a) 10, (b) 63%
2. (a) 2, (b) 28.6%

13.1.6
1. 63 g mol^{-1}
2. 39, KOH

13.1.7
1. (a) 0.85, (b) 0.0025 mol
2. (a) 1.1×10^{21}, (b) 4.7 g

13.1.8
1. 0.055 dm^3
2. 0.14 dm^3

FURTHER PRACTICE PROBLEMS
1. 0.30 mol dm^{-3}
2. 0.24 mol dm^{-3}
3. 22 cm^3
4. (a) 0.084 mol dm^{-3}, (b) 0.013 mol
5. (a) 0.19 mol dm^{-3}, (b) 0.019 mol/100 g H$_2$O
6. (a) 0.11 mol, (b) 2.2 mol dm^{-3}, (c) 230 g dm^{-3}
7. (a) 0.0064 mol, (b) 0.0032 mol, (c) 2:1
8. (a) 0.0038 mol, (b) 0.0038 mol, (c) 1:1
9. 14.5%
10. (a) 13 g, (b) 80%
11. 52.5%
12. (a) 7, (b) 54%
13. (a) 1, (b) 15%
14. (a) 0.10 mol dm^{-3}, (b) 98 g mol^{-1}
15. (a) 0.084 mol dm^{-3}, (b) 56 g mol^{-1}, (c) 39, potassium hydroxide
16. (a) 0.13 mol dm^{-3}, (b) 1.9×10^{21}, (c) 0.27 g
17. (a) 0.15 mol dm^{-3}, (b) 3.0×10^{21}, (c) 0.84 (d) 0.44 g
18. (a) 0.28 mol dm^{-3}, (b) 0.13 dm^3
19. (a) 0.25 mol dm^{-3}, (b) 0.59, (c) 0.0076 dm^3 (76 cm^3).

CHAPTER 14

PRACTICE PROBLEMS
14.2
1. 0.30 g min^{-1}
2. 0.0083 cm^3 s^{-1}

Answers

14.3

1. $-\dfrac{1}{2}\dfrac{\Delta[SO_2]}{\Delta t} = -\dfrac{\Delta[O_2]}{\Delta t} = \dfrac{1}{2}\dfrac{\Delta[SO_3]}{\Delta t}$

2. $H_2 = 0.0152$ mol dm^{-3} s^{-1}, $N_2 = 0.00507$ mol dm^{-3} s^{-1}, $NH_3 = 0.0101$ mol dm^{-3} s^{-1}

14.5

(a) $r = k[H_2O_2]$
(b) 1
(c) 2.0 s^{-1}

14.7.1

0.020 cm^3 s^{-1}

14.7.2

1. (a) 3.08×10^{-4} s^{-1} (b) 0.110 mol dm^{-3}

2. (a)

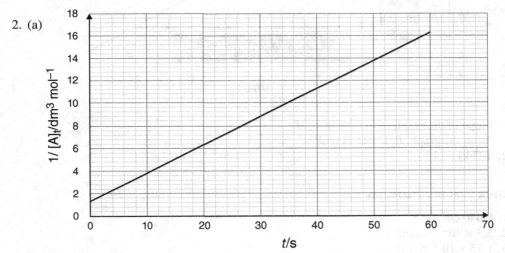

The graph of $1/[A]_t$ against t
The graph of $1/[A]_t$ against t is a straight line; hence, the reaction is second-order.

(b)

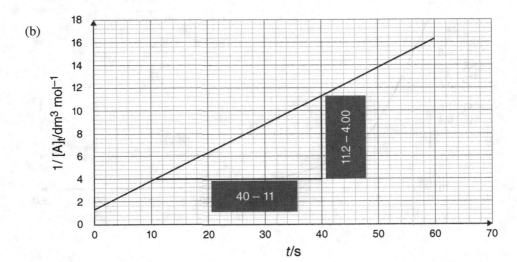

Slope of graph of the graph of 1/[A]$_t$ against t
From the graph, $k = 0.248$ dm^3 mol^{-1} s^{-1}
(c) 5.38 s

14.8

1.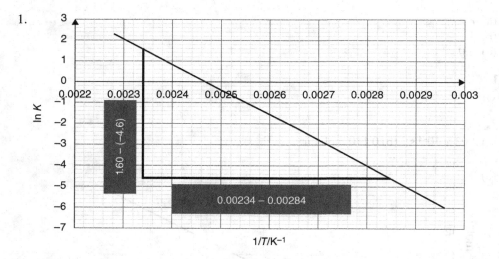

    Slope of the graph of ln k against $1/T$
    From the graph, $E_a = 103$ kJ mol^{-1}
2. 107 kJ mol^{-1}

FURTHER PRACTICE PROBLEMS

1. 0.040 cm^3 s^{-1}
2. 3.3 × 10^{-4} mol dm^{-3} s^{-1}
3. 1.75 × 10^{-4} mol s^{-1}
5. (a) 3.00 × 10^{-4} mol dm^{-3} s^{-1} (b) 6.00 × 10^{-4} mol dm^{-3} s^{-1}

6. (a)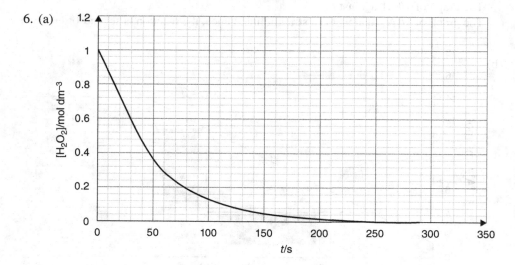

The rate curve for the decomposition of H_2O_2

(b)
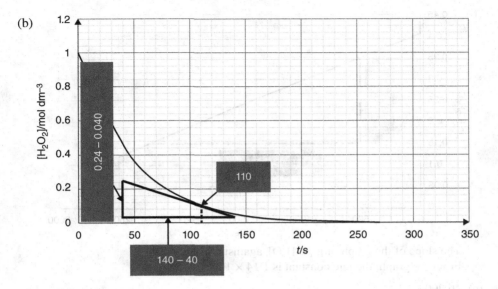

The slope of the tangent to the rate curve of for the decomposition of H_2O_2 at 110 s. From the graph, the instantaneous rate of reaction is 0.0020 mol dm^{-3} s^{-1}.

7. (a) The reaction is first-order in both ICl and H_2 (b) 2 (c) 0.400 dm^3 mol^{-1} s^{-1}.
8. (a) The reaction is second-order in NO_2 and zero-order in CO, (b) 2, (c) 6.50 dm^3 mol^{-1} s^{-1}.
9. (a) 0.0170 dm^3 mol^{-1} s^{-1}, (b) 3030 s.
10. (a) 1500 s

11. (a)

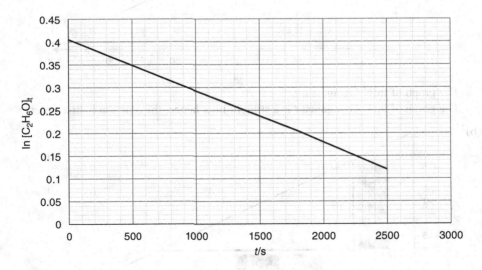

The graph of $\ln[C_2H_6O]_t$ against t

The graph of $\ln[C_2H_6O]_t$ against t is a straight line; hence, the reaction is first-order.

(b)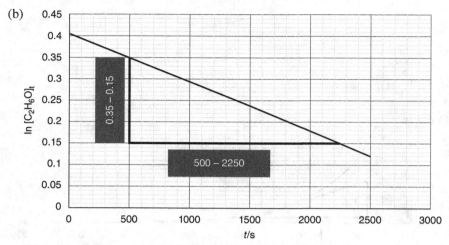

The slope of the graph of $\ln[C_2H_6O]_t$ against t
From the graph, the rate constant is 1.14×10^{-4} s^{-1}.

(c) 20,200 s

12. (a)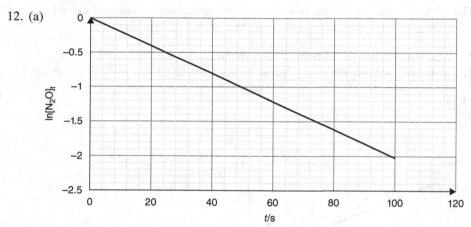

The graph of $\ln[N_2O]_t$ against t
The graph of $\ln[N_2O]_t$ against t is a straight line; hence, the reaction is first-order.

(b)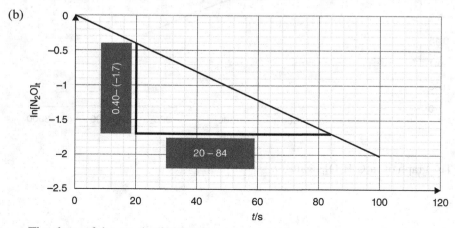

The slope of the graph of $\ln[N_2O]_t$ against t

Answers

From the graph, the rate constant is 0.0328 s^{-1}.
   (c) 21.1 s
13. 86.2%
14. 544 kJ mol^{-1}
15. 7.55 × 10^{18} dm^3 mol^{-1} s^{-1}
16. 1.40 × 10^{-3} s^{-1}

17.

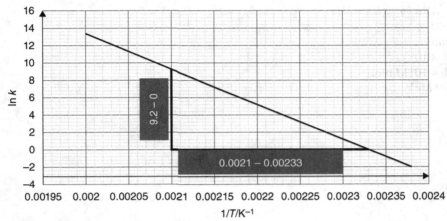

The slope of the graph of ln$k$ against $1/T$
From the graph, E_a = 333 kJ mol^{-1}

18. 75.0 kJ mol^{-1}
19. 1.19 s^{-1}
20. (a) 120 kJ mol^{-1}, (b) 696 K

CHAPTER 15

Practice Problems

15.1.1

1. $^{234}_{92}$U → $^{230}_{90}$Th + $^{4}_{2}$He
2. $^{214}_{83}$Bi → $^{210}_{81}$Tl + $^{4}_{2}$He
3. $^{206}_{80}$Hg → $^{206}_{81}$Tl + $^{0}_{-1}$e
4. $^{222}_{86}$Rn → $^{218}_{84}$Po + $^{4}_{2}$He
5. $^{234}_{91}$Pa → $^{234}_{92}$U + $^{0}_{-1}$e

15.1.2

1. 2.31 s
2. 0.00357 yr^{-1}
3. 24.2 days
4. 146
5. 0.182 days

15.2

1. 93 × 10^8 kJ

15.3

1. 6.32×10^8 kJ mol^{-1}
2. 2.21 MeV

FURTHER PRACTICE PROBLEMS

1. (a) $^{210}_{83}\text{Bi} \rightarrow {}^{210}_{84}\text{Po} + {}^{0}_{-1}\text{e}$
   (b) $^{222}_{86}\text{Rn} \rightarrow {}^{218}_{84}\text{Po} + {}^{4}_{2}\text{He}$
   (c) $^{226}_{88}\text{Ra} \rightarrow {}^{222}_{86}\text{Rn} + {}^{4}_{2}\text{He}$
2. 0.0625 g
3. 88 min
4. 4.51×10^8 kJ
5. 8.61×10^9 kJ mol^{-1}
6. 0.930 MeV

Index

Pages in *italics* refer to figures and pages in **bold** refer to tables.

A

acid and base pairs
    Arrhenius theory, 187–188
    Brønsted-Lowry theory, 188, **188**
    Lewis theory, 189
    pH and hydrogen ion concentration, 189–193, *190*
acid-base titration, 292, **293**
activation energy, 332
actual pressure, of gas, 140
actual yield, 106
addition, 43, 48–49
addition of oxygen, 113
algebraic method, 86
α-decay, 358–359
alpha particles, 357, *357*
alums, *see* double salts
amorphous solids, 5
analyte, 292
approximations
    decimal numbers, 40–42, *41*
    significant figures, 42–45
Arrhenius acid, 187–188
Arrhenius base, 188
Arrhenius equation, 348
Arrhenius, Svente, 187
Arrhenius theory, 187–188
arrow (→), 86
ascending paper chromatography, 22, *23*
atom, 23–27, **25**, *25*, **26**
atomic mass, 24–25
atomic number, 25, 133, 358
atomic pile, 366
atomic structure
    mass number, 133–135
    orbitals and electronic configuration, 129–133
    relative atomic mass calculation, 135–136
Aufbau or building up principle, 131
average titre value, 292
Avogadro's law, 103, 147
Avogadro's number, 63

B

balanced nuclear equation
    lead-214, 359
    uranium-238, 359
balancing chemical equations, 86–89
balancing redox equations, 124–128
Berzelius, Jons Jacob, 1, 11
β-decay, 358
beta particles, 357–358, *357*
binary ionic compounds, 119, **120**
binary molecular compounds, 119–120, **120**
binding energy, 369–370
boiling point, **35**

bond energies, 217–220
Born-Haber cycle, 203, 208, *209*, 211, *212*
Boyle, Robert, 1
Boyle's law, 141–143, *142*
Brønsted, Johannes Nicolaus, 188
Brønsted-Lowry theory, 188, **188**
butane, 104

C

calcium carbonate, 92, 97
calcium oxide, 92, 97
calorimetry, 204
carbon, 113
catalyst, 333, *334*
cell diagram, 267
cell potential, 268–269
    and concentration, 274–277
    and equilibrium constant, 272–274
    and free energy, 269–272
centrifugation, 19, *20*
Charles' law, 143–145, *144*
chemical bonds, 28–30, *29*, **30**
    covalent bonding, 30–33, *31–34*
    covalent compounds, 35–36
    electrovalent bonding, 30, *31*
    electrovalent compounds, 35–36
    formation, 34–36, **35–36**
chemical equations, 86–90
    balancing, 86–89
    information from, 89–90
chemical formula
    empirical formula, 76, **77**, 81
    of ionic compounds, 50
    molecular formula, 81
    types, 76
chemical symbols, 11–12, **13–14**
chlorine, 114
chromatography, 22, *23*
coefficient, 86
collision theory, 331–332
common-ion effect, 258–260
compounds, 12
concentration, 164–170, 333–335
condensation, 8–9
coordinate covalent bonding, 28, 32–33, *33–34*
copper-copper ion half-cell, *265–266*
covalent bonding, 30–33, *31–34*
covalent compounds, 35–36
covalent solids, 6
critical mass, 366
crystal lattice, 5
crystalline solids, 5
crystallisation, 22

411

D

Dalton, John, 1, 24
Dalton's law of partial pressures, 139–141
Daniell cell, 267, *267*
    equilibrium constant, 273–274
    free energy, 270
    standard potential, 268
daughter nuclide, 358
decantation, 17–18, *19*
decay series of uranium-238, *356*, 356–357
decimal fraction, 40
decimal numbers, 40–42, *41*
decimal point, 40
Democritus, 24
denominator, 40
deposition, 9
diatomic molecules, 27
dilution law, 170–172
dissolution equilibrium, 252, 254
distillation, 19, *21*
division, 46–48
d orbitals, *131*
double salts, 123, **124**
dumb bell shape, orbital, 130

E

effective collisions, 331
Einstein's mass-energy equation, 367
electrochemical cell, 265–267, **266**, *266–267*
electrochemical series, 265, **266**
electrode potential, 264–265
electrolysis, 277–278
    Faraday's first law, 279–286
    Faraday's second law, 286–291
    mechanism, 278–279
electrolytes, 36
electrolytic cell, 277–278, *278*
electromotive force, 268
electron acceptor, 114
electron donor, 114
electronegative elements, 114
electronic configurations, 132–133
electrons, 129
electron transfer, 114
electrovalent bonding, 30, *31*
electrovalent compounds, 35–36
elements, 9–11
    chemical symbols, 11–12, **13–14**
    compound, 12
    mixture, 12, 14–16, **15**, *15*, **16**
empirical formula, 76, **77**, 81
endothermic reaction, *332*
endothermic reactions, 200, *201*
energy levels, 129
enthalpy change, 200, 215–216
entropy, 220–224
equation of state, 148
equilibrium constant, 261
    cell potential, 272–274
    and concentrations, 236–246
    of Daniell cell, 273–274
    and temperature, 249–251

equivalence point, of titration, 292
ethene, combustion of, 103
evaporation, 8–9, 16
excitation, 129
excited state, electron, 129
exothermic reaction, *332*
exothermic reactions, 200, *201*

F

Faraday's first law, of electrolysis, 279–286
Faraday's second law, of electrolysis, 286–291
filling orbitals with electrons, principles for, 131
    Aufbau or building up principle, 131
    Hund's rule, 131
    Paul exclusion principle, 131
filtration, 17, *18*
first-order reaction, 342
fissionable nucleus, 366
flotation, 18
formula unit, *see* chemical formula
fractional crystallisation, 22
fractional distillation, 20–21
free energy, 224–228
    Daniell cell, 270
    galvanic cell, 269–270
freezing, 8
freezing point, 8

G

galvanic cell, 265, 267, 269–270
gamma rays, 357, *357*
gases, 7–8, *8*
gas laws
    Boyle's law, 141–143, *142*
    Charles' law, 143–145, *144*
    Dalton's law of partial pressures, 139–141
    general (combined), 145–147
    Graham's law of diffusion or effusion, 154–161
    ideal gas equation, 147–152
    mole fraction, 137–139
    relative molecular mass, 152–154
    vapour density, 152–154
gas (volume-volume) stoichiometry, 103–106
Gay-Lussac, Frenchman Joseph Louis, 1
general (combined) gas law, 145–147
Gibbs, Josiah Willard, 224
Graham's law, 154, 156–157
    of diffusion/effusion, 154–161
ground state, electron, 129

H

Haber equilibrium, 234, 249, 262
half-life, 342–343, 347, *360*, 360–365
heat of reaction, 200
    calculation, from bond energies, 217–220
    calorimetry, 204
    chemical and physical changes, 200–201
    combustion, 201, 214–215
    condensation, 202, 212–213
    deposition, 203
    experimental determination, 204–208, 213–214

Index

formation, 202, 210–211
fusion, 202
Hess's Law and, 208–217, *209*
lattice energy, 203
neutralisation, 201, 206–207
solidification, 202
solution, 204
solvation energy, 203
sublimation, 203, 211
types, 201–204
vaporisation, 202, 209–210, 212–213
Henry's Law, 179
Hess's Law, 203, 208–217, *209*, 220, 225
Hund's rule, 131
hydrated salts, 123, **124**, 193
hydration energy, *see* solvation energy
hydrochloric acid solution, 97
hydrogen bonds, 36
hydrogen halides, aqueous solutions of, 123, **123**
hydrogen peroxide, 90
hydrohalic acids, 123

I

ideal gas equation or law, 147–152
improper fraction, 40
indicator, 293
instantaneous rate
determination, 339–341
integrated rate law, 341–348
of reaction, 339
integrated rate laws, 341
first-order reaction, 342
half-life, 342–343, 347–348
second-order reaction, 342
zero-order reaction, 341, 343
International Union of Pure and Applied Chemists (IUPAC), 11, 119
cell diagram rule, 267
nomenclature, 117, 119–120, 122
oxyanions naming convention, 121
ion, 27, 316
ionic solids, 6
ionic theory, 278
ionisation, 278
ionising radiations, 357
ion product, 255–258
isotopes, 25–27

K

kinetic energy, 4
KLMN notation, 28–29

L

lattice energy, 203
lattice point, 5
Lavoisier, Antoine, 1
law of conservation of mass, 108
law of conservation of matter, 86
law of definite (constant) proportions, 108
law of multiple proportions, 108
law of radioactive decay, 359–365
laws of chemical combination, 108–110
lead-214 decays, 359
Le Châtelier's principle, 234–236
left hand side (LHS), 86
Leucippus, 24
Lewis, Gilbert, 189
Lewis structure, 32, *33*
Lewis theory, 189
light, 333
limiting reagent, 99–103
liquids, 7, *7*
Lowry, Thomas Martin, 188

M

magnetic quantum number, 130
magnetic separation, 17
mass, 60, 86, 173–174
of oxygen, 90
percent of copper, 109
of salt, 316
of sodium sulfate, 95
mass composition, 54
mass concentration, 164–166, 172–173
mass defect, 367
mass number, 133–135
mass percent composition, 54
matter, 36
change of state, *8*, 8–9
chemical changes, 4, **4**
classification, 9, 9–23, *10*, **11**, 13–15, *15*, **16**, *17–21*, 23
defined, 3–4
elements, 9–11
chemical symbols, 11–12
compound, 12
mixture, 12–16
separation, 16–22
particles
atom, 23–27, **25**, *25*, **26**
ion, 27
molecule, 27
particulate theory, 4
physical changes, 4, **4**
physical states, 4
gases, 7–8, *8*
liquids, 7, *7*
solids, *5*, 5–7
properties, 4
purification, 23
purity test, 23
melting, 8
melting point, 8, **35**
metal ion-metal system, 264
cell potential, 268–269
and concentration, 274–277
and equilibrium constant, 272–274
and free energy, 269–272
electrochemical cell, 265–267
electrode potential, 264–265
metallic solids, 6, *7*
metalloids, 10
metals, 10, **11**, 101
methane, 94
molar concentration, 164–167, 172–173, 252–254

molar entropy, 220
molar mass, 51, 313
  of gas, 155–156
  of unknown gas, 157
molar volume of gases at STP, 72
mole, 59, *59*
  mass, 60
  molar volume of gases at STP, 72
  number of particles, 63
  volume, 68
molecular formula, 81
molecular solids, 6
molecule, 27
mole fraction, 137–139
  of argon, 137, 139
  of helium, 137–138
  of neon, 138
mole ratio, 301
multiplication, 43, 46–48

N

natural science, 1, 2
nature of reactants, 333
Nernst equation, 274–277
neutrons, 366
nitrogen monoxide, 100
noble gases electronic configuration, *29*, 29–30, **30**
non-metal elements, 10, **11**
nuclear chain reaction, 366, *366*
nuclear chemistry, 356–371
  binding energy, 369
  nuclear fission and fusion, 366–369
  radioactivity, 356–366
nuclear equations, 358–359
nuclear fission, 366–369
nuclear fusion, 367
nuclear radiations
  penetrating powers, 358, *358*
  properties, **357**
nuclear reactors, 366
number of molecules of water of crystallisation, 307–308
number of particles, 63–64
numerator, 40

O

opposite and complimentary processes, 113
opposite directions, 86
orbital diagrams, 132–133, *133*
orbitals and electronic configuration, 129–133
  electronic configurations and orbital diagrams, 132–133
  magnetic quantum number (m_l), 130
  principal quantum number (n), 129
  principles, 131
  spin quantum number (M_s), 130
  subsidiary, azimuthal or angular momentum quantum number (l), 129
order of reaction, 334
ordinary covalent bonding, 28
overall order of reaction, 334
oxidation and reduction concept

electropositive/electronegative elements, 114
  hydrogen transfer, 113
  modern definitions of, 114, *115*
  oxygen transfer, 113
oxidation, definition, 114, 118
oxidation number, 115–118
  of chromium in $Cr_2O_7^{2-}$, 116–117
  in defining and identifying redox reactions, 118–119
  increase and decrease, 118
  of manganese in $KMnO_4$, 116
  of NH_3 in $[Pt(NH_3)_2Cl_4]$, 116–117
  of sulfur in H_2SO_4, 116
oxidising agent, 113–114
oxoacids, 122–123, **123**
oxyanions, 120–122
oxygen, 106
  addition of, 113
  mass of, 90
  removal of, 113

P

parent nuclide, 357–358
partial pressure, 139, 141
particles
  atom, 23–27, **25**, *25*, **26**
  ion, 27
  molecule, 27
particulate theory, 4
Paul exclusion principle, 131
penetrating powers of nuclear radiations, 358, *358*
percentage yield, 106–108
percent purity, 303
Periodic Table, 11–12, *12*
pH, 179, 189–193, 316, 319
pH indicator, 388
phosphorus, 100
polyatomic ions, 27, 120, **121**
polyatomic ions and inorganic compounds nomenclature, 119, 122
  binary ionic compounds, 119
  binary molecular compounds, 119–120, **120**
  double salts, 123–124, **124**
  hydrated salts, 123, **124**
  hydrogen halides, 123
  IUPAC names, **121–122**
  oxoacids, 122–123, **123**
  oxyanions, 120–122
p orbital, *130*
potassium chlorate, 106
precipitation, 19
pressure, 333
Priestley, Joseph, 1
principal quantum number, 129
propane, 104
proper fraction, 40
purity, 303

R

radioactive decay series, 356
radioactive isotope, 356

Index

radioactivity, 356–357
   law of radioactive decay and half-life, 359–365
   nuclear equations, 358–359
   types, 357
radioisotope, 356
rate curve, *328*
rate law, 334–335
rate of diffusion/effusion, 154–155
   of ammonia, 159
   of chlorine, 160
   of nitrogen, 160
reaction quotient, 246–249
reaction rates
   chemical reactions, 329
   factors affecting, 332–333
   measuring, 327
   and temperature, 348
redox equations, balancing, 124–128
redox reactions, 113
redox titrations, 293, **294**
reducing agent, 113–114
relative abundance, 25
relative atomic mass, 25–27, **26**, 135–136
relative formula mass, 51
relative molecular mass, 152–154
reversible reactions, 86, 234, *235*, 260
right hand side (RHS), 86
rough value, 292

S

saturated solution, 179, 298
scientific method, 1–3, *3*
second-order reaction, 342
self-indicating titrations, 293
separation technique, 16, *17*
   centrifugation, 19, *20*
   chromatography, 22, *23*
   crystallisation, 22
   decantation, 17–18, *19*
   distillation, 19, *21*
   evaporation, 16
   filtration, 17, *18*
   flotation, 18
   fractional crystallisation, *21*
   fractional distillation, 20–21
   funnel, 16–17
   magnetic separation, 17
   precipitation, 19
   sieving, 16
   sublimation, 18
shells, 129
sieving, 16
significant figures, 42–45
sodium iodide, 114
sodium sulfate, 95
solids, *5*, 5–7
solubility, 179–187, *180*, *186–187*, 298
solubility curve, 180, *180*, *198*
solubility equilibrium, 258–260
solubility product, 251–255

solution
   of acids and bases
      Arrhenius theory, 187–188
      Brønsted-Lowry theory, 188, **188**
      Lewis theory, 189
      pH and hydrogen ion concentration, 189–193, *190*
   concentration, 164–170
   dilution law, 170–172
   molarity, 165–167
   preparation, 172–179
   solubility, 179–187, *180*, *186–187*
   volume, 170–171
   water of crystallisation, 193–197
solvation energy, 203
s orbital, *130*
Sørensen, Søren Peter Lauritz, 189
spherically symmetrical shape, orbital, 130
spin quantum number (M_s), 130
standard enthalpy change, 201
standard form, 45
   addition in, 48–49
   division in, 46–48
   expressing numbers in, 45–46
   multiplication in, 46–48
   subtraction in, 48–49
standardisation, 293–294
standard molar entropy, 220
standard solution preparation, 172–179
stoichiometric coefficient, 86
stoichiometry, 90–98
   gas (volume-volume), 103–106
   the laws of chemical combination, 108–110
   limiting reagent, 99–103
   percentage yield, 106–108
sublimation, 9, 18
subsidiary (azimuthal or angular momentum) quantum number *(l)*, 129
subtraction, 43, 48–49
surface area of reactants, 333
systematic name, 119

T

temperature, 332
theoretical yield, 106
thermodynamics first law, 228–233
titrant, 292
titration, *292*, 292–293, **293–294**
trial-and-error approach, 86

U

unit cell, 5
universal gas constant (R), 147, **148**
uranium-238, *356*, 356–357

V

valency, 27–28, **28**
van't Hoff Equation, 249–250
vapour density, 152–154

velocity constant, 334
voltmeter, 264
volume, 68
volume of gas evolved, 321
volumetric analysis, 292

W

water of crystallisation, 193–197

Z

zero-order reaction, 341
zinc, 114
zinc oxide, 113
zinc-zinc ion half-cell, *265*

Printed in the United States
by Baker & Taylor Publisher Services